1

Maths

Teacher's Guide

Ray Huntley
Tony Cotton

Caroline Clissold
Linda Glithro
Cherri Moseley
Janet Rees

OXFORD

OXFORD
UNIVERSITY PRESS

Great Clarendon Street, Oxford, OX2 6DP, United Kingdom

Oxford University Press is a department of the University of Oxford. It furthers the University's objective of excellence in research, scholarship, and education by publishing worldwide. Oxford is a registered trade mark of Oxford University Press in the UK and in certain other countries.

© Ray Huntley, Tony Cotton and Janet Rees 2021

The moral rights of the author have been asserted.

First published in 2014

All rights reserved. No part of this publication may be reproduced, stored in a retrieval system, or transmitted, in any form or by any means, without the prior permission in writing of Oxford University Press, or as expressly permitted by law, by licence or under terms agreed with the appropriate reprographics rights organization. Enquiries concerning reproduction outside the scope of the above should be sent to the Rights Department, Oxford University Press, at the address above.

You must not circulate this work in any other form and you must impose this same condition on any acquirer.

British Library Cataloguing in Publication Data

Data available

ISBN 9781382017268

12

Paper used in the production of this book is a natural, recyclable product made from wood grown in sustainable forests. The manufacturing process conforms to the environmental regulations of the country of origin.

Printed and bound by CPI Group (UK) Ltd, Croydon, CR0 4YY

Acknowledgements

The publisher and authors would like to thank the following for permission to use photographs and other copyright material:

Cover artwork by Peskimo. **Photos: pv(t):** foto-bee/Alamy Stock Photo; **pv(m):** Alistair McDonald/Shutterstock; **pv(b):** 2R fotografia/Shutterstock; **px:** Monkey Business Images/Shutterstock; **pxviii:** monkeybusinessimages/iStockphoto; **p139(t):** Intellson/Shutterstock; **p147:** Awesome_art_Creation/Shutterstock; **p217:** Maria Averburg/Shutterstock.

Artwork by Q2A Media Services Pvt. Ltd and Oxford University Press ANZ.

Every effort has been made to contact copyright holders of material reproduced in this book. Any omissions will be rectified in subsequent printings if notice is given to the publisher.

The manufacturer's authorised representative in the EU for product safety is Oxford University Press España S.A. of El Parque Empresarial San Fernando de Henares, Avenida de Castilla, 2 – 28830 Madrid (www.oup.es/en or product.safety@oup.com). OUP España S.A. also acts as importer into Spain of products made by the manufacturer.

Contents

Introduction

The joy of learning maths

We are living in an ever-changing world, where the way we work, live, learn, communicate and relate to one another is constantly shifting. In this climate, we need to instill in our learners the skills to equip them for every eventuality so they are able to overcome challenges, adapt to change and have the best chance of success. To do this, we need to evolve beyond traditional teaching approaches and foster an environment where students can start to build lifelong learning skills for success. Students need to learn how to learn, how to problem solve, be agile and work flexibly. Going hand-in-hand with this is the development of self-awareness and mindfulness through the promotion of wellbeing to ensure students learn the socio-emotional skills to succeed.

With *Oxford International Primary Maths*, students develop lifelong learning skills as well as mathematical skills. The course promotes the development of real-world skills including financial literacy. The activities in the Student Books and Practice Books offer numerous opportunities to think creatively and develop interpersonal skills. Fundamentally, *Oxford International Primary Maths* promotes students' self-development as critical thinking and motivation are at the heart of the problem-solving approach in the course.

This series is based on the English National Curriculum Programme of Study for Primary Maths. The books for each stage meet all the learning objectives. Each lesson includes the learning objectives from the curriculum and summary of the key teaching points. A full mapping grid identifying the unit and lesson where each objective can be found is available online at https://www.oxfordowl. co.uk/

Oxford International Primary Maths: a problem-solving approach

In this second edition of *Oxford International Primary Maths*, there is a strong focus on using a problem-solving approach. Whilst mathematical facts are important, it is unlikely that simply giving students the information they need will result in them understanding the mathematics and being able to apply their learning in new problem-solving situations. This is often described as a move from 'surface learning' to 'deep learning'.

Many people remember mathematics lessons as places where the teacher stood at the front of the class writing on the board. The students copied the information down, maybe worked through a couple of examples with the teacher and then proceeded to complete a series of exercises to practise the skill that they had been taught. This can be described as a *didactic* approach and it relies on the idea that direct instruction is the appropriate strategy to adopt. The authors of this series would argue that *heuristic* strategies encourage students to explore the mathematics for themselves supported by the teacher. 'Heuristic' derives from the Greek word meaning to discover, and in mathematics learning, heuristic strategies are ones where the student engages in exploration and discovery to solve a problem. Heuristic strategies include making a visual representation of a problem, making a calculated guess or estimate, simplifying a problem or following a known method. This results in a deeper understanding for the student.

When faced with any problem in mathematics, there are recognised stages to go through in order to solve the problem, and these have been developed and agreed by many researchers. One version that summaries the problem-solving process comes from Georg Polya:

1. Understand the problem

2. Devise a plan

3. Carry out the plan

4. Check the reasoning

In following these stages, students will engage in a number of skills which support problem solving such as trial and improvement, working systematically, pattern spotting, visualising, conjecturing and generalising.

Embedding a mastery approach

In recent years, the term 'mastery' has been used in conjunction with mathematics learning. It has been drawn from teaching approaches in countries where mathematics performance is deemed to be very high. The essence of mastery is to produce students who have deep conceptual understanding and procedural fluency through learning in a collaborative and problem-solving context. Mastery learning incorporates use of manipulatives, exposure to different methods of solving a problem, dialogue and explanation.

Following a Concrete Pictorial Abstract (CPA) approach

One of the more successful approaches to learning was provided by Jerome Bruner in his model of enactive, iconic and symbolic modes. This has been developed in recent years to form the CPA approach, which stands for concrete, pictorial and abstract, each of which aligns with Bruner's modes. The concrete phase is about making use of physical manipulatives to help understand the learning, before moving to record the learning in pictorial form as an individual student. As the learning develops, students will begin to recognise how to record their learning in a more general and abstract way. The CPA approach is not necessarily sequential, and students might move between the different modes as they work through a problem.

Oxford International Primary Maths and the use of manipulatives

Throughout the series, students are encouraged to use manipulatives, or concrete objects, to model addition, subtraction, multiplication and division. These manipulatives include:

- base-ten equipment (ones-cubes, tens-rods, hundreds-flats and thousands-cubes)

- place-value counters

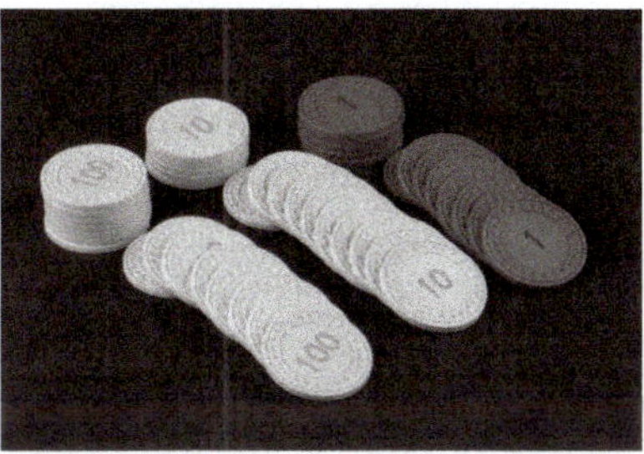

- number rods

Such manipulatives are used to explain to students how the written methods 'work', for example, by modelling exchanging 10 ones cubes for 1 tens rods in an addition.

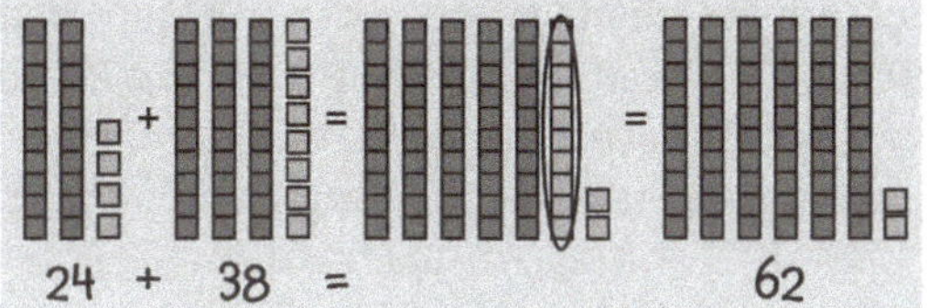

Differentiation

There are several ways that you can differentiate learning in the classroom. These include:

- Differentiation by task
- Differentiation by outcome
- Differentiation by support
- Differentiation by grouping.

It has been traditional in some schools to offer up to three different levels of tasks for each lesson. This is differentiation by task. It is important that all students are exploring the same area of mathematics as they can collaborate and discuss their mathematics in a way that is not possible if students are engaged on different activities. This approach has been extensively researched and published by Jo Boaler of Stanford University, California. In her book 'The Elephant in the Classroom' (2nd Edition, 2015, Souvenir Press) she outlines projects which gave students in different schools either a differentiated approach in lessons, or lessons where everyone worked on the same task. Where all abilities worked on the same task, every student made and sustained 'better than expected' progress, and performed better on statutory tests and exams. The Education Endowment Foundation teacher's toolkit suggests that collaborative learning can result in a five-month acceleration in student learning. (See https:// educationendowmentfoundation.org.uk/resources/ teaching-learning-toolkit .)

The expectation in this series is that all students will be offered the same starting point. The activities are carefully designed to be accessible to all students in your class and the teacher's notes for the activity offer differentiated outcomes for students. It is also important that you offer differentiated support to different students. You will mainly do this through the sort of questioning that you engage in and support you offer. You will ask challenging questions and supporting questions to help all students access the task. For example, when engaging in a simple counting activity with some students you may model the action of counting by placing a finger on each object as you count and emphasise the last number you say to model that the last number you say gives the number of objects in the set. For other students engaged in the same activity you may ask them to compare two sets, or to find one more or one less than the set they are counting.

Grouping students to promote a growth mindset

When engaging in learning mathematics, it is expected that you will use a variety of student groupings. This may be a change for some teachers who have previously grouped students by prior attainment in their classroom. Research has shown that grouping students 'by ability' which usually means grouping students using test results, can have a negative impact on their future attainment. It is more effective to use a range of ways of grouping students. You will decide on the most appropriate way of grouping students depending on the activity. You are also given advice in the teacher's notes. It is important that the teacher is active in deciding which form of grouping is appropriate. It is also important that students learn how to operate in a range of different groups and with a range of different students so that they get used to working in a variety of ways and with different people.

There are three main ways of grouping students:

- Friendship
- Ability/Prior experience
- Mixed attainment.

Friendship groups, are most appropriate for activities in which the students have been given some element of choice. Perhaps they are carrying out some research for a data handling project or exploring data on animals to develop their understanding of measurement. This grouping is the default if teachers do not actively group students.

Ability groups, or groups based on the prior experience of students, may be helpful if the lesson requires a very specific prior knowledge. You can group students who you know have this knowledge together as they can then work with minimal teacher guidance which then allows you to focus on groups who need additional support.

Mixed-attainment groups are the grouping that is encouraged for the majority of the activities. This is also the form of grouping favoured by those following a mastery approach. Working in collaborative, all-attainment groups also supports students well-being and promotes a growth mindset. , as described in research by Carol Dweck. She found that students who were put in ability groupings tended to stay in those groupings throughout their school life, and regard themselves as having a fixed ability that could not be changed. This has dire consequences for students in middle or lower sets. By using mixed ability groupings, all students can develop a growth mindset which enables them to believe they can learn and improve, whatever their starting point. (Dweck, C., 2007, The Perils and Promise of Praise, Educational Leadership, October 2007, 65(2),, 34-39). A growth mindset is promoted when students do not feel that their future success is predicated on prior achievement. This kind of grouping is particularly helpful for students new to English. Mixed-attainment groups allow students who are less confident in English to hear their more confident peers using mathematical vocabulary. Research has shown that mixed-attainment groups benefit both high attainers, who become more secure in their mathematics knowledge through explaining their thinking to peers, and to those less secure in their mathematical knowledge as peer teaching has been shown to be effective.

Whatever form of grouping you choose, it is helpful to assign roles to individuals in the group. Some teachers use 'role cards' to remind members of the group of the role they should play. Examples of these roles are:

- Leader: You should make sure everyone has a chance to speak and focus the discussion around the task.
- Time keeper: You should encourage the group to stay on task. Announce when the time is half way through and when time is nearly up.
- Recorder: You should write down the group member's ideas or draw a collective graphic. You will write on the board during the presentation.
- Presenter: You will present the groups findings to the whole class at the end of the session.
- Resource organiser: You will make sure the group has all the resources they need during the task.

Assessment is the process of establishing how each student is progressing and what they have achieved, or a means of measuring their learning. Assessment is usually carried out in two main ways – assessment of learning and assessment for learning.

Assessment of learning is sometimes called summative assessment, and takes place at the end of a lesson, a unit, a term or even a year. It measures what students know at that point as a summary of their learning to that point. In *Oxford International Primary Maths*, summative assessment opportunities are provided in the Review lesson at the end of each unit in the Student Book, whilst half-termly summative assessment opportunities are provided through printable resources, available online.

Assessment for learning is an approach brought to prominence by Paul Black and Dylan Wiliam and is based on the notion that students have a full, clear sense of what they are learning, where they have reached in their learning and what they need to do to improve further. It is carried out during lessons and gives teachers continuous data on each student's learning, as well as allowing students to track their own learning, which provides greater motivation. (Black, P., Harrison, C., Lee, C., Marshall, B., and Wiliam, D. (2004) Inside the Black Box: Assessment for learning in the classroom. Phi Delta Kappan, Vol. 86 No. 1 pp8-21)

It is suggested that there are five key strategies for assessment for learning. These are outlined below with suggestions of how you can do this in your classroom.

1 Being clear about learning objectives and success criteria with the students.

Each activity has at least one learning objective. At the beginning of a lesson, share the activity's learning objective with the students. This should be more than simply stating the objective. You should make sure that students understand the objective and how you will measure success. For example, you might say: *I know that you can all count 10 objects and all count to 10 as a class.* Then you point to '20' on a number line and ask: *Does anyone know what this number is?* If a student knows it is 20 praise them, if no-one knows, tell them it is twenty and say: *By the end of the lesson I will be able to listen to you count to twenty.*

2 Planning student discussions that give you evidence of their learning.

Every activity plan in the Teacher's Guide offers the opportunity for small-group or whole-class discussion. There are also examples of probing questions that you can ask to assess the student's current understanding. For example, if a group has been counting two sets of objects you can ask: *Were there more or less in the second group? How do you know?*

3 Giving students feedback that helps them move forward.

This allows students to know whether or not they are meeting the success criteria and what they can do next to move their learning on. Developing the example above, if a group has been comparing two sets and understands the concept of 'more' and 'less' you could ask them to make sets that are 'one more' and 'one less' or even 'two more, and, two less'.

4   Activating students to act as instructional resources for each other.

Collaborative group work in mixed-attainment groups, as described by Jo Boaler in her research (see under Differentiation earlier), gives students the opportunity to operate both as learners and teachers, with peer learning being highly effective. Not only is understanding of the mathematics enhanced, but students can support each other in assessing their progress.

5   Activating students as owners of their own learning.

The key point here is to listen carefully to the students and adapt your questioning to support individual development and to follow individual interests.

Questioning is key

The most skilled mathematics teachers can ask open questions to elicit students' current understandings. Skilful open questioning also allows students to articulate their current understanding carefully and though this process either consolidate their understanding or come to realise where they have made a mistake. The list below offers a series of open questions that can be used whatever mathematics you are teaching:

- *How are these the same/different?*
- *About how many/how long/many more …. do you think there will be?*
- *What would happen if …?*
- *How else could you have done that?*
- *Why did you ….?*
- *How did you …?*
- *How do you know that is correct?*

If you want students to check their solutions and consolidate their learning it is helpful to ask them to explain how they reached their solution to a friend. Similarly, to support students in reflecting on their learning you might ask:

- *What mathematics did you use to solve the problem?*
- *What new mathematics did you learn?*
- *What key words did you use?*
- *What was the most challenging part of the activity?*
- *What did you do when you got stuck?*
- *What other questions could you ask?*
- *Did this remind you of any other areas of mathematics?*

In *Oxford International Primary Maths*, there is an opportunity to ask these reflective questions, and for students to reflect on their learning, at the end of each unit in the Review lesson of the Practice Book.

Word problems

Word problems are useful as an assessment of children's understanding of the correct mathematics to use in any given situation. In *Oxford International Primary Maths* word problems are included throughout the units and on every Student Book Review page as part of the end-of-unit assessment. Many teachers find teaching word problems a challenge. This area is particularly challenging for students with a limited English vocabulary as word problems are tightly bound to linguistic ability. We have to decode and understand what the problem is asking us to do before we can begin to apply our mathematical knowledge. Some teachers have found the following acronym helpful when working with students on solving word problems.

R: Read the problem carefully.

U: Understand what the problem is asking you to do.

C: Choose the mathematics or arithmetical operations that you need to use to solve the problem.

S: Solve the problem.

A: Answer the problem.

C: Check the answer is accurate and reasonable.

It is often helpful for students to underline key facts and write down the operations they are going to use before they solve the problem. For example:

> Tony rode his bicycle 7 miles to school with his friend. On his way home he took a short cut which was only 5 miles. How far did he cycle altogether?
>
> *This will be an addition calculation.*

It is a useful activity for students to annotate word problems and write down the operation(s) they will use without carrying out the calculation as this focuses on the skill of understanding the problem and choosing the operations appropriately.

Another activity which helps students becomes skilled at solving word problems is asking them to write their own word problems based on a picture or a set of objects. For example:

- How many black cubes are there? (3)
- Two friends took three cubes each. How many were left? (2)
- If I take out the black cubes, how many are left? (5)
- If I share the cubes equally between two people, how many do they each get? (4)

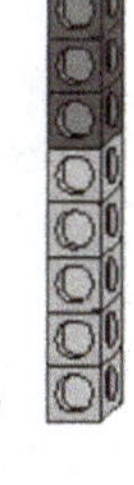

Wellbeing and *Oxford International Primary Maths*

It is thought that children learn more and feel more connected to their learning when they are active in their lessons. OIPM has active learning at its heart. Most lessons start with a whole-class session which usually includes a range of physical or active activites. You will see this signified by a star jump icon in the Teacher Guide.

Many adults and children have felt anxious about their learning of mathematics at some stage. This anxiety is reduced by working collaboratively in all-attainment groups. There is also a reflective session at the end of each lesson and the formative assessment activity in the Practice Book asks students to reflect on their learning across the unit.

Well-being is also supported by effective questioning to support and stretch students and by planning group work carefully. These areas have already been discussed above.

Language Support

The challenges

Ministries of Education at both local and national level are increasingly adopting the policy of English Medium Instruction (EMI), for either one or two subjects or across the whole curriculum. The rationale for doing so varies according to the local context, but improving the levels of achievement in English is an important factor.

In international schools an additional reason is likely to be that students do not share a mother tongue with each other or perhaps the teacher. English is, therefore, chosen as the medium for instruction so that all students are in the same position and to provide the opportunity to develop proficiency in an international language.

This does not mean that the mathematics teacher is now being asked to replace the English teacher, or to have the same skills or knowledge of English (though in many primary schools one teacher may indeed teach both). What it does mean, however, is that the mathematics teacher has to view his/her role differently: he/she has to become much more language aware. It is this recognition of the need to ensure that the delivery of the content is not negatively impacted by the use of the second language that informs the planning and methodology of EMI.

This raises significant challenges, including:

- the teacher's knowledge of English
- students' level of English (which may vary considerably in international schools)
- resources which provide appropriate language support
- assessment tools which ensure that it is the content and not the language which is being tested
- differentiation which acknowledges different levels of proficiency in both language and content.

Meeting the challenges positively

Perhaps lack of confidence in their own English proficiency is one of the most common concerns among teachers. However, while it is a factor, success in EMI is not necessarily linked to the teachers' proficiency in English. Teachers who have English as their mother tongue may well lack the sensitivity to, or awareness of,

the language that a non-native speaker has acquired through learning and studying the second language. Developing this awareness and demonstrating it in both materials and method is the key to effective EMI.

Classroom language/Teacher Talk

Often non-native-speaker teachers are more concerned about their ability to run and manage the whole class in English than they are about the actual teaching of the mathematics concepts, as the resources or textbook should help them with the latter. However, this use of English in the class is very important as it provides exposure to the second language, which plays a valuable role in language acquisition. It is also true that the teacher talk for purposes such as checking attendance and collecting homework does not have to be totally accurate or accessible to the students. When teaching the mathematics concepts, however, it is essential that the Teacher Talk is comprehensible. Some basic strategies to ensure this include:

- simplification of your language
- use short, simple sentences and project your voice
- paraphrase (say in a different way) as necessary
- use visuals, write or draw on the board, gestures and body language to clarify meaning
- repeat as necessary
- plan before the lesson
- prepare clear, simple instructions and check understanding.

Creating a language-rich environment

Primary teachers often excel at providing a colourful and engaging physical environment for students. In the EMI classroom, this becomes even more important. Posters, 'Word walls', lists of key structures, students' work, English signs and notices all provide a backdrop which provides the opportunity for language exposure and language acquisition.

Planning

When planning, look carefully at each stage of the unit and identify what the language demands are. This means thinking about what language students will need to understand or produce, and deciding how best to scaffold the learning to ensure that language does not become an obstacle to understanding the concept. This involves providing language support and goes beyond the familiar strategy of identifying key vocabulary.

Support for listening and reading

Listening and reading are receptive skills, requiring understanding rather than production of language. If you are asking your students to listen to or read texts in English, ask yourself the following questions when you are planning the unit:

- Do I need to teach any vocabulary before they listen/read?
- How can I prepare them for the content of the text so that they are not listening 'cold'?

- Can I provide visual support to help them understand the key content?

- How many times should I ask them to read/listen?

- What simple question can I set before they listen/read for the first time to focus their attention?

- How can I check more detailed understanding of the text? Can I use a graphic organiser (e.g. tables, charts and diagrams) or gap-fill task to reduce the language demands?

- Do I need to differentiate the task for those students who find reading/listening difficult?

- Could I make the tasks interactive (e.g. jigsaw reading, when students access different information before coming together, and information share)?

- How am I going to check their answers and give feedback?

Support for speaking and writing

Speaking and writing are productive skills because students doing these need to produce language. They are different to the receptive skills of listening and reading where students receive language from other sources. These skills may require more input from the teacher.

When you plan to use a task which requires students to *produce* English (speak or write), you need to think about how to help them do this.

This means that you have to think in detail about what language the task requires (Language Demands, LD) and what strategies you will use to help them use English to perform the task (Language Support, LS).

You need to ask yourself the following questions:

- What *vocabulary* does the task require? (LD)

- Do I need to teach this before they start? How? (LS)

- What *phrases/sentences* will they need? Think about the language for learning mathematics, e.g. predicting and comparing. What structures do they need for these language functions? (LD)

- Will they be able to produce these sentences or should I provide some *scaffolding* [e.g. sentence starters/sentence frames/gapped sentences (see below)]? (LS)

  A square has _____ sides.

  A triangle has _____ sides.

  A quadrilateral has _____ sides.

  A pentagon has _____ sides.

- While I am *monitoring* this task is there any way I can provide further support for their use of English (especially for the less-confident students)? (LS)

- What language will students need to use at the *feedback* stage (e.g. when they present their task)? Do I need to scaffold this? (LD, LS)

Teaching vocabulary and structures

Vocabulary

Learning the key mathematics vocabulary is central to EMI and 'learning' means more than simply understanding the meaning. Knowing a word also involves being able to *pronounce* it accurately and *use* it appropriately. Below is a list of strategies which could be useful:

- Avoid writing the list of vocabulary on the board at the start of the unit and 'explaining' it. The vocabulary should be introduced as and when it arises in the unit. Word boxes are provided on each page of the Student Books and Practice Books with the key words for the lesson. This helps students associate the word or phrase with the concept and context.

- Record the vocabulary clearly on the board when you first introduce it in the lesson, and check that you are confident with the pronunciation and spelling. before the lesson. If you think students may struggle to pronounce words, decide how best to model this pronunciation.

- Give students a chance to say the word once they have understood it. The most efficient way to do this is through repetition drilling.

- Use visuals whenever possible to reinforce students' understanding of the word.

- Ensure students are recording the vocabulary systematically in their glossaries, at the back of the Student Books, and, if possible, use a 'Word wall' which lists the vocabulary under unit/topic headings.

- Remember to use and revise the vocabulary.

Structures

In order for students to talk or write about their mathematics, they will need to go beyond vocabulary: they will also need to use those phrases and sentence frames which a particular task requires.

For example, they may need the following expressions in mathematics:

> *X is the same as Y.*
>
> *The sides are the same length.*
>
> *The next number in the sequence.*
>
> *I predict that X will happen.*
>
> *If X happens, then Y happens.*
>
> *The next step is …*

You need to build up these banks of common mathematics phrases and encourage students to record them. This is an important part of identifying the language demands and providing the necessary support. The teacher does not have to focus on grammar here as the language can be taught as phrases rather than specific grammatical structures.

Every unit of the Teacher's Guide begins with some useful background information. This includes:

The Big idea: The main mathematical concept covered in the unit.

Look out for: Tricky concepts that may need explaining prior to any learning taking place.

Common misconceptions: Common errors that students make or misunderstandings that students have. This section offers advice on how to deal with these misconceptions.

Key vocabulary: The key mathematical words used in the unit.

Coverage in lessons: The English National Curriculum objectives covered in the unit.

Every lesson in the Student Book and Practice Book has corresponding lesson notes in the Teacher's Guide. Each comprehensive set of lesson notes includes:

A mini reproduction: The relevant pages from the Student Book.

Global skills: These are the skills that aim to foster a classroom environment where students develop the skills for success. The skills are: *Creative skills* where students are problem solving, investigating or exploring new maths content; *Real-world skills* where students are taking part in research, or presenting and interpreting information, or if they are dealing with money and developing their financial literacy; *Interpersonal skills* where students are practising their teamwork and communication, often through working in pairs or larger groups; and *Self-development skills* where students have the opportunity to reflect on their learning and talk about what went well and what they are still uncertain about.

The key vocabulary and resources: A list of key vocabulary used in the lesson and the concrete resources required for the activity.

Language support: A range of strategies, including card sorts and card games, word walls, team games to define or explain words, use of similar words to explain meaning and exploration of the origins of words.

The key principles underpinning the language support are:

Words should be introduced and explained carefully.

The word should be explained in context.

Repetition is vital.

Words should be linked to pictures or actions.

Students should develop their own glossaries.

The learning of mathematics vocabulary should be fun.

Language should not be a barrier to effective learning of mathematics.

Detailed lesson notes: Comprehensive lesson notes, including an introduction activity and main activity.

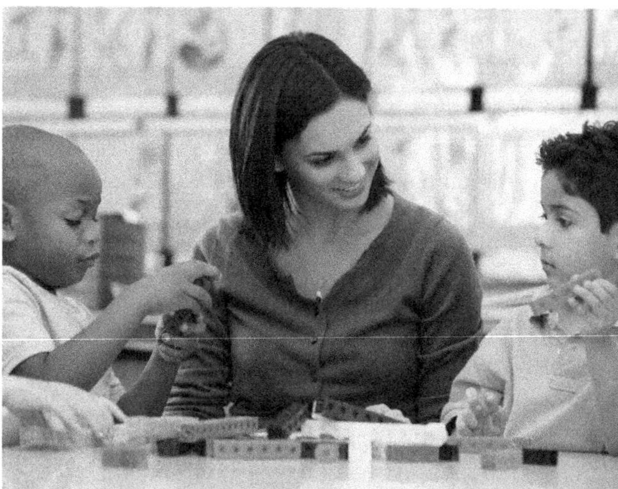

These notes refer to the Student Book and Practice Book, where relevant. The notes include probing questions for formative assessment, which are italicised. Icons are used to suggest the groupings that should be used at each point of the activity (whole class, small group, pairs, individual). A separate 'star-jump' icon indicates that the activities give students an opportunity for physical movement (standing up, jumping, moving around) rather than doing activities sitting down.

Differentiation: The Teacher's Guide offers strategies for you to *support* those students who may have difficulty accessing the task; to *consolidate* the learning for those students who need a little more practice; and to *extend* the learning for those who need more challenge.

The Teacher's Guide also offers differentiated outcomes. These outcomes are listed in the form of:

All students

Most students

Some students

Stretch zone: Each activity in both the Student Book and the Practice Book has a stretch zone question to support deeper learning. The Teacher's Guide provides additional notes on these activities.

Reflection time: Suggestions on how to bring the class back together to reflect on the learning and share ideas.

Answers: Answers to all the Student Book and Practice Book activities are provided.

Review pages: The Teacher's Guide provides notes on the Review pages of the Student Book (summative assessment), including answers to the assessment questions, and the Practice Book (a formative, reflective review).

Digital resources: Where it is appropriate to use digital resources in a lesson, such as sharing the interactive Student eBook page on an interactive whiteboard (IWB), suggestions are embedded in the lesson plan.

Resources sheets: these photocopiable resources can be used with some of the main activities. They are referenced in the resources section of the lesson plan and are available from the *Oxford International Primary Maths* page on the Oxford Owl website (www.oxfordowl.co.uk).

Tour of a typical unit

The 'Big question' provides a discussion stimulus about the key idea of the unit.

1 Numbers and counting

?

How do we use numbers?

In this unit you will:

- count, read and write numbers to 100
- count in twos, fives and tens
- know and make numbers using objects and pictures
- use words such as equal to, more than, less than (fewer), most, least
- read and write numbers from 1 to 20 in words.

Learning objectives are stated clearly at the beginning of every unit.

Engage

Which numbers can you see in the classroom?

Which numbers can you see on your way to school?

What is the biggest number you have ever seen?

Further questions allow students to develop communication skills.

6

The Engage spread is bright and colourful, with artwork or photos to spark interest in young students and provide discussion points.

Student Book Discover and Explore

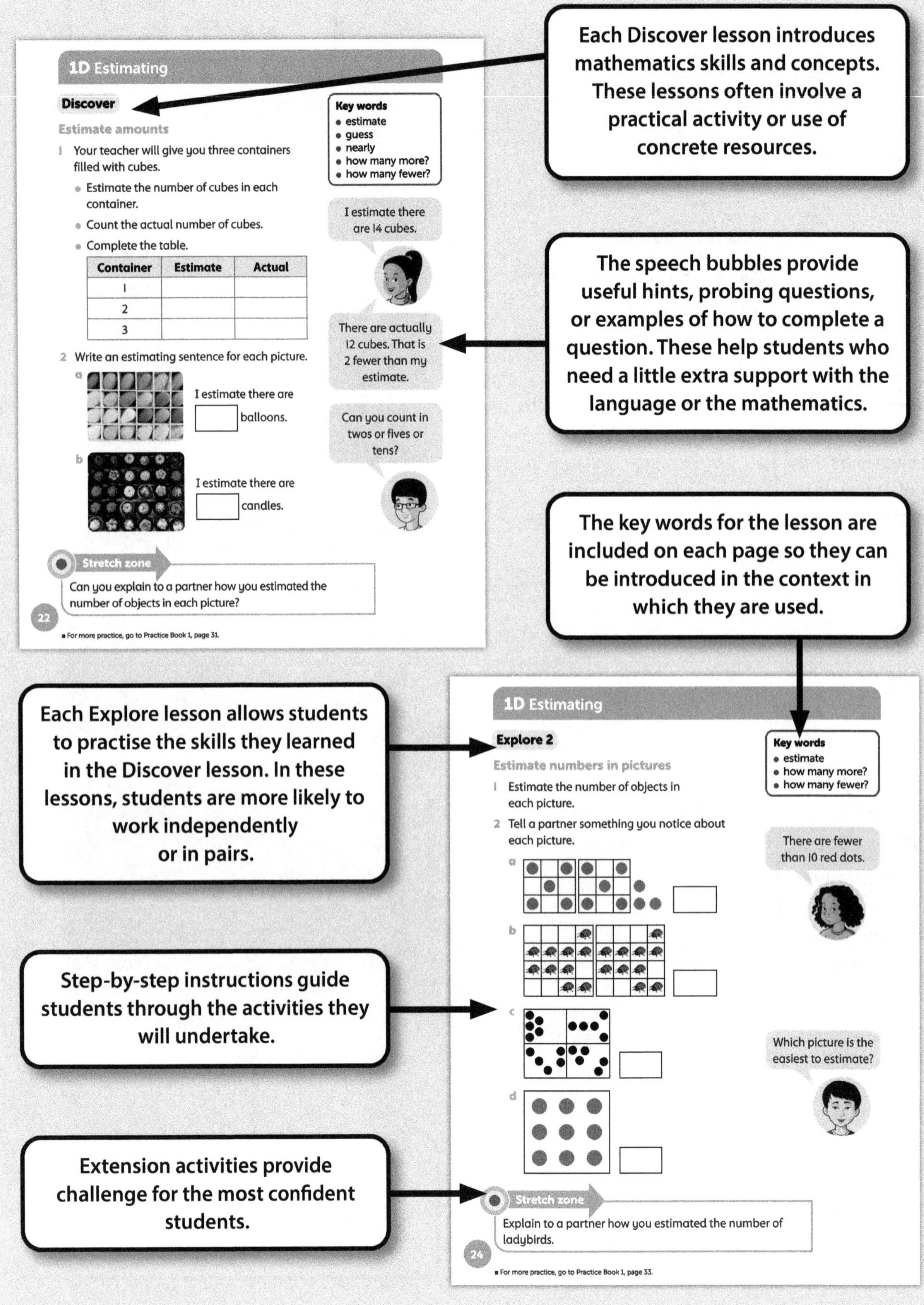

Each Discover lesson introduces mathematics skills and concepts. These lessons often involve a practical activity or use of concrete resources.

The speech bubbles provide useful hints, probing questions, or examples of how to complete a question. These help students who need a little extra support with the language or the mathematics.

The key words for the lesson are included on each page so they can be introduced in the context in which they are used.

Each Explore lesson allows students to practise the skills they learned in the Discover lesson. In these lessons, students are more likely to work independently or in pairs.

Step-by-step instructions guide students through the activities they will undertake.

Extension activities provide challenge for the most confident students.

The Connect lesson makes [links] between the [diff]erent areas of [ma]thematics in the unit.

1 Numbers and counting

Connect

Make a number poster

Work as a group.

1 Collect some magazines. Talk about which magazines might have numbers in them. What do the numbers tell us?

We use numbers to count or to say how many of something there are.

The 'Big idea' sums up what students have discovered in the unit. It answers the Big question on the Engage page.

[Con]nect activities [ar]e often set in [rea]l-life contexts [t]o make the [li]nk between [mat]hematics and [th]e real world.

2 Cut out pictures that have numbers.

What is the biggest number on your poster?

What is the smallest number on your poster?

3 Make a poster to display in class.

4 Talk in your group about the numbers you have found.

Stretch zone

Take photographs of numbers on the way home from school. What job are the numbers doing? Explain your ideas to a partner.

A further extension activity provides a challenge for the most confident students.

1 Numbers and counting

Review

1 Draw the beads and write the numbers in the spaces.

Beads	Numbers	Words
	5	
		sixteen
		three
	12	
		nineteen
	1	
		four
	14	
		twenty

2 Samir has a bracelet with 19 beads. Lina's bracelet has one more bead than Samir's. How many beads are on Lina's bracelet?

Celine's bracelet has 10 more beads than Lina's. How many beads are on Celine's bracelet?

26

Practice Book Discover and Explore

Practice Book activities can be completed in the school lesson or as homework.

Most of the Discover and Explore lessons in the Student Book have a corresponding page in the Practice Book. These activities provide opportunities for students to consolidate and deepen their learning.

Step-by-step instructions guide students through the activities they will undertake.

If students require concrete resources, these are listed in a box at the top of the page. This is particularly useful if students are completing the activities at home.

Extension activities provide challenge for the most confident students.

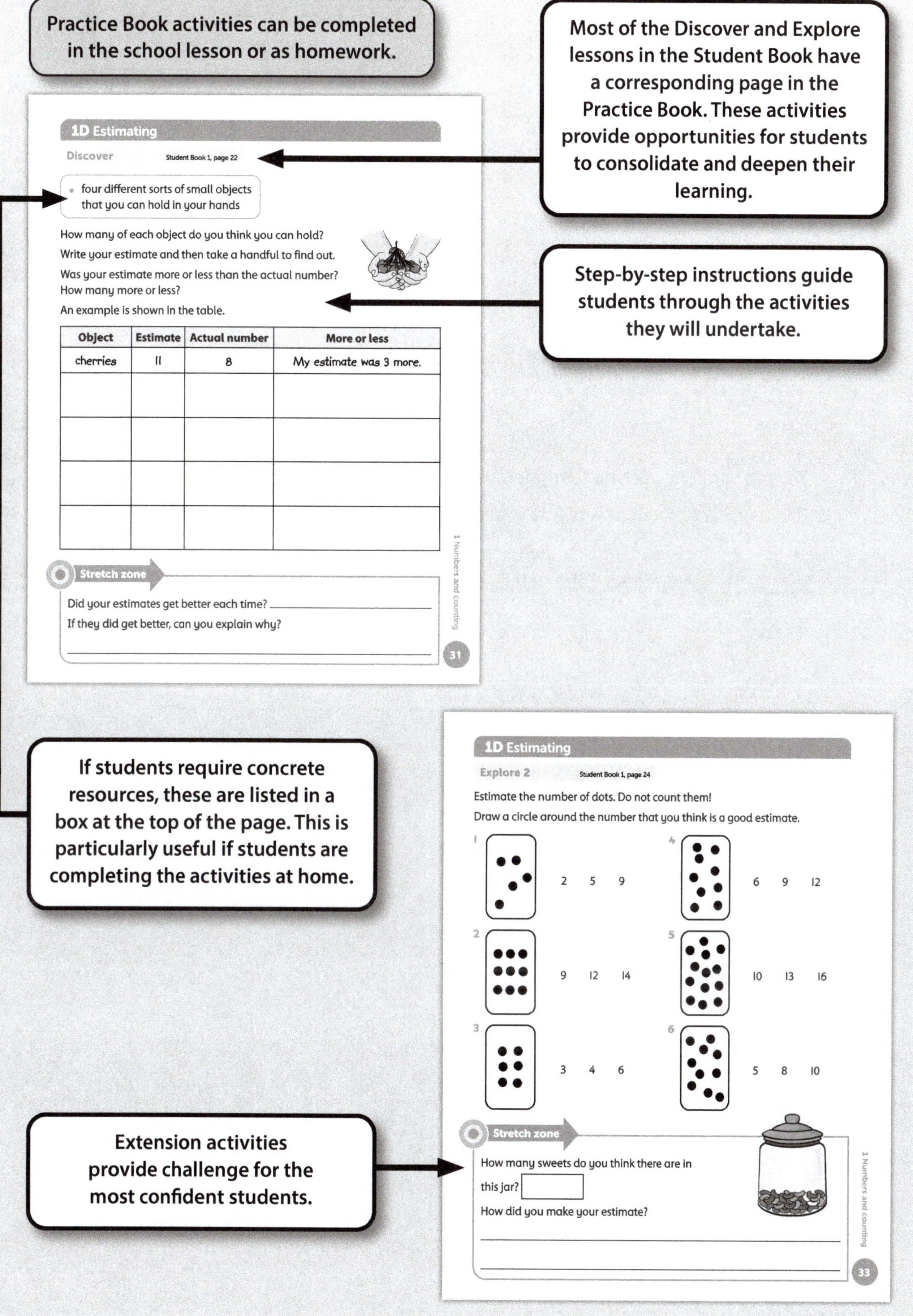

Each Review page in the Practice Book includes a reminder of all the topics learned in the unit.

1 Numbers and counting

Review

 1 Draw a face next to each bubble to show how you feel about your learning.

counting objects

reading and writing numbers

counting in twos, fives and tens

estimating quantities

 2 Tell a partner about one thing you did really well in this unit.

3 Draw or write about things you found easy, challenging or really hard.

What work did you feel confident doing?

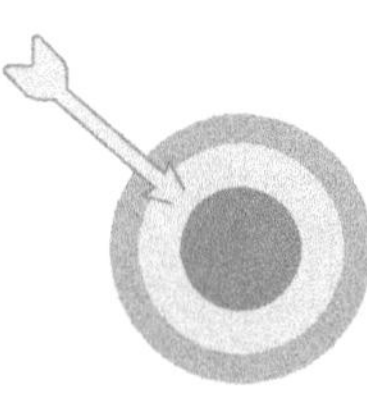

What work was challenging?

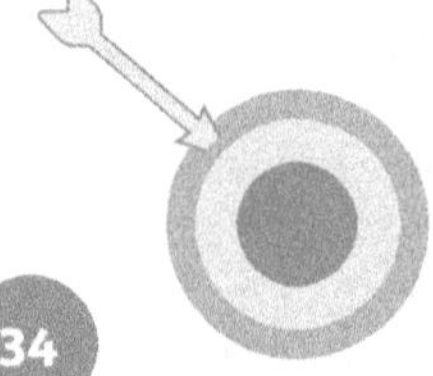

Is there any work you might need some extra help with?

34

Self-assessment activities help students to reflect on their learning.

The Student Books

The Student Books are write-in textbooks for students to read and use. There are six Student Books: one for each school year at primary school. The Student Books introduce learning through a mixture of practical, discussion and independent activities.

Student Book	Typical student age range
Student Book 1	Age 5–6
Student Book 2	Age 6–7
Student Book 3	Age 7–8
Student Book 4	Age 8–9
Student Book 5	Age 9–10
Student Book 6	Age 10–11

  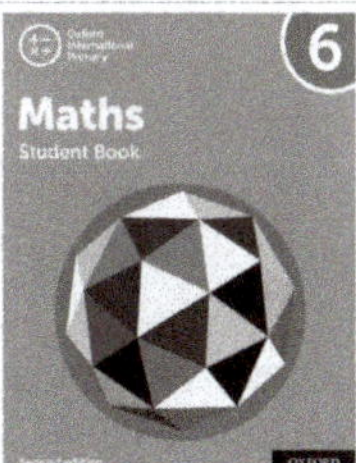

The Practice Books

The Practice Books are write-in workbooks for students to read and use. There are six Practice Books: one for each school year at primary school. The Practice Books provide deeper learning opportunities through a range of independent activities, which can be completed in school or at home.

Practice Book	Typical student age range
Practice Book 1	Age 5–6
Practice Book 2	Age 6–7
Practice Book 3	Age 7–8
Practice Book 4	Age 8–9
Practice Book 5	Age 9–10
Practice Book 6	Age 10–11

  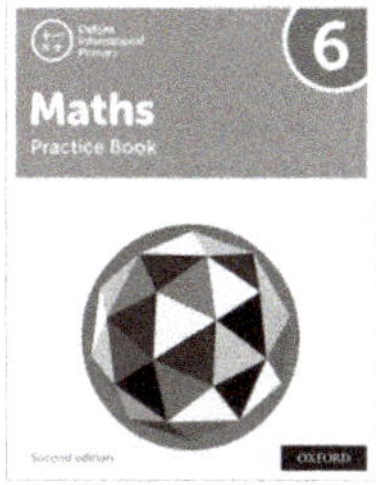

The Teacher's Guides

There are six Teacher's Guides: one for each school year at primary school. Each Teacher's Guide includes:

- An introduction with advice about delivering mathematics in primary schools using *Oxford International Primary Mathematics*.
- A unit overview, giving advice on teaching each unit, including common misconceptions and how to deal with them.
- A lesson plan for every lesson in the Student Book and corresponding pages in the Practice Book.
- Model answers to each question in the Student Book and Practice Book.

Interactive eBooks

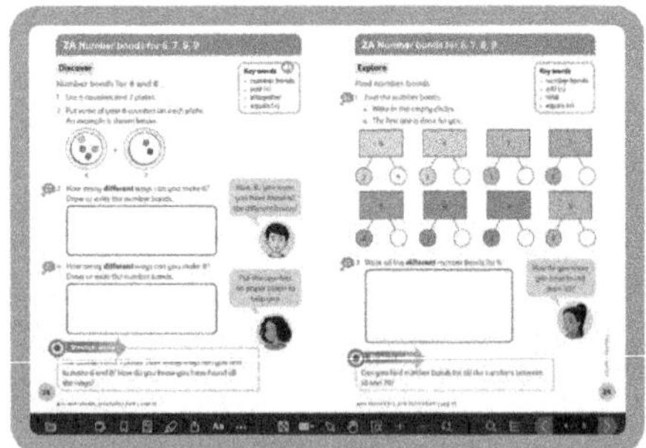

For the teacher

Teachers can access the Student Books, Practice Books and Teacher's Guides online in eBook format, on the Oxford Owl website (www.oxfordowl.co.uk).

The enhanced eBooks show the course content on screen, making it easier for teachers to deliver engaging lessons.

For the students

Teachers can allocate an eBook version of the Students Books to the students for use at home. The Student eBooks include interactive activities, worksheets and audio of all the key vocabulary,

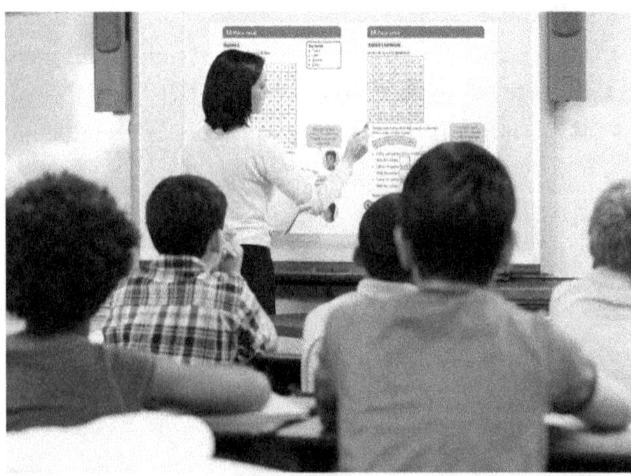

Assessment resources

The downloadable assessment materials offer you additional opportunities to assess students' progress. The materials include:

- end-of-unit summative assessment
- end-of-year summative assessment.

Every test comes with everything you need to assess and record progress including:

- Answers
- Mark schemes and guidance on assessment

Oxford Primary Illustrated Maths Dictionary

The *Oxford Primary Illustrated Maths Dictionary* gives comprehensive coverage of the key maths terminology children use in the course. Entries are in alphabetical order, and each includes a clear and straightforward definition along with a fun and informative colour illustration or diagram to help explain the meaning. The dictionary is suitable for Students with English as an Additional Language.

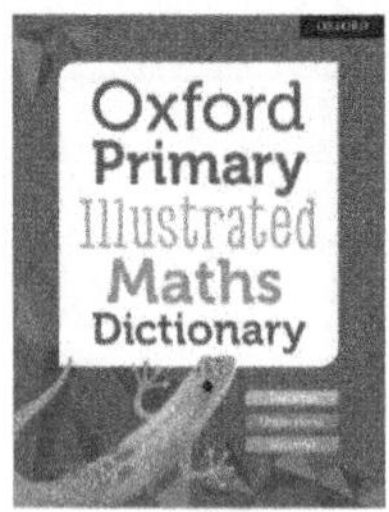

The curriculum

The Oxford International Curriculum offers a new approach to teaching and learning focused on wellbeing, which places joy at the heart of the curriculum and develops the global skills students need for their future academic, personal and career success.

Through six subjects – English, Maths, Science, Computing, Wellbeing and Global Skills Projects – the Oxford International Curriculum offers a coherent and holistic approach to ensure continuity and progression across every student's educational journey, equipping them with the skills to shape their own future. Through this approach, we can help your students discover the joy of learning and develop the global skills they need to thrive in a changing world.

1 Numbers and counting

Overview

Big idea

Number sense refers to a person's general understanding of number and operations. In order to develop a good sense of numbers, there are three key elements, which need to be mastered:

- Place value – an understanding of numbers expressed in a base system, usually base 10, and an ability to think of them in more than one way.
- Concepts that involve understanding the operations ($+$, $-$, $\times$ and $\div$), what each of the operations are and what they do to the numbers, and an ability to write and interpret symbolic expressions, using the operation symbols, e.g. $5 + 9$ or $5 - 3$.
- Knowing number facts (mental strategies) using addition and subtraction to at least 20.

Use these strategies to enable students to develop a good sense of number:

- Speaking: encourage students to count aloud, explain their solutions and use correct mathematical language.
- Listening: listen for patterns when counting to and from larger numbers, for example skip counting (e.g. leaving out all even numbers when counting).
- Reading: read numbers in different contexts and estimate how many without counting.
- Writing: begin with informal, leading to more formal writing of numbers as numerals and in words.

Look out for

- **Students who are confused if the first number is called 'zero', the second 'one', the third 'two' and so on.** Show numbers on a number line that has illustrations showing the value of the number. Zero will have none, 1 has one and so on. Students then have a visual model of each quantity increasing by 1. Don't say to students that 1 is the first number.
- **Students who are confused between a number track used for ordering numbers and a number line used for calculating.** A number track has the number in a space. A number line has the number at its own unique place on the line.
- **Students who do not know that the last number in a count is the total.** Count using practical materials such as counters or cubes. When the count has finished, ask *How many altogether?*
- **Students who have difficulty with one-to-one matching,** e.g. saying '1, 2, 3, 4, 5' when there are only three objects present. As objects are counted, drop them in a bowl or bucket so that they cannot be seen. Slow the counting down to match the objects.

Possible misconceptions

- **When counting along a number line or track, students may count the number/space that they are already on.** Use a floor number line that involves students counting along the line by making jumps backwards and forwards.

Key vocabulary

- number, zero, one, two, three, …, twenty, count
- ones, tens, cubes, tens-rods, total
- more, less, fewer, more than, less than, fewer than, most, fewest, smallest, largest, 1/2/5/10 more, 1/2/5/10 less
- count on, count back, order, steps, multiple
- estimate, guess, actual, predict, pattern

Coverage in lessons

Learning focus	Learning outcomes (the ENC objectives)
Counting objects	Count to and across 100, forwards and backwards, beginning with 0 or 1, or from any given number.
Reading and writing numbers	Count, read and write numbers to 100 in numerals.
Counting on and back	Given a number, identify 1 more and 1 less. Count in multiples of 2s and 10s.
Estimating	Identify and represent numbers using objects and pictorial representations including the number line, and use the language of: equal to, more than, less than (fewer), most, least.

1 Numbers and counting

Engage Student Book page 6

Big question

- How do we use numbers?

Global skills

- **Creative skills:** exploring
- **Real-world skills:** interpreting information
- **Interpersonal skills:** communication / teamwork
- **Self-development skills:** reflecting on learning

Key vocabulary

- number, zero, one, two, three, …, twenty

Resources

- a range of pictures that contain numbers (see Student Book page 1 for ideas)
- number cards 1–20
- mini whiteboards and pens

Language support

Check that students can use the correct number names. Repeat number names to model the correct pronunciation, for example, emphasising the difference between teens and multiples of ten, e.g. four, fourteen and forty. Carry out repeated activities that involve counting out loud as a group.

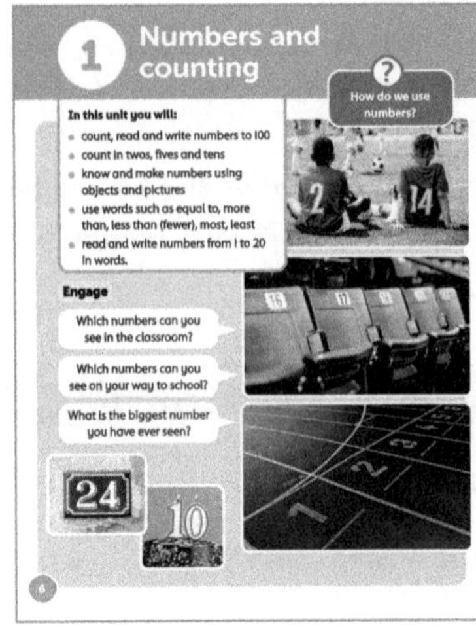

Introductory activity

If you have access to an interactive whiteboard (IWB), you could display the Student Book Engage page and discuss the Big idea of the unit with students. Ask students to talk about the pictures in the Student Book and to relate them to any experiences of numbers they have had at home, at school or when out with the family.

Main activity

Ask students to share some of their ideas with the rest of the class. To encourage discussion use questions such as: *What numbers do you see when you are at home? What do you see when you go out? What is the same? What is different?*

Choose one of the pictures to discuss. *Where do you think this is? Have you seen numbers like these?*

Choose a picture. Describe it to a partner. Can they find the picture you chose?

Encourage specific mathematical language and vocabulary during discussions, e.g. rather than 'I can see a door' students should say 'I can see a door with a number 24'. Support the vocabulary with pictures, diagrams or whole-class participation in a physical activity to demonstrate the words/numbers (e.g. *hold up four fingers, point to two eyes, show me two feet*).

Ask students which numbers they can count up to. Model counting up to 100 with the whole class and counting backwards from 20.

Differentiation

Supporting: Some students may need support by counting with you – place their finger on each object as you count. Ask them to repeat the numbers after you say them. Remind students of the number names for single-digit numbers.

Consolidating: Allow students to count on their own, only counting with them if they forget numbers or miscount. If they make an error in counting, check with them by counting together. Remind students of the number names for 2-digit numbers.

Extending: Ask students to count forwards and backwards starting at 10 and 20.

Stretch zone: *Clapping maths:* Place a set of number cards 1–20 face down. Ask students to take turns to turn over and clap the number they have. Other students count in their heads and write the number on a whiteboard.

 Reflection time

Give pairs a number picture that you have brought into the class. Use the think/pair/share technique. That is, each student thinks quietly for one minute, shares their ideas with a talk partner for two minutes and then some pairs share their discussion with the whole class.

1A Counting objects

Discover
Student Book page 7 • Practice Book page 16

Specific learning focus
- Count up to 20 objects, understanding that the total number of objects does not change when the objects are rearranged.

Global skills
- **Creative skills:** problem solving / exploring
- **Real-world skills:** presenting information
- **Interpersonal skills:** communication / teamwork

Key vocabulary
- count, more, less, fewer

Resources
- materials for counting: cubes, coins, buttons, beads
- tray, cloth, small bucket, small open boxes, ten coins

Language support
Work with individuals during the activity to model the correct pronunciation of the numbers. Say the number first and ask the student to repeat the number.

 **Introductory activity**

Put a number of different objects under a cloth on a tray.

Ask a student to feel under the cloth and guess how many objects there are. Remove the cloth and **count** together as a class by picking up each object as you say its number name and dropping it into a small bucket.

Put a different number of objects under the cloth and choose a different student to repeat the activity. Repeat this activity five times. After each count, ask the group: *Were there **more** or **fewer** objects that time?* For the last two repeats of the activity make the objects all the same.

 **Main activity**

Give each student a small box. Ask them to collect a set of their favourite counting things to put in it. Ask them to count how many they have. Then ask them to take some of the items out. *How many do you have on your table? How many do you still have in your box? If you put them all back in the box again, how many will you have?*

Ask students to replace some of the items again Repeat the questions above. Finally, replace all the items and ask them to count again. *Are there the same number now as there were at the beginning?*

Count ten coins as you drop them into a bucket. Ask how many there are to begin with, then how many there are

if you take one out and put it on the table. Always refer back to the total number of coins in the bucket and on the table (always ten).

When students have completed this practical activity, they can complete the activities on page 7 in the Student Book. Put students in pairs and give them a selection of three different colours of cubes and beads (10 of each: ideally, blue, red and yellow to match the Student Book). Model how to take a handful and sort the cubes by colour, reviewing each colour name. Each student should count and sort their cubes out loud for their partner to check.

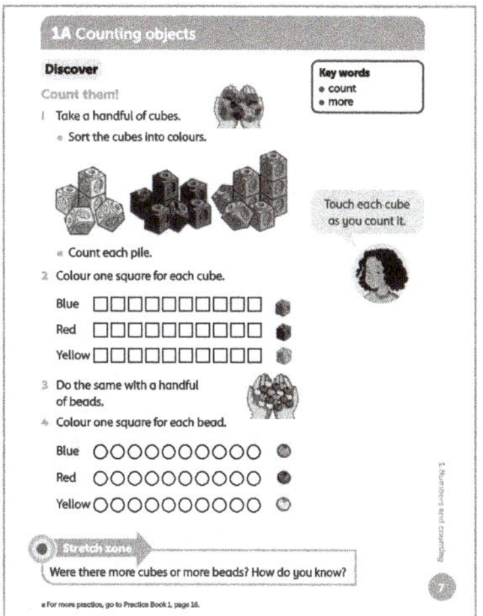

Differentiation
Supporting: Ask students to count with you – place their finger on each object as you count.

Consolidating: Allow students to count on their own, only counting with them if they forget numbers or miscount. If they make an error in counting, check with them by counting together.

Extending: Ask students to count how many more there are in one set by comparing two sets. Line the sets up with objects next to each other so that students can see the difference.

Stretch zone: Ask students to compare the numbers of cubes and beads from the Student Book activity. How many more ...? Ask students to explain how they know if there were more cubes or more beads.

 **Reflection time**

Put a row of buttons on the table. Ask the class to count with you as you point to each button in turn. Ask a student to cover each button with a cube. *We had [say the number] buttons. How many cubes do you think we have?* (same number)

Repeat several times, changing the number of cubes and buttons, always asking questions such as: *How many cubes are there now? How many buttons? Are there more buttons than cubes?*

Practice Book: Students can complete Practice Book page 16. This can be done directly after the main activity, as homework, or as the focus of a separate mathematics session to help students consolidate their learning and build fluency.

<table>
<tr><td colspan="2">Differentiated outcomes</td></tr>
<tr><td>All students</td><td>should be able to count to 20 accurately.</td></tr>
<tr><td>Most students</td><td>will say whether there are more or less in one group of objects compared to another.</td></tr>
<tr><td>Some students</td><td>may say how many more or less there are in one group of objects compared to another.</td></tr>
</table>

Answers

Student Book page 7

Students will have different numbers of coloured cubes and beads. Check that students have coloured the correct numbers of squares for each colour.

Practice Book page 16

For each number, check that students have coloured in the correct number of squares and numbered them.

1 3 squares coloured, with numbers 1–3 written correctly beneath the squares.

2 4 squares coloured, with numbers 1–4 written correctly beneath the squares.

3 1 square coloured, with number 1 written correctly beneath the square.

4 8 squares coloured, with numbers 1–8 written correctly beneath the square.

5 4 squares coloured, with numbers 1–4 written correctly beneath the squares.

Stretch zone: 3

1A Counting objects

Explore 1   Student Book page 8 · Practice Book page 17

Specific learning focus

- Count up to 20, understanding conservation of number.

Global skills

- **Creative skills:** exploring / investigating
- **Real-world skills:** presenting information
- **Interpersonal skills:** communication

Key vocabulary

- count, most, fewest, number

Resources

- 0–20 number line, marbles, counters, beads, coins, small bowls

Language support

For individuals who need to work on the vocabulary of counting, count for them and ask them to repeat the number names after you have said them. Learners developing their vocabulary should work with more confident language learners so they can hear peer modelling of the vocabulary.

 Introductory activity

Count up to 10 along a number line at the front of the class. As you count, miss a number. *Which **number** was missing? How do you know?* Repeat the activity several times missing out a different number.

Main activity

Tell this story, using marbles to act out the story: *Max likes to play marbles. Every day Max puts his marbles into his pocket.* Count out five marbles.

Sometimes he puts them in this pocket (demonstrate) *and sometimes he puts them in this pocket* (choose your other pocket). *I wonder how many different ways he can put five marbles in his pockets?*

Draw on the board the two pockets and the five marbles.

Show all the marbles in one pocket and then all in the other pocket. *What if he puts one marble in this pocket? How many will he have to put in the other pocket?* Count out the five marbles. Put one of them in one pocket and count out the rest. *One is in this pocket and so 1...2...3...4 have to go in this pocket. Does he still have five marbles?* (yes) Count them all together: *1...2...3...4...5.*

Ask students to complete page 8 in the Student Book individually. Give out bowls holding a number (between 20 and 30) of 'coins'. Explain that the bowls are their pots and, if using counters or marbles, that these are their 'coins'. Ask individual students as they work, *Which bag has the **fewest** coins so far? Which has the **most**?*

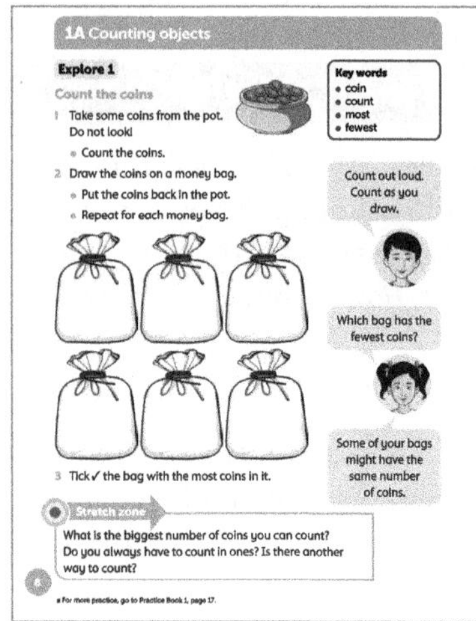

Differentiation

Supporting: Some students may need support in counting. Touch each marble or counter as you count it and say the number name. Ask them to repeat your actions.

Consolidating: Ask students to count forwards and backwards from 1 to 20 without the use of concrete objects to support them.

Extending: Encourage students to count beyond 20.

Stretch zone: Ask children what is the biggest number they know. Encourage them to count as far as they can, counting in larger steps, for example, tens or hundreds, if they know how.

 **Reflection time**

Bring the whole class back together.

What if Max has ten marbles?

He puts two in this pocket and eight in that pocket. How many will he have altogether? (ten). Repeat with other pairs of numbers, always with a total of ten.

Practice Book: Students can complete page 17 of the Practice Book. This can be done directly after the main activity, as homework, or as the focus of a separate mathematics session to help students consolidate their learning and build fluency. Support students to read each number aloud and count out the corresponding number of counters (or another concrete resource). They can then draw this number of counters.

Differentiated outcomes	
All students	should count to 20 saying the number names with support.
Most students	will count to 20 confidently knowing that the total is given by the last number in the count.
Some students	may count beyond 20.

Answers

Student Book page 8

You should try to mark this activity as you walk around the classroom. As you look at students' books, ask students to tell you how many coins they have drawn in each bag. Some students can write the numbers next to the bag. You can write the correct numbers for those students who are not yet confident in writing numbers.

Practice Book page 17

1 Check that 7 objects have been drawn.

2 Check that 13 objects have been drawn.

3 Check that 14 objects have been drawn.

4 Check that 10 objects have been drawn.

5 Check that 16 objects have been drawn.

6 Check that 8 objects have been drawn.

Stretch zone: Students should have drawn a tick next to the set of 16.

Students should have drawn a cross next to the set of 7.

There are 9 more counters in the box with the most than in the box with the fewest. (If students include the worked example, they may tick that box, with 3 counters, as the 'fewest' and give this final answer as 13, which is also acceptable.)

1A Counting objects

Explore 2 Student Book page 9 • Practice Book page 18

Specific learning focus

- Count two sets of objects and find the total.

Global skills

- **Creative skills:** problem solving / exploring / investigating
- **Real-world skills:** presenting information
- **Interpersonal skills:** communication / teamwork

Key vocabulary

- number, zero, one, two, three, …, twenty, more than, fewer than, altogether, total, largest, smallest

Resources

- cubes, counters, beads or similar

Language support

For students who need to work on the vocabulary of counting, count for them and ask them to repeat the number names after you have said them. Learners developing their vocabulary should work with more confident language learners so they can hear peer modelling of the vocabulary.

 **Introductory activity**

Ask a student to come forward and pick up one handful of cubes and place them in a pile. Then ask another student to come forward and pick up a handful of cubes and place them beside the first pile.

Help students count the number of cubes in each pile by moving each cube away from the pile, counting as each one is moved, until they have all moved to a new pile. Ask them which pile has more cubes. Now put the piles together and count the **total**.

Main activity

Students work in pairs to complete page 9 of the Student Book. Ask them to take turns to pick up a handful of objects (cubes, counters, beads, etc.). When they have each taken a handful, ask them to count how many they each picked up. Then ask them to count to find the total number (how many **altogether**). As students are working, watch them as they count and listen to how they are saying the numbers. Help them count correctly if you hear them making counting errors. Using the example at the end of question 2, show them how to record each handful of cubes and then the total number of both handfuls of cubes in a number sentence. Addition will be introduced formally in Unit 2 so do not spend too much time making this link.

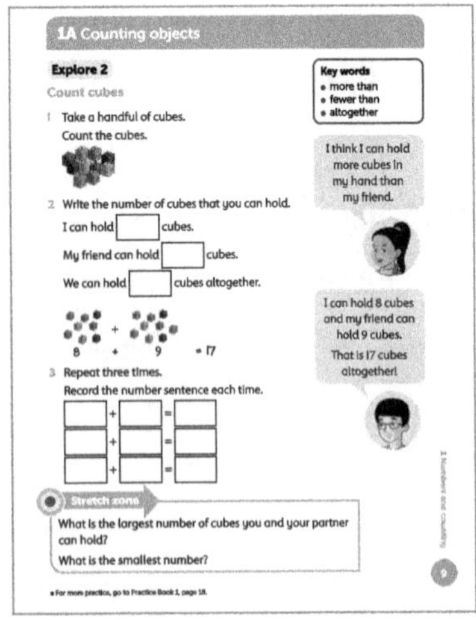

Differentiation

Supporting: Some students may need support in counting. Touch each cube as you count it and say the number name. Ask students to repeat your actions.

Consolidating: Ask students if they are sure that they have counted the total number of cubes from both piles correctly. Ask them how they know and how they can check.

Extending: Use larger numbers of cubes – three or four handfuls.

Stretch zone: *What is the **largest** number of cubes you and your partner can hold? What is the **smallest** number?*

Ask students to think about how they can count the cubes in their handfuls. Discuss why the number might change each time they pick some cubes up. Do they count ones that are dropped? What is the smallest number they can pick up? Why?

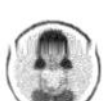

Reflection time

Ask each pair of students to describe how they counted their piles of objects. Ask a series of questions to find out who has picked up the most cubes for their pile. *Who has picked up **more than** 5 cubes? More than 6? More than 7?* Continue until you establish who picked up the most. Ask which pair had the biggest total of objects by using similar questioning. *Which pair picked up more than 10 cubes in total? More than 11?* and so on. Discuss what they did when they were counting to avoid losing their place.

Practice Book: Students can complete page 18 of the Practice Book. This can be done directly after the main activity, as homework, or as the focus of a separate mathematics session to help students consolidate their learning and build fluency. Support students to read each number aloud and count out the corresponding number of beads (or another concrete resource). They can then draw this number of beads on each 'string'.

Differentiated outcomes	
All students	should pick up a handful of objects and count them.
Most students	will count to find the total in two piles.
Some students	may count how many more in one of the piles.

Answers

Student Book page 9

Observe as students count their handfuls of cubes and help them count correctly where they are making mistakes. Try to mark this activity as you walk around the classroom. As you look at their books, ask students to tell you how many cubes they held. Check that they have written the correct numbers in the addition sentences. Use the example in question 2 to remind students how to record the total from two piles of cubes using the '+' and '=' signs. You can write the correct numbers for those students who are not yet confident in writing numbers.

Practice Book page 18

1 Check that 12 beads have been drawn.

2 Check that 7 beads have been drawn.

3 Check that 18 beads have been drawn.

4 Check that 13 beads have been drawn.

5 Check that 4 beads have been drawn.

6 Check that 15 beads have been drawn.

7 Check that 11 beads have been drawn

Students should have drawn a tick next to the string of 18 beads.

Stretch zone: Check that there are 3 more red beads than blue beads.

1B Reading and writing numbers

Discover 1
Student Book page 10 · Practice Book page 19

Specific learning focus
- Read and write numerals from 0 to 20.

Global skills
- **Creative skills:** problem solving / investigating
- **Real-world skills:** presenting information
- **Interpersonal skills:** communication

Key vocabulary
- number, zero, one, two, three, …, twenty, 1 less, 1 more

Resources
- display of pictures showing different uses of numerals: buses with numbers on, house numbers, telephone numbers, car number plates and so on. You can use the pictures prepared for the 'Engage' activity.
- number cards 1–20, wooden or plastic numerals, bag, counters
- number word cards for 1–20, e.g. one, ten
- mini whiteboards and pens (enough for one between two students)

Language support
Count small amounts (less than 5) as a class and increase by one or two each time. Use an illustrated number line to make a connection between the words that you say and the numbers on the line. Display a list of the number words from 0 to 20 for students when they complete the activities in the Student Book.

 ## Introductory activity

Choose two students to come to the front of the class. Ask one student to select a number card. They write this number with their finger on the back of the other student. Ask the second student to try to guess the number. Repeat several times using a range of students.

 ## Main activity

Put wooden or plastic numerals in a bag and ask a student to come to the front of the class to identify them by touch. The student should not say the number aloud, but should draw it in the air. The student needs to have their back to the class, otherwise the drawing will be reversed. Ask the other students to guess what the number is. After a few turns, begin to say after each number, e.g. *This number was 10. One less than 10 is 9. One more than 10 is 11.* Ask students to also say the number that is **1 less** and **1 more** than the number that was drawn.

Repeat the activity with a different student each time. Continue until the bag is empty.

Ask students to complete the activity on page 10 of the Student Book individually with sets of number cards and number word cards 1–20.

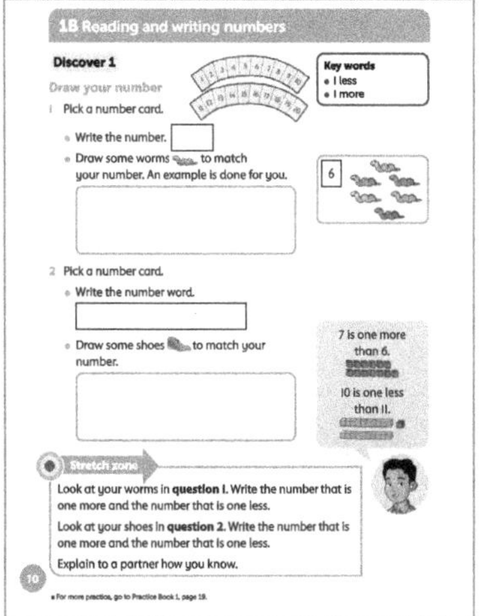

Differentiation
Supporting: Some students will need to use number word cards to help with the last question in the Student Book.

Consolidating: Ask students to use the 2-digit numbers in the Student Book.

Extending: Some students may be confident to use numbers up to 20.

Stretch zone: *Look at your worms in question 1. Write the number that is one more and the number that is one less.*

Look at your shoes in question 2. Write the number that is one more and the number that is one less. Explain to a partner how you know.

Ask students to identify the numbers one more and one less than those on the cards they pick for the Student Book activity. They may need to use a number line to help with this. Encourage them to explain to a partner how they found the numbers that are one more and one less than the number on the card.

 ## Reflection time

Bring the whole class together. Arrange students in pairs so that there is a student who can write the numbers 1–20 in each pair. Give each pair a mini whiteboard and pen.

- *Show me a number that is more than 7.*
- *Show me a number that is more than 7 and less than 10.* (8 or 9)
- *Show me number that is next to 15.* (16 or 14)
- *Show me a number that is more than 16 and less than 18.* (17)

Select students to ask similar questions for the rest of the class to answer.

Practice Book: Students can complete Practice Book page 19. This can be done directly after the main activity, as homework, or as the focus of a separate mathematics session to help students consolidate their learning and build fluency. Students should continue to work with number word cards and concrete resources to support counting.

Differentiated outcomes	
All students	should identify the numbers up to 10.
Most students	will identify the numbers up to 20.
Some students	may be confident with counting up to 20.

1B Reading and writing numbers

Discover 2 Student Book page 11 · Practice Book page 20

Specific learning focus
- Read and write numerals from 0 to 100.

Global skills
- **Creative skills:** problem solving
- **Real-world skills:** interpreting information
- **Interpersonal skills:** communication

Key vocabulary
- number, zero, one, two, three, …, twenty, ones, tens, cubes, tens-rods

Resources
- base-ten equipment: tens-rods and ones-cubes

Language support
Help students to use the correct words for the cubes (ones) and the tens-rods (tens). Encourage them to say the tens and then the ones when they are counting.

Answers

Student Book page 10
1 Check that students have drawn the correct number of worms to match the number they have written.

2 Check that students have drawn the correct number of shoes to match the number they have written.

Practice Book page 19
Check that students have drawn the correct number of objects for each question and have written the correct word for each number.

Stretch zone: 6. Check that students have drawn 6 cats.

 Introductory activity

Show students an arrangement of a **ten-rod** and ten 'ones'-**cubes**. Count together to see how many 'ones' there are, then count the parts along the ten-rod to see that they have the same number.

 Main activity

Show students a collection of 2 tens-rods and 5 cubes. Explain that you are going to count how much the total is, remembering that the rod is the same value as ten cubes. Point to a ten-rod and ask how many cubes this represents. Students should recall it is 10. Help them count the parts on the rod to check. Now ask the value of 2 tens-rods together (20). Explain to students that 'twenty' is the number we use for 2 tens-rods. Count cubes to get from 20 to 25, e.g. 20, 21, 22, 23, 24, 25. Show students how this is written and point out that the 2 tens-rods show how many **tens** (2) and the 5 cubes show how many **ones** (5).

Repeat this using 1 ten-rod and 7 cubes, and again with 2 tens-rods and 8 cubes to count 17 and 28 respectively. Emphasise the link between the numbers of tens-rods and cubes, the number of tens and ones, and the digits in the written numbers.

Ask students to complete the activity on page 11 of the Student Book in pairs. Provide each pair with tens-rods and cubes so that they can physically count each rod and cube. Support pairs to count and record the numbers throughout the activity.

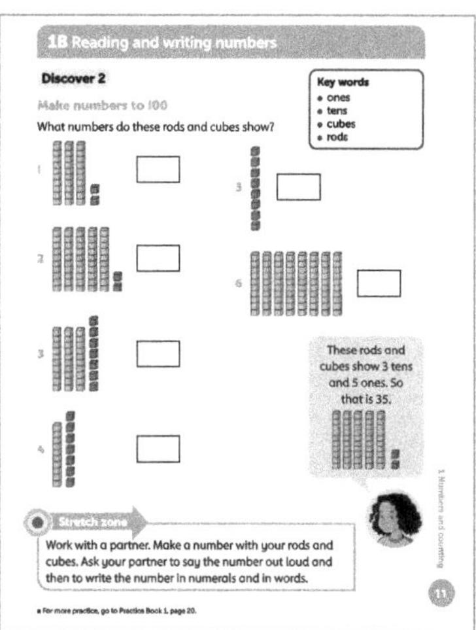

Differentiation

Supporting: Some students will need help with recalling that the ten-rod has the same value as 10 cubes.

Consolidating: Ask students to use the 2-digit numbers in the Student Book.

Extending: Ask students to represent numbers up to 100.

Stretch zone: *Work with a partner. Make a number with your tens-rods and cubes. Ask your partner to say the number out loud and then to write the number in numerals and in words.*

Working in pairs, students should be able to count the number represented by the tens-rods and cubes held by their partner. Some students may know the number words up to 100 and can make larger numbers with confidence.

 **Reflection time**

Ask students to explain to a partner how they counted the tens-rods and cubes to make their number. Did they count correctly? Did they remember to count each rod as ten? Were they able to write down the correct number in digits to represent the tens-rods and cubes?

If students made errors with counting, show an example, e.g. 3 tens-rods and 5 cubes. Ask the class to count with you as you touch each rod and then each cube in turn to get to the correct number. If students wrote the digits incorrectly, show them again how the tens and ones digits represent the tens-rods and cubes.

Practice Book: Students can complete page 20 of the Practice Book. This can be done directly after the main activity, as homework, or as the focus of a separate mathematics session to help students consolidate their learning and build fluency.

Differentiated outcomes	
All students	should count cubes as ones.
Most students	will count cubes as ones and tens-rods as tens.
Some students	may count tens-rods and cubes and write the number.

Answers

Student Book page 11

1 32

2 52

3 37

4 17

5 8

6 80

Practice Book page 20

Check that students have drawn the correct number of tens-rods and cubes for each number.

1 31   3 tens-rods, 1 cube

2 36   3 tens-rods, 6 cubes

3 26   2 tens-rods, 6 cubes

4 25   2 tens-rods, 5 cubes

5 43   4 tens-rods 3 cubes

6 47   4 tens-rods, 7 cubes

Stretch zone: Check that the numbers and drawings match for the stretch zone question.

1B Reading and writing numbers

Explore 1   Student Book page 12 · Practice Book page 21

Specific learning focus

- Read and write numerals to 20.

Global skills

- **Creative skills:** problem solving
- **Real-world skills:** interpreting information
- **Interpersonal skills:** communication

Key vocabulary

- number, zero, one, two, three, …, twenty, count on, count back, order, fewer, fewest, more, most

Resources

- 0–20 number line, small plant pots and seeds (or a realistic substitute), sticky labels for numbering

Language support

Help students to recite the number names correctly up to twenty. Check that they are matching the number names with the numbers on the number line as you point to them.

 **Introductory activity**

Count together as a class from 0 to 20. Use a number line to help. Then start from a number other than 0 and **count on** to 20 from that number. Choose a different number and repeat.

Using the number line to help, **count back** from 20 to 0.

 **Main activity**

Explain that you are going to plant a different number of seeds in several pots and see how many seeds grow.

Plant one seed in the first pot, two in the second, three in the third and then stop. Ask students, *How can I remember how many seeds I planted in each pot?* Encourage students to suggest labelling each pot with the number of seeds planted in it and to carry on planting up to ten pots.

Ask students to help you label each of the pots and plant the right number of seeds in them.

Mix the pots up and ask students to put them in **order**, starting with the smallest number of seeds. *Which pot has the fewest seeds in it? Which pot has the most seeds? Which pot will be first? Which pot will be last?*

Ask students to complete the activity on page 12 of the Student Book individually. Ask them to mentally count along the number line to 20. You may also want to ask them to do this aloud with a partner to check the accuracy of their counting. Point to the line drawn from the first piece of the jigsaw to the frame below and explain that the first one is done for them. Encourage them to focus on the last number of the first piece to help them 'build up' the jigsaw in order.

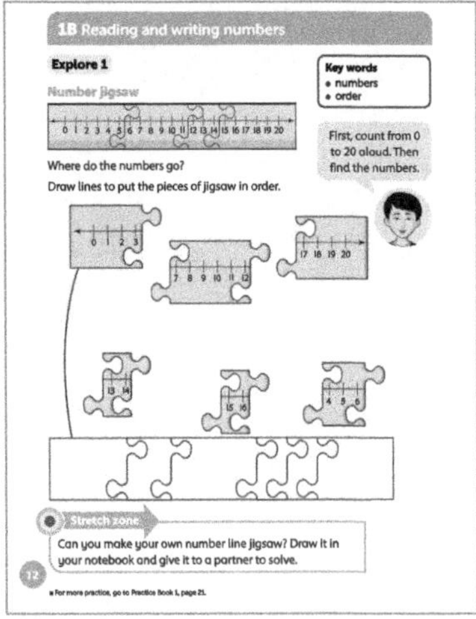

Differentiation

Supporting: Some students will need help with keeping their place as they count by placing their finger on the starting number and moving it one place at a time along the number line until they reach the answer.

Consolidating: Encourage students to count on from different numbers.

Extending: Ask students to try to count backwards from 20 without the number line.

Stretch zone: *Can you make your own number-line jigsaw? Draw it in your notebook and give it to a partner to solve.*

Check that students have drawn a number line correctly and then made it into a puzzle for a partner to solve by drawing out the number line and then cutting it into several pieces using cuts other than straight lines.

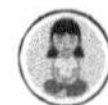 **Reflection time**

Ask students to explain to a partner how they were able to order the plant pots. *How did they know which pot came first? How did they decide for each pot? Did they move the pots around and check the numbers were in order? Did they use 'before' or 'after' as they were sorting?*

Practice Book: Students can now complete the activities on page 21 of the Practice Book. This can be done directly after the main activity, as homework, or as the focus of a separate mathematics session to help students consolidate their learning and build fluency. Model the first question for them, read each instruction aloud. *I start at 3. This means I need to find 3 on the number line. I add 2. This means I count on 2 from 3. 3 … 4 … 5. Mark 5 on the number line. I write '5' in the box here.* Point to the box. Repeat for question 1, asking students to take you through the steps to finding the answer.

Differentiated outcomes	
All students	should say the number names to 20 and match them to a numeral
Most students	will count on from any number to 20.
Some students	may count backwards from 20.

Answers

Student Book page 12

Check that students have marked the pieces of the number line in the correct order.

Practice Book page 21

1 12

2 11

3 10

4 14

5 4

6 18

7 17

8 16

Stretch zone: Check that students have written three number sentences correctly that total 19.

1B Reading and writing numbers

Explore 2   Student Book page 13 · Practice Book page 22

Specific learning focus

- Read and write numerals to 20.

Global skills

- **Creative skills:** problem solving / investigating
- **Real-world skills:** representing information / interpreting information
- **Interpersonal skills:** communication

Key vocabulary

- number, zero, one, two, three, … , twenty, count on, count back, comes after, comes before, is between, column, row.

Resources

- set of large number cards 1–20
- 0–20 number lines: a large one for the front of the classroom and small ones for each group
- a large caterpillar with body segments labelled 1–20 (drawn on the board or on a large piece of paper so that all students can see it)

Language support

Encourage counting out loud as a group so that all students hear the correct pronunciation of the numbers. Work with less confident individuals by saying a number and asking them to repeat it. Over pronounce the 'teens numbers' in preparation for contrasting between multiples of 10 from 30 to 90. For example, emphasise fifTEEN in contrast to fifTY.

 **Introductory activity**

Ask a student to come to the front of the class and to select one of the number cards. They count on to 20 in ones starting from the selected number. Repeat, with a different student and number, but ask them to count back to **zero**. For less-confident students, ask the class to start the count and the student to join in. Use a large number line to help students count correctly, pointing to each number as you say it.

 **Main activity**

Display the number caterpillar at the front of the class. Tell students that you're going to cover one of the numbers with a piece of paper. Do this without the students seeing which number you choose.

Which number have I covered?
How do you know?

Listen to the explanations. Model the phrases **comes after**, **comes before**, **is between**.

Repeat several times with different numbers, each time asking *How do you know?* Encourage students to use the phrases above as part of their explanation.

Repeat but cover more than one number in sequence, such as 4, 5, 6 or 15, 16, 17, or swap numbers around, for example 14 and 15, to encourage students to use the correct language.

What are the numbers I have covered?
How did you know?

Students work on the activities on page 13 of the Student Book individually. Check that they have filled in the missing numbers in the caterpillar correctly before they move on to completing the grid. Explain that the ticket numbers are the green numbers. As students work, ask question such as *Are there some ticket numbers that could be easier to place in the grid? Why? Can you use the numbers that are already in the grid? What will the first number in the grid be? What about the last? How do you know? What number goes between 17 and 19?*

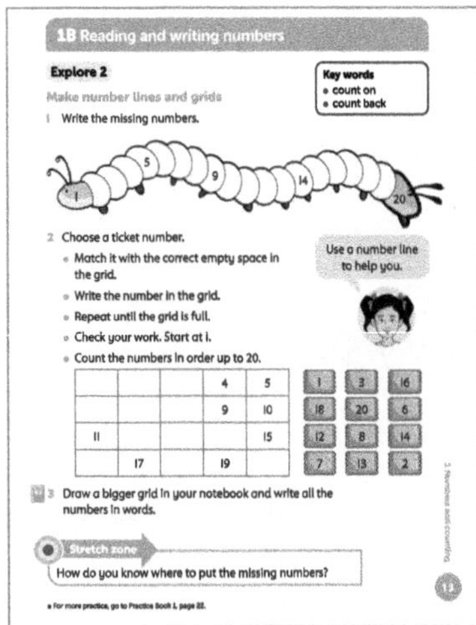

Differentiation

Supporting: Students should use 1–20 number lines to help identify missing numbers. You can help write the answers for less confident writers.

Consolidating: Ask students to explain how they know where to place numbers in the grid.

Extending: Ask students to look for patterns in the grid, e.g. the last digit alternates in each column (1, 5, 11, 15 and 4, 9, 14, 19).

Stretch zone: *How do you know where to put the missing numbers?*

Students should say the numbers aloud as they count to work out the missing numbers. They should explain to you or a partner why they placed a number where they did, and they might notice patterns in the columns of the grid, for example, the ones digits in a column make a repeating pattern. (The column with 4 at the top has 4, 9, 4, 9 as its 'ones' digits reading down.)

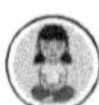 **Reflection time**

At the end of the lesson ask students to feed back on the activities from the Student Book.

What helped you put the numbers in order?

Did you put all the tickets in order first time, or did you have to change some?

*Look at the **columns** and **rows**. Can you see any patterns?* (The numbers in each column go up by 5 in each row.)

Ask further questions to help students notice patterns on a diagonal (e.g. 5, 9, 13, 17). *Count from 5 to 9, how many is that?* (4) *Do the same for 9 to 13. Now 13 to 17. What do you notice?*

Practice Book: Students can complete Practice Book page 22. This can be done directly after the main activity, as homework, or as the focus of a separate mathematics session to help students consolidate their learning and build fluency. Provide students with number cards or encourage them to look at the number line on page 21 of the Practice Book to help them write the numerals correctly.

Differentiated outcomes	
All students	should identify missing single numbers up to 10.
Most students	will identify missing single numbers up to 20.
Some students	may identify missing sequences of numbers.

Answers

Student Book page 13

1 Numbers should be written in order from 1 to 20.

2 Check that students have written the correct numbers in their grid

1	2	3	4	5
6	7	8	9	10
11	12	13	14	15
16	17	18	19	20

3 Check that students have written the correct number names for the ticket numbers in their grid.

Practice Book page 22

Numbers should be written in order from 1 to 20.

Stretch zone: Check students have written thirteen, fourteen, fifteen, sixteen, seventeen, eighteen and nineteen.

1B Reading and writing numbers

Explore 3   Student Book page 14 • Practice Book page 23

Specific learning focus

- Read and write numbers to 100.

Global skills

- **Creative skills:** problem solving / exploring
- **Real-world skills:** presenting information
- **Interpersonal skills:** communication / teamwork

Key vocabulary

- ones, tens, smallest, largest

Resources

- base-ten equipment

Language support

Help students to recall that the tens-rods are worth 10, and so they need the number names for ten, twenty, thirty, …, hundred. Listen to make sure that students are not confusing, for example, thirTEEN with thirTY, so emphasise the difference when saying the numbers.

Introductory activity

Count together as a class in tens from 0 to 100 and back to 0 as a reminder of that sequence. Show students two numbers using tens-rods and cubes. In one pile place 4 tens-rods and 2 cubes, and in the other pile 2 tens-rods and 9 cubes. Ask students which pile shows the larger number. Some students may say the second pile because it has more pieces, but remind them what the tens-rods represent, and then help them count the number shown by each pile. *If the tens-rods have a greater value, then what is more important, the number of pieces or the number of tens-rods?* (If one pile has more tens-rods, it will be the larger number.)

Main activity

Students work in pairs. Each student should pick up a handful of tens-rods and cubes. Tell them to count the tens-rods and cubes and write down the number they have picked up. Then they can compare to see who has the larger number. Remind them to compare the numbers by looking at the number of tens-rods first. The student with more tens-rods has the larger number. If they both have the same number of tens-rods, then count the cubes to see who has more.

Does the student with the larger number always have the most pieces? For example, if one student has recorded 61 and the other has recorded 48, ask *Who has the larger number? Who has more pieces?* Emphasise that it is not the number of pieces that matters, but the value of the pieces.

Students can repeat this with another pile of tens-rods and cubes each.

Students continue to work with ones-cubes and tens-rods to complete the activities on page 14 of the Student Book individually. Students may find it easier to draw the rods and cubes if they trace around them.

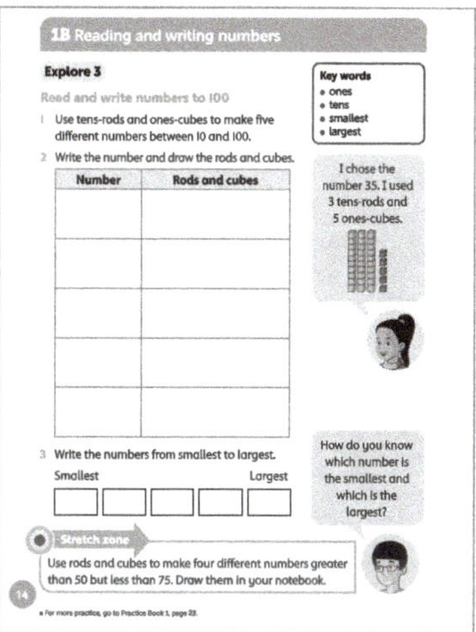

Differentiation

Supporting: Remind students that the tens-rods are worth 10 when counting. Help them by counting up in tens and then count the cubes in ones. You can help write the answers for less confident writers.

Consolidating: Ask students to explain how they know how much their piles of tens-rods and cubes are worth.

Extending: Ask students to identify which number is larger by comparing the tens, looking at the ones only when the number of tens is the same.

Stretch zone: *Use tens-rods and cubes to make four different numbers greater than 50 but less than 75. Draw them in your notebook.*

Students should ask themselves the following questions: *Can I draw four numbers that all use 5 tens-rods? What could they be? What is the largest number I can make? What is the smallest? Do all my numbers need to have a mix of cubes and tens-rods? Which do not?*

 Reflection time

Ask *Who made the largest number? How many tens-rods and cubes did they have? If I have 3 tens-rods and 2 cubes and my partner has 2 tens-rods and 3 cubes, whose pile shows the larger number?*

Practice Book: Students can complete Practice Book page 23. This can be done directly after the main activity, as homework, or as the focus of a separate mathematics session to help students consolidate their learning and build fluency. Look at the first example together and ask questions to get students thinking about smaller than and larger than, e.g. *23 is larger than 20. How do we know? Can you tell me another number larger than 20? What about a number smaller than 20?*

Differentiated outcomes	
All students	should make a number using tens-rods and cubes.
Most students	will write the number made from their pile of tens-rods and cubes.
Some students	may compare the numbers by counting the tens first.

1C Counting on and back

Discover 1 — Student Book page 15 • Practice Book page 24

Specific learning focus

- Find 2 more or less than a number to 20, recording the jumps on a number line.

Global skills

- **Creative skills:** problem solving / exploring
- **Real-world skills:** presenting information
- **Interpersonal skills:** communication

Key vocabulary

- 2 more, 2 less, count on, count back

Resources

- number cards 0–20
- washing line or string, pegs
- 0–30 number line: a large one for class work and small ones for each table
- drum, markers in two different colours

Language support

Make sure that students know the counting words. For example, saying 'twelve' when they see 12. Count forwards and backwards so that they can see, hear and touch the pattern of counting. This is especially important when counting in steps.

Answers

Student Book page 14

Check that students have chosen five different numbers between 10 and 100, and then written them and drawn the tens-rods and cubes correctly for each number. Check they have written the numbers in the correct order from smallest to largest.

Practice Book page 23

Check that students have correctly identified a number smaller or larger than the given number and have drawn tens-rods and cubes to match. There are different possible answers for each question.

 **Introductory activity**

Use a washing line or string and pegs to hang number cards. Hang the '10' number card in the middle of the line. Give the other number cards to students. Ask them to peg a number on the line according to instructions. For example, *Can you find the number that is **2 less** than 10? Can you find the number that is 1 more than 5?* Numbers can be moved as more numbers are added. Ask students, *Can you choose a number and explain why you chose to place the card there?*

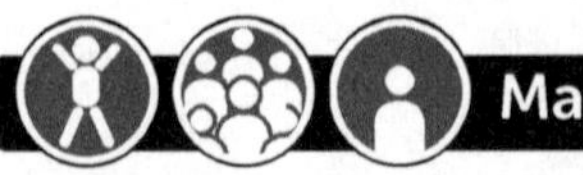 **Main activity**

Choose two students to come to the front of the class. Ask the first student to bang the drum eight times and to record each beat by marking a jump on the 0–30 class number line, starting at 0. Circle 8 on the number line. Ask the second student to slowly clap their hands twice, and to mark the two jumps on the number line, starting at 8. Circle the new end number. (10)

How many drumbeats? (Point to 8.) *How many claps?* (Point to the 2 jumps) *8 and **2 more** is 10.*

Repeat the activity using different start numbers, counting on 2 each time. For example, starting at 4 and counting on in twos will give 6, 8, 10, 12, and so on.

Show the class how to jump back in twos from a starting number, by marking the drumbeats and counting back the number of claps. For example, if you start on 9 and count back in twos, you land on 7, 5, 3, 1.

Ask students to look at page 15 in the Student Book. *Look at the first frog and the arrow. In what direction is Frog jumping?* (forward/on) *How many jumps did it make?* (2) *Where did it start from?* (0) *Where did it stop?* (4) *What size were the jumps?* (jumps of 2) *What about the next frog?*

Students complete the activities individually.

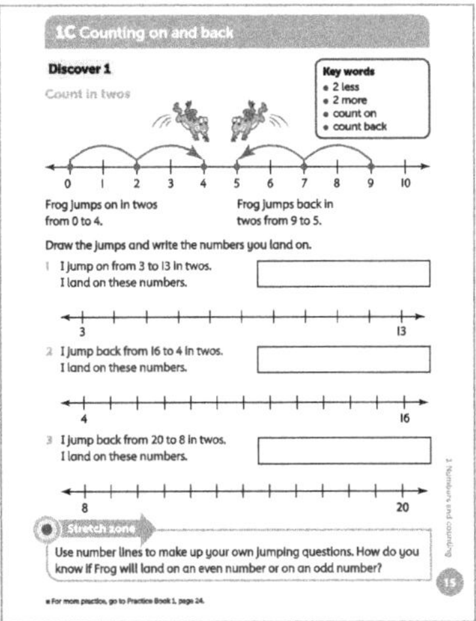

Differentiation

Supporting: Ask students to count on in twos from different starting numbers by pointing to a number line and saying the numbers aloud after each jump of 2.

Consolidating: Ask students to predict the numbers they will land on.

Extending: Ask students to describe the patterns they see when jumping in twos, such as: they are all even numbers or they are all odd numbers.

Stretch zone: *Use number lines to make up your own jumping questions. How do you know if Frog will land on an even number or on an odd number?*

Encourage students to make some 'jumping questions' which involve jumping 2 places along the number line. They might begin to notice that if they start on an odd number, they land on an odd number. Similarly, starting on an even number means they land on an even number.

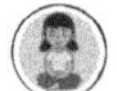 Reflection time

Choose a number on the class number line, for example 3. Then ask students, *What is the last number I will land on if I count on in twos three times? Which numbers will I land on?*

Choose a larger number, for example 17, and ask, *What is the last number I will land on if I count back in twos five times? Which numbers will I land on?* Can they predict the numbers and where they will finish before counting on the number line?

Practice Book: Students can complete Practice Book page 24. This can be done directly after the main activity, as homework, or as the focus of a separate mathematics session to help students consolidate their learning and build fluency. Encourage students to form each number carefully.

Differentiated outcomes	
All students	should count on and back on the number line successfully.
Most students	will be able to jump two spaces without counting.
Some students	may start to recognise the patterns when jumping in twos.

Answers

Student Book page 15

1 3, 5, 7, 9, 11, 13

2 16, 14, 12, 10, 8, 6, 4

3 20, 18, 16, 14, 12, 10, 8

Practice Book page 24

1 6, 8, 10, 12, 14, 16, 18, 20

2 17, 15, 13, 11, 9, 7, 5, 3, 1

3 14, 12, 10, 8, 6, 4, 2, 0

1C Counting on and back

Discover 2
Student Book page 16 • Practice Book page 25

Specific learning focus

- Find 10 more or less than a number to 100, recording the jumps on a 100-square.
- Within the range 0 to 100, say the number that is 1 or 10 more or less than any given number.

Global skills

- **Creative skills:** problem solving / exploring / investigating
- **Real-world skills:** presenting information
- **Interpersonal skills:** communication / teamwork

Key vocabulary

- 10 more, 10 less, count on, count back

Resources

- number cards 0–100
- washing line or string, pegs
- 100-square: a large one for the front of the classroom and small ones for each table
- drum, markers in two different colours

Language support

Make sure that students know the counting words, for example, saying twenty when they see 20. Count forwards and backwards so that they can see, hear and touch the pattern of counting. This is especially important when counting in steps.

Introductory activity

Use a washing line or string and pegs to hang number cards. Hang the '26' number card in the middle of the line. Give the other number cards to students. Ask them to hang a number on the line according to your instructions. For example, *Can you find the number that is **10 less** than 60? Can you find the number that is **10 more** than 30?* Move numbers where necessary as more numbers are added. Ask students, *Can you choose a number and explain why you chose to place the card there?*

Main activity

Choose two students to come to the front of the class. Circle '10' on the large 100-square. The first student bangs the drum 4 times, and each time counts down 1 row, finishing at 50. Circle this number in a dfferent colour. Write on the board the numbers that are landed on (20, 30, 40, 50). Ask the second student to slowly clap their hands twice, and each time count down 1 row, starting at 50. Write on the board the numbers that are landed on (60, 70). Circle the final number. (70)

Where did we start? (10) *How many drumbeats?* (4) *Where did we move to?* (50)

How many claps? (2) *50 and 20 more makes 70.*

How much is 1 jump of 10 if you start from 0? (10) *How much is 2 jumps of 10?* (20) Write on the board 10 + 10 = 20. Now ask, *How much is another jump of 10?* Write down 10 + 10 + 10 = 30 and tell students that each jump of 10 is adding 10 more.

Now ask, *How much is 1 jump of 10 back from 80?* (70) *How much are 2 jumps of 10 back from 80?* (60)

Ask students to look at the number lines on page 16 in the Student Book. *What do you notice about these number lines?* (The first has only 0 and 30 labelled and the second has only 0 and 20 labelled.) Explain that labelling the number lines will help students to answer the rest of the questions. Check, or organise students in pairs to check, that number lines have been labelled correctly before they begin answering the remaining questions individually.

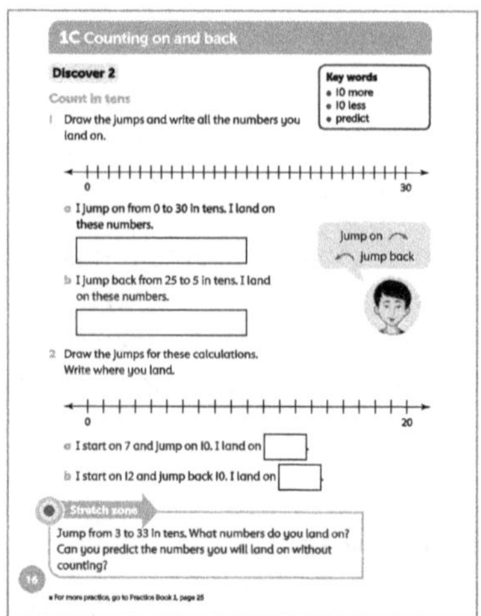

Differentiation

Supporting: Ask student to count on in tens from different starting numbers by pointing to a 100-square and saying the numbers aloud after each jump of 10.

Consolidating: Ask students to predict the numbers they will land on.

Extending: Ask students to describe the patterns they see when jumping in tens, such as all the numbers ending in the same digit.

Stretch zone: *Jump from 3 to 33 in tens. What numbers do you land on? Can you predict the numbers you will land on without counting?*

Students should be encouraged to think about the numbers they land on when jumping in tens. Help them to focus on what changes (the tens digit) and use this to help them make predictions about the number sequences.

 **Reflection time**

Start at 45 on the 100-square and jump back 10. *What number do you get?* (35) Ask students, *What is the same about 45 and 35?* (the ones) *What is different?* (the tens)

Repeat this, starting with 68 and jumping back 10. *What number do you get?* (58) *What is the same and different about 58 and 68?*

Ask students to notice what happens to the tens and ones digits when you jump back 10.

Practice Book: Students can complete Practice Book page 25. This can be done directly after the main activity, as homework, or as the focus of a separate mathematics session to help students consolidate their learning and build fluency. Draw students' attention to the use of 'jump back' and 'jump on' in each question to tell them whether they should count on or back.

Differentiated outcomes	
All students	should count on and back on the 100-square.
Most students	will be able to jump on or back 10 using the 100-square.
Some students	may start to recognise the patterns when jumping in tens.

Answers

Student Book page 16

1 a 10, 20, 30 **b** 15, 5

2 a 17 **b** 2

Stretch zone: 13, 23, 33

Practice Book page 25

1 12, 22

2 19, 9

3 18, 28

4 13, 3

5 17, 27

Stretch zone: 8 + 10 = 18, 23 − 10 = 13

18 + 10 = 28, 13 + 10 = 23

1C Counting on and back

Discover 3 Student Book page 17 • Practice Book page 26

Specific learning focus

- Find 5 more or less than a number to 100, recording the jumps on a 100-square or number line.
- Within the range 0 to 100, say the number that is 5 more or less than any given number.

Global skills

- **Creative skills:** exploring / investigating
- **Real-world skills:** interpreting information
- **Interpersonal skills:** communication / teamwork

Key vocabulary

- 5 more, 5 less, count on, count back

Resources

- number cards 0–100
- washing line or string, pegs
- 0–30 number line – a large one for the front of the classroom
- drum, markers in two different colours
- 100-square – a large one for the front of the classroom

Language support

Make sure that students know the counting words, for example, saying 'twelve' when they see 12. Count forwards and backwards so that they can see, hear and touch the pattern of counting. This is especially important when counting in steps.

 Introductory activity

Use the washing line or string to peg number cards on to. Hang '33' in the middle of the line. Give the other number cards to students. Ask them to peg a number on the line according to instructions such as *Show me 5 more than 33*, *Show me 5 less than 33*. Move numbers as more numbers are added, when necessary.

 Main activity

Choose two students to come to the front of the class. Mark '3' on the large 0–30 number line. The first student bangs the drum 2 times, and each time counts on 5 places. Circle the end number. (13) Write on the board the numbers that are landed on. (8, 13) Ask the second student to slowly clap their hands 4 times, each time counting on 5 places, marking the numbers landed on. (18, 23, 28, 33) Circle the end number. (33)

Where did we start? (3) *How many drumbeats?* (2) *Where did we move to?* (13)

How many claps? (4) *So 13 and 20 more makes 33.*

How much is 1 jump of 5 if you start from 0? (5). *How much is 2 jumps of 5?* (10) Write on the board 5 + 5 = 10. Now ask, *How much is another jump of 5?* (15) Write down 5 + 5 + 5 = 15 and tell the students that each jump of 5 is adding 5 more.

Now ask students, *How much is 1 jump of 5 back from 40?* (35) *What about 2 jumps of 5 back from 40?* (30)

Ask students to complete the activities on page 17 in the Student Book with a partner. They should work on activities 1 and 2 individually and then check with a partner by saying in tandem all the numbers they coloured. They should work on question 3 together. For students completing the stretch zone question, encourage them to copy the phrasing in the previous questions.

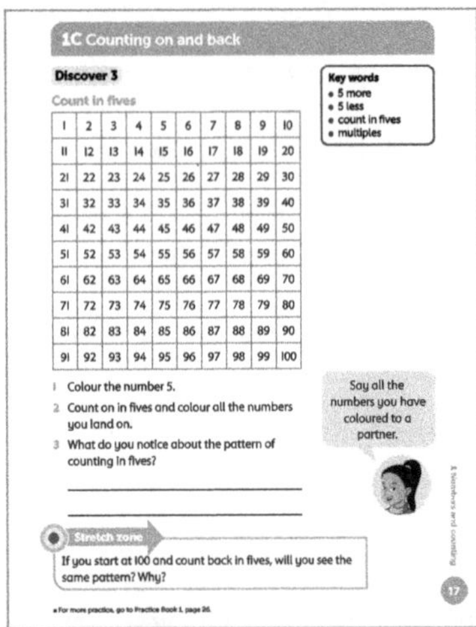

Differentiation

Supporting: Ask students to count on in fives from different starting numbers, by pointing to a 100-square and saying the numbers aloud after each jump of 5.

Consolidating: Ask students to predict the numbers they will land on.

Extending: Ask students to describe the patterns they see when jumping in fives, for example, that the units digits alternate.

Stretch zone: *If you start at 100 and count back in fives, will you see the same pattern? Why?*

Encourage students to look at the pattern on the number square and notice how the numbers that are counted are all in two columns, the ones ending in 5 and the ones ending in 0.

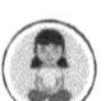

Reflection time

Cover one or more numbers on a large 100-square. Ask students what numbers are covered and how they know. Ask them to say how much is 5 more or 5 less than a given number.

Practice Book: Students can complete Practice Book page 26. This can be done directly after the main activity, as homework, or as the focus of a separate mathematics session to help students consolidate their learning and build fluency. Draw students' attention to the use of 'jump back' and 'jump on' in each question to tell them whether they should count on or back.

Differentiated outcomes	
All students	should count on and back on the number line successfully.
Most students	will be able to count on and back 5 spaces on the number square.
Some students	may start to recognise the patterns when jumping in fives.

Answers

Student Book page 17

5, 10, 15, 20, 25, 30, 35, 40, 45, 50, 55, 60, 65, 70, 75, 80, 85, 90, 95, 100

Practice Book page 26

1 7, 12, 17, 22, 27

2 4, 9, 14, 19, 24, 29

3 21, 16, 11, 6, 1

4 22, 17, 12, 7, 2

Stretch zone: Check that students have written a sentence to describe the jumps they chose.

1C Counting on and back

Explore 1
Student Book page 18 • Practice Book page 27

Specific learning focus

- Find 2 more or less than a number to 20, recording the jumps on a number line.
- Within the range 0 to 30, say the number that is 2 more or less than any given number.
- Begin to recognise numbers that are 2 more or 2 less than any given number.

Global skills

- **Creative skills:** exploring
- **Real-world skills:** presenting information / interpreting information
- **Interpersonal skills:** communication / teamwork

Key vocabulary

- 2 more, 2 less, count on, count back

Resources

- large empty 0–30 number line for display
- 0–30 number line for each pair
- number cards 1–9

Language support

Diagnostic questioning will help to reveal any difficulties. Do this in a small group or individually, through mathematics or language. Observing students' work and listening to their explanations will help.

For example, *Can you tell me how you did that? Can you draw the steps on the number line?*

What number comes before/after? Look at this number. I want this to be my answer. What is the question?

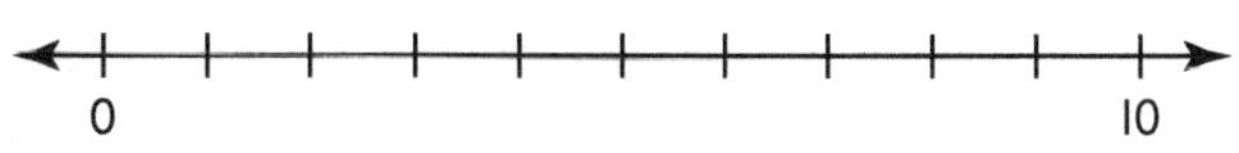 Introductory activity

Draw or display an empty number line on the board. Write 0 and 10 and the marks in between.

Start on 0 and ask, *What number will be 2 more than 0?* Show the jump and write 2. Continue until all the even numbers to 10 are written.

Draw or display a new number line and repeat, counting back from 10. Ask students to tell you the missing numbers and write them on the line.

Main activity

Draw some jumps on an empty 0–20 number line on the board and write the numbers where you land. For example, start on 8, make the jump (+ 2) and write 10, make the jump (+ 2) and mark 12.

Ask students to predict the next number (14) and tell you how they know (they are counting on in twos).

Work backwards along the line. Start on 20 and mark jumps of 2 asking students for the rule (counting back in twos).

Give each pair a 0–30 number line. Ask students to work with a partner making forward or backward jumps on the number line for the partner to guess the rule. Each student has four turns.

Ask students to complete the activities on page 18 of the Student Book individually. After they have completed the questions, ask them to join with a partner to discuss the patterns they notice. Move between pairs asking them to share with you what they have noticed.

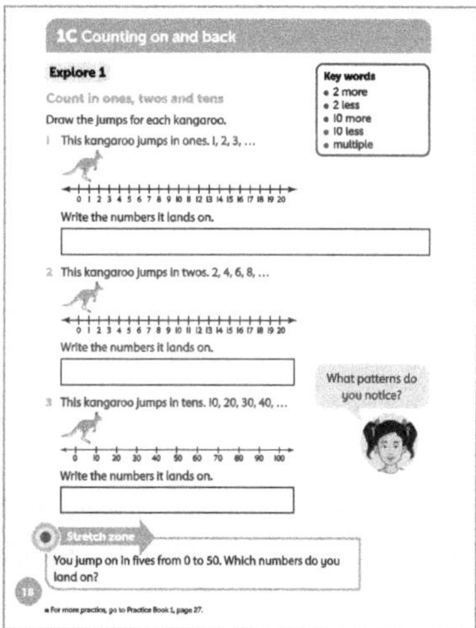

Differentiation

Supporting: Ordering number cards and filling in numbers on the number line as well as correction of naming of the written numerals will help less-confident students.

Consolidating: Ask students to predict the numbers they will land on.

Extending: Ask students to describe the patterns they see when jumping in twos.

Stretch zone: *You jump on in fives from 0 to 50. Which numbers do you land on?*

Students can use a number line to track the numbers as they jump forward 5 each time, starting from 0. They can record the numbers they land on, and they might begin to notice that the ones digits alternate between 0 and 5.

 **Reflection time**

Give each pair a 0–30 number line and number cards 1–9 (face down). They each start at 0 on the number line and take turns to pick a number card. If the number is even, they jump forward 2. If the number is odd, they jump forward 2 and then another 2. Always jump on from the last number. The first person to get to, or go past, 30 is the winner.

Practice Book: Students can complete Practice Book page 27. This can be done directly after the main activity, as homework, or as the focus of a separate mathematics session to help students consolidate their learning and build fluency. Encourage students to look at the numbers they record for each question. Can they see a pattern and describe it to someone?

Differentiated outcomes	
All students	should count on and back on the number line successfully.
Most students	will be able to jump two spaces without counting.
Some students	may start to recognise the patterns when jumping in twos.

1C Counting on and back

Explore 2 Student Book page 19 · Practice Book page 28

Specific learning focus

- Count forward or back in twos, fives or tens from a given number.

Global skills

- **Creative skills:** exploring
- **Real-world skills:** presenting information
- **Interpersonal skills:** communication / teamwork

Key vocabulary

- more than, less than, predict

Resources

- 0–30 empty number line: a large one for the front of the classroom and a small one for each pair
- 0–50 number line for each student

Language support

Ensure that students are able to use the correct counting words when doing steps of 2, 5 or 10. Help them to understand that 'more than' refers to the larger of two numbers and 'less than' is for the smaller of two numbers.

Answers

Student Book page 18

Check that the correct jumps and numbers have been written on each number line.

1 1, 2, 3, 4, 5, 6, 7, 8, 9, 10, 11, 12, 13, 14, 15, 16, 17, 18, 19, 20

2 2, 4, 6, 8, 10, 12, 14, 16, 18, 20

3 10, 20, 30, 40, 50, 60, 70, 80, 90, 100

Stretch zone: 5, 10, 15, 20, 25, 30, 35, 40, 45, 50

Practice Book page 27

1 3, 5, 7, 9, 11, 13, 15, 17, 19, 21, 23, 25, 27, 29

2 13, 23

3 6, 8, 10, 12, 14, 16, 18, 20, 22, 24, 26, 28, 30

4 14, 24

5 15, 25

Stretch zone: 6, 11, 16, 21, 26.

Check students' explanations.

 **Introductory activity**

Draw or display an empty number line on the board. Write 0 and 30 and the marks in between. Start on 0 and ask, *What number will be 2 more than 0?* Show the jump and write 2. Continue until all the even numbers to 10 are written.

Use a new line and count on from 0 in jumps of 5. Ask students to tell you the missing numbers and write them on the line.

Finally repeat this, counting on from 0 in jumps of 10.

Main activity

Draw some jumps on an empty number line and write the numbers where you land, for example, start on 5, make the jump (+2) and write 7, make the jump (+2) and write 9.

Ask students to **predict** the next number (11) and tell you how they know (they are counting on in twos, and 11 is 2 more than 9).

Draw some jumps on an empty number line and write the numbers where you land, for example, start on 9, make the jump (+5) and write 14, make the jump (+5) and mark 19.

Ask students to predict the next number (24) and tell you how they know (they are counting on in fives, and 5 more than 19 is 24).

Ask students to work with a partner to make forward or backward jumps of 2, 5 or 10 on a 0–30 number line for their partner to guess the rule. Each student has four turns.

Ask students to complete the activities in the Student Book individually. Look at question 1 together. *If I need to find the number 2 more than 10, do I need to count on or back?* (count on) *What size of jump?* (2) *And, to find 2 less, do I count on or back?* (count back) *Will the number be smaller or bigger than 10?* (smaller)

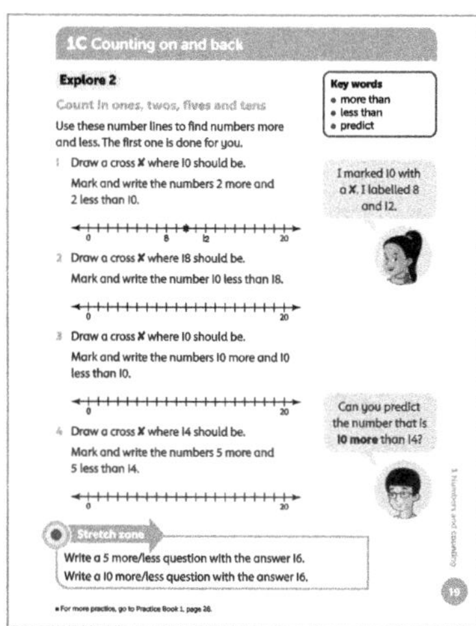

Differentiation

Supporting: Help students to make jumps of 2, 5 and 10 rather than counting up, '1 and 2' for a step of 2.

Consolidating: Ask students to predict the numbers they will land on.

Extending: Ask students to describe the patterns they see when jumping in twos, fives or tens.

Stretch zone: *Write a 5 more/less question with the answer 16.*

Write a 10 more/less question with the answer 16.

For this stretch zone, students are being challenged to finish at a given answer after finding 5 more or 5 less than a given number, or 10 more and 10 less. They will need to understand that, to do this, they are working backwards from the answer to the starting number. Similarly, to end up on a given number after finding 5 less, they have to count 5 more than that answer.

Check that students have composed questions to match the task, and perhaps see if some students can form further questions of a similar style, e.g. *Write a question for a partner that requires them to find 5 less than a number and finish on 23.*

 **Reflection time**

Show students a number sequence that goes up in jumps of 5, for example, 3, 8, 13, 18, 23, 28,… but with one or more numbers covered. Ask students to say what the missing numbers are and how they know. *Can you find missing numbers in sequences with jumps of 2? Jumps of 10?*

Practice Book: Students can complete Practice Book page 28. This can be done directly after the main activity, as homework, or as the focus of a separate mathematics session to help students consolidate their learning and build fluency. Provide students with a 0–50 number line to help them or they can use the number lines on page 27 of the Practice Book to support them for most questions.

Differentiated outcomes	
All students	should count on a number line and use the correct number names.
Most students	will count in twos, fives or tens from one number to another.
Some students	may recognise the patterns of counting in twos, fives or tens.

Answers

Student Book page 19

1 8 and 18 should be marked.

2 0, 10 and 20 should be marked.

3 9, 14, 19 should be marked.

Practice Book page 28

1 11, 13, 15	**4** 19, 24, 29	**7** 14, 24, 34
2 18, 20, 22	**5** 9, 14, 19	**8** 4, 14, 24
3 20, 25, 30	**6** 15, 25, 35	

1C Counting on and back

Explore 3 — Student Book page 20 • Practice Book page 29

Specific learning focus

- Recognise number patterns that occur in series that jump by 2, 5 or 10.

Global skills

- **Creative skills:** exploring
- **Real-world skills:** presenting information / interpreting information
- **Interpersonal skills:** communication / teamwork

Key vocabulary

- multiple, count on, count back, steps, patterns

Resources

- 100-square – a large one for class use and a small one for each pair
- coloured pencils to mark numbers on the 100-square, counters
- number cards 2, 2, 5, 5, 10, 10 (for each pair)

Language support

Reinforce the use of vocabulary relating to patterns and counting in steps. Relate this to everyday situations such as going up and down stairs.

 **Introductory activity**

Display a large 100-square at the front of the classroom. Ask one student to choose a single-digit number in the top line of the square. Ask another student to choose a number to use for counting on – either 2, 5 or 10. Count together as a class from the starting number up in **steps** of the chosen number.

 **Main activity**

Tell students they are going to explore number **patterns** on their 100-squares. Show on the large 100-square how to find their starting number by choosing a number card (2, 5 or 10). Then choose another number card to find which number to count in, for example 5. Explain that they will mark each number they land on using a coloured pencil to create a trail of numbers.

Ask students to work with a partner making forward jumps of 2, 5 or 10 on their 100-square. Repeat with a different starting number and a new counting number, marking this trail in a different colour.

Ask students to complete the activities on page 20 of the Student Book individually. Encourage them to look at

the numbers they colour. *How are they the same? How are they different?*

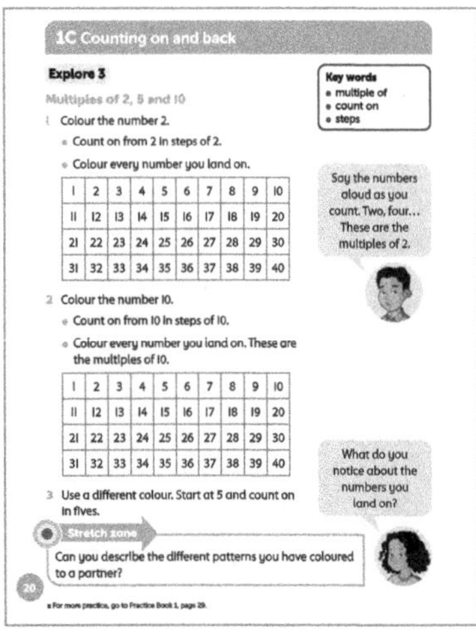

Differentiation

Supporting: Ask students to use counters to mark the jumps before colouring.

Consolidating: Ask students to predict the numbers they will land on.

Extending: Ask students to describe the patterns they see when jumping in twos, fives or tens.

Stretch zone: *Can you describe the different patterns you have coloured to a partner?*

Encourage students to describe their patterns using observations about the ones digits, for example, counting on in twos from 13, the ones digits are 5, 7, 9, 1, 3 and then back to 5 again.

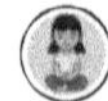 **Reflection time**

Ask if any students counted on in the same size jump as their starting number.

Ask students to share examples of this, or show some if no pairs come up with any:

2, 4, 6, 8, 10, 12, 14, …

5, 10, 15, 20, 25, …

10, 20, 30, 40, 50, …

Explain to students that these patterns of numbers are called **multiples**. Ask students to tell you some more multiples of 2, 5 and 10.

What do you notice about the numbers you landed on? Encourage them to notice, for example, that 10 is a multiple of 2 and of 5.

Practice Book: Ask students to complete the activities on Practice Book page 29 individually. This can be done directly after the main activity, as homework, or as the focus of a separate mathematics session to help students consolidate their learning and build fluency. Encourage students to describe the pattern they shaded for each question to another child or an adult.

Differentiated outcomes	
All students	should count in jumps of 2.
Most students	will jump in fives and tens.
Some students	may predict the pattern when jumping in twos, fives or tens.

Answers

Student Book page 20

1 2, 4, 6, 8, 10, 12, 14, 16, 18, 20, 22, 24, 26, 28, 30, 32, 34, 36, 38, 40

2 10, 20, 30, 40

3 5, 10, 15, 20, 25, 30, 35, 40

Practice Book page 29

1 25, 27, 29, 31, 33, 35, 37, 39

2 38, 36, 34, 32, 30, 28, 26, 24, 22, 20, 18, 16, 14, 12, 10, 8, 6, 4, 2, 0

3 5, 10, 15, 20, 25, 30, 35, 40

4 50, 40, 30, 20, 10

1C Counting on and back

Explore 4 Student Book page 21 • Practice Book page 30

Specific learning focus

- Relate counting on and back in tens to finding 10 more/less than a number (< 100).

Global skills

- **Creative skills:** exploring / investigating
- **Real-world skills:** presenting information / interpreting information
- **Interpersonal skills:** communication

Key vocabulary

- 10 more, 10 less, count on, count back, column, row

Resources

- 100-square – a large one for class use and a small one for each student
- set of two different-coloured cubes or counters per student

Language support

When you use words such as 'above' or 'below', relate them to classroom situations such as the clock is above the shelf. Display simple diagrams highlighting the prepositions above, below, next to, before.

Introductory activity

Show the class a large 100-square. Ask students what they can see and to discuss it with their partner. Take any suggestions and prompt them, if necessary, by saying for example, *Look at the last digit. Find all the numbers that end with 7.* Leave time for discussion. *What do you notice?* They should notice that the ones digit is the same all the way down each column, and that only the tens digit is changing.

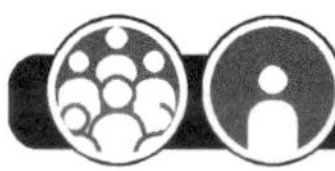

Main activity

Give each student a small 100-square, and ask them to put a red cube on 2, then count on 10 to find 10 more and put a blue cube on the new number. Ask what they notice about the position of the two cubes. (The blue cube is below the red cube.) Repeat using other start numbers and ask if the same thing always happens. Explain that if 10 more is added to a number, the answer is always the number in the same column in the row below on the 100-square.

Using the same strategy, ask students to use the 100-square and cubes to find 10 less by counting back and landing on the number above. Repeat with different start numbers. *To find 10 more, we go to the number below. To find 10 less, we go to the number above.*

Play this game. *I am going to say, '10 more' or '10 less' and you move your finger to the next number. Put your finger on 14. Find 10 more. Tell me the new number.* Repeat for different start numbers.

Put your finger on 23. Find 10 less. Tell me the new number. Repeat for different start numbers.

This time I am going to say '10 more' and '10 less'. Put your finger on 10. Find 10 more, 10 more, 10 less. What is your number? Repeat several times using both addition and subtraction.

Display the large 100-square. Cover a number with paper or card and ask students which number is covered and how they know. Encourage vocabulary such as 'it is 10 more than' or 'it is 10 less than'. Repeat with other covered start numbers.

Ask students to complete the activities on page 21 in the Student Book individually. After they have finished each question, ask them to describe to a partner what they notice about their start and finish numbers.

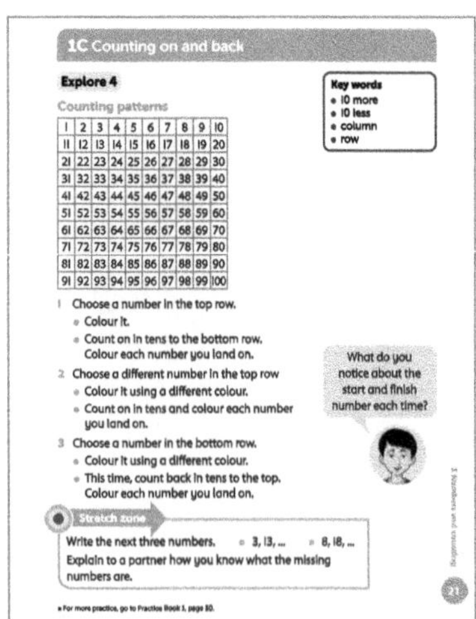

Differentiation

Supporting: Support students in placing counters accurately on the 100-square.

Consolidating: Ask students to predict the numbers in the pattern.

Extending: Ask students to explain the patterns in the 100-square.

Stretch zone: *Write the next three numbers.*

3, 13, …

8, 18, …

Explain to a partner how you know what the missing numbers are.

Students can use a 100-square to notice what jump takes them from 3 to 13, and they can then repeat that jump to find the next numbers in the sequence… 23, 33, 43.

Similarly, they can jump from 8 to 18 and use this to get 28, 38, 48.

Students should be able to describe to a partner how they know the next numbers in each sequence.

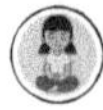 Reflection time

Ask students to close their eyes and imagine number 5 on the 100-square. *What numbers are either side of 5?* (4 and 6) Repeat with numbers above and below (adding and subtracting 10).

Ask students to work with a partner. Each chooses a start number and tells the other the numbers that are below and above it and either side of it. The partner has to say the number.

Practice Book: Students can complete Practice Book page 30. This can be done directly after the main activity, as homework, or as the focus of a separate mathematics session to help students consolidate their learning and build fluency. Encourage students to record each number carefully.

Differentiated outcomes	
All students	should be able to use the 100-square to support their calculations.
Most students	will notice the patterns in the 100-square.
Some students	may be able to explain the patterns in the 100-square.

Answers

Student Book page 21

Observe students as they work through the activity, and check that they are adding 10 correctly each time. Ask them what patterns they notice (the numbers are in the same column and the ones digits are the same).

Stretch zone: 3, 13, 23, 33

8, 18, 28, 38, 48

The last digit is the same each time.

Practice Book page 30

10 less	Number	10 more
14	24	34
37	47	57
9	19	29
32	42	52
48	58	68
78	88	98
80	90	100
35	45	55
61	71	81
18	28	38
26	36	46

1D Estimating

Specific learning focus

- Give a sensible estimate of some objects that can be checked by counting, e.g. to 30.

Global skills

- **Creative skills:** exploring / investigating
- **Real-world skills:** research / presenting information / interpreting information
- **Interpersonal skills:** communication / teamwork

Key vocabulary

- more than, fewer than, estimate, guess, nearly, how many more, how many fewer, actual

Resources

- red and blue cubes, counters, transparent pots (three per group), sticky notes

Language support

Model the key language such as more than …, less/ fewer than …, how many, estimate, and actual. Create a poster showing an estimation game with these phrases as sentence starters.

 ### Introductory activity

Take a handful of cubes. Show them to the class and ask them to guess how many there are and to make statements such as, *There are **more than** … cubes. There are **fewer than** … cubes*. Ask several students to give you statements that they are sure are true. Count the cubes to see who is right. Then ask students to work with a partner. They take a handful of cubes each and repeat the activity.

 ### Main activity

Throughout this activity students should use the phrases 'more than' and 'fewer than' to get a sense of 'range'. Tip a box of cubes onto the floor and ask students to **estimate** how many red cubes there are. Write the estimates on the board. Repeat for the blue cubes. Choose two students to count the cubes of each colour. Write the **actual** amounts next to the estimates. Discuss whether the estimates were **nearly** the same as the actual numbers.

Ask students to work in small groups. Give each group three pots with different numbers of cubes (or counters). Ask students to estimate how many cubes are in each pot on their table and to write the estimate in the table on page 22 of the Student Book. Then ask them to take

the cubes out of each pot one at a time and count them, and write the actual number in the table. At the end of the activity, ask students how they made their estimates. For example, did they **guess** or count some and guess the rest?

Ask students to complete question 2 on page 22 of the Student Book individually. Encourage them to count in twos, fives or tens to check the actual number of candles and balloons.

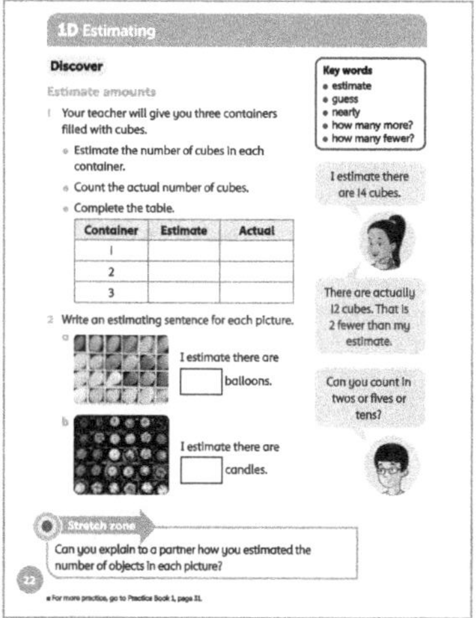

Differentiation

Supporting: Use smaller amounts for students who are still at the early stages of counting. Make the items large so that there aren't as many to count.

Consolidating: Ask students to explain how they decided on their estimate.

Extending: Challenge students to work with increasingly large numbers of items, which could also be smaller objects.

Stretch zone: *Can you explain to a partner how you estimated the number of objects in each picture?*

Students should be encouraged to describe the way they estimated the number of objects.

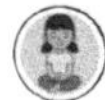 ### Reflection time

Pour some counters onto the table or floor. Ask students, *We need to put these counters away. How many are there do you think? About 10? More than 10? Could there be 30?* Ask students to write their estimate on a sticky note and put it on the board. Count the counters, putting each one back in a tub as you count. Part way through counting, you may want to ask if any students want to change their estimate. Ask them why.

When the counters have been counted, discuss the estimates on the board.

Practice Book: Students can complete Practice Book page 31. This can be done directly after the main activity, as homework, or as the focus of a separate mathematics session to help students consolidate their learning and build fluency. If students are unsure of what the object they are counting is called or how to write it, say that they can draw a picture of it instead.

Differentiated outcomes	
All students	should make sensible estimates.
Most students	will know how many more or how many fewer than their estimate there are.
Some students	may work with numbers over 30.

Answers

Student Book page 22

Observe students while they work and ask them to explain how they worked out their estimates. For example, they might compare the look of the set with a number of objects they've counted previously.

Practice Book page 31

Check that students do estimate before they count. Observe whether their estimates get better as they try handfuls of the different objects.

1D Estimating

Explore 1 Student Book page 23 • Practice Book page 32

Specific learning focus

- Give a sensible estimate of some objects that can be checked by counting, e.g. to 30.

Global skills

- **Creative skills:** exploring / investigating
- **Real-world skills:** research / presenting information / interpreting information
- **Interpersonal skills:** communication

Key vocabulary

- estimate, how many more, how many fewer?

Resources

- photographs or pictures of objects for estimating, sticky notes
- jar of dried beans (up to 50) and a scoop for class work, three transparent jars with different numbers of beans or cubes
- other small items that can be held in the hand

Language support

Use the language of estimation (guess how many, estimate, nearly, about, close to, about the same, just over, just under, too many, too few) during everyday activities, relating the words to real-life experiences. Repetition of these words will allow students to become more familiar with the vocabulary.

Introductory activity

Show photos or pictures of different numbers of objects and ask students to estimate how many objects there are. As in the last lesson, they should use the phrases 'there are more than …' and 'there are fewer than …'.

Main activity

Show the jar of beans to the class. Ask students to use the phrases 'more than' and 'fewer than' to estimate the number of beans. Discuss the range of answers. Discuss when it is appropriate to make an estimate (how many beans in the jar) and when it is not (measuring medicine).

Go back to the bean jar. Take one scoop of beans out and count them. Ask questions such as, *About how many more scoops of beans are in the jar? About how many beans are in the jar?*

Show the class the three jars of beans or cubes and arrange them in a line. Either during or at the end of the lesson, ask students to write their estimates in the table on page 23 of the Student Book. They can then complete question 2 individually.

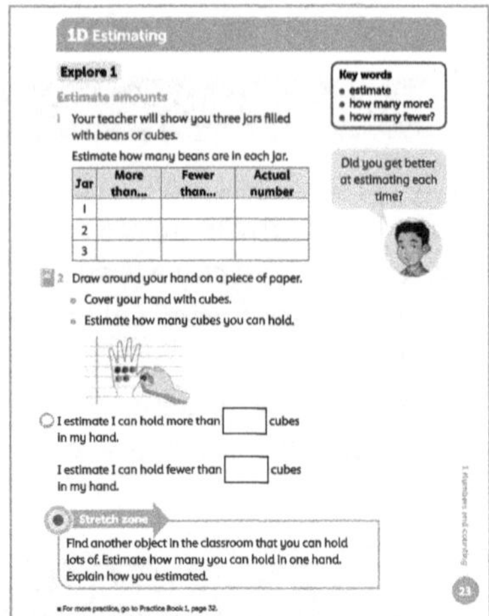

Differentiation

Supporting: Use resources to estimate up to 30, checking by counting.

Consolidating: Use resources to estimate up to 50, checking by counting.

Extending: Use resources to estimate beyond 50 checking by counting.

Stretch zone: *Find another object in the classroom that you can hold lots of. Estimate how many you can hold in one hand. Explain how you estimated.*

Students might describe their estimate compared with something else they held. For example, if they could hold 10 cubes and a cube is about the same size as 2 beans, then they might estimate that they can hold 20 beans.

 **Reflection time**

Show photos or pictures of different numbers of objects and ask students to estimate how many objects there are. Use the pictures as a display with the student estimates on sticky notes next to them.

Practice Book: Students can complete Practice Book page 32. This can be done directly after the main activity, as homework, or as the focus of a separate mathematics session to help students consolidate their learning and build fluency. If students are completing this activity in class, organise students in small groups. If they are completing it at home, tell them to ask an adult for help to set the activity up.

Differentiated outcomes	
All students	should count a number of objects.
Most students	will accurately estimate using previous experience.
Some students	may estimate and check larger numbers.

1D Estimating

Explore 2 Student Book page 24 • Practice Book page 33

Specific learning focus

- Give a sensible estimate of some objects that can be checked by counting, e.g. to 30.

Global skills

- **Creative skills:** exploring / investigating
- **Interpersonal skills:** communication / teamwork

Key vocabulary

- estimate, how many more, how many fewer?

Resources

- photographs or pictures of objects for estimating, sticky notes
- sets of different items (e.g. toy cars, cotton wool balls, pencils, cubes, counters), sheets to cover the items

Answers

Student Book page 23

Observe students while they work and ask them to explain how they worked out their estimates. For example, they might compare the look of one set of objects with a number of objects they've counted previously.

Practice Book page 32

Check that students estimate before they count. Observe whether their estimates get better as they try the different containers.

Language support

Use the language of estimation (guess how many, estimate, nearly, about, close to, about the same, just over, just under, too many, too few) during everyday activities, relating the words to real-life experiences. Repetition of these words will allow students to become more familiar with the vocabulary.

 **Introductory activity**

Show the class an image of a set of items (people, animals, vehicles) and ask them to estimate how many there are. Show the image for a short time and then remove it. Ask students to describe how they arrived at their estimate. For example, they may have looked at all of the items and guessed, or they may have grouped them in some way, perhaps in twos or fives and estimated how many groups. Share these strategies with students if they did not offer them.

 **Main activity**

Place a number of objects under a sheet. Tell students you are going to remove the sheet for a short time and they are to estimate the number of objects without counting.

After showing the objects, allow the students time to discuss in pairs, then ask for some estimates. Ask students to explain how they estimated.

The students will then play a game in pairs. One student takes a handful of objects and hides them under a sheet, then their partner has to have a quick look and estimate how many. If they are close (within 2 either way), they score a point. Then swap and repeat. The winner is the one who scored most points in the time allowed.

The students should then complete the activity on page 24 in the Student Book in pairs. Ask all pairs to consider the question in the margin, 'Which picture is the easiest to estimate?'. Can they explain their thinking? For each question they should make comparison statements, e.g. there are the same number of purple dots in each row.

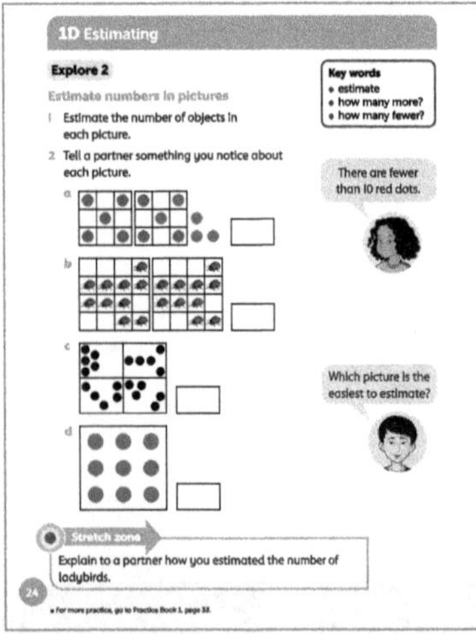

Differentiation

Supporting: Use a small number of objects 'hidden' in an arranged layout under a sheet for students to estimate.

Consolidating: Use other known patterns like spots on a domino for students to estimate.

Extending: Encourage students to use strategies such as grouping to get more accurate estimates.

Stretch zone: *Explain to a partner how you estimated the number of ladybirds.*

Some students may have estimated the overall group, others may have estimated or counted one group and added the same amount again.

 **Reflection time**

Have some objects (different numbers) hidden under two different sheets. Show each set of objects briefly. Ask students to estimate which set had more objects. Can they explain why? Also try this using the same number of objects under each sheet, but more spaced out under one sheet. See if students think there are more objects in the spaced out group.

Practice Book: Students can then complete Practice Book page 33. This can be done directly after the main activity, as homework, or as the focus of a separate mathematics session to help students consolidate their learning and build fluency. Before students begin, remind them that they are meant to estimate the number of dots for each, so they must not count them but reason which of the three numbers is likely to be closest to the actual number of dots.

Differentiated outcomes	
All students	should count a number of objects.
Most students	will accurately estimate using previous experience.
Some students	may estimate and check larger numbers.

Answers

Student Book page 24

Students should provide reasonable estimates for the number of objects in each picture. The numbers shown are:

a	13	**b**	20	**c**	20	**d**	9

Practice Book page 33

Students should provide reasonable estimates for the number of objects in each picture. The numbers shown are:

1	5	**3**	6	**5**	13
2	9	**4**	9	**6**	10

1 Numbers and counting

Connect Student Book page 25

Big idea

- We use numbers to count or say how many of something there are.

Global skills

- **Creative skills:** exploring / investigating
- **Real-world skills:** research / presenting information
- **Interpersonal skills:** communication / teamwork

Key vocabulary

- number, zero, one, two, three, ..., twenty

Resources

- magazines and newspapers, large sheets of paper, scissors, glue and pens

Language support

Encourage students to read the numbers aloud. Ask more confident peers to model the correct pronunciation of numbers.

 ### Introductory activity

Put students in mixed-attainment groups of 4 to 6 students. Give out the magazines and newspapers to the groups. Each group should find one feature in the magazine which contains numbers and share this with the whole class.

 ### Main activity

Students work as a group to create a 'Numbers' poster to be displayed. Use the instructions on page 25 of the Student Book to take students through the steps of the activity.

They cut out all of the different numbers that they find in pictures in the magazines.

They annotate the posters by writing 'We can use numbers to…'

They choose some of the numbers that they display to write in words, and add these to their poster.

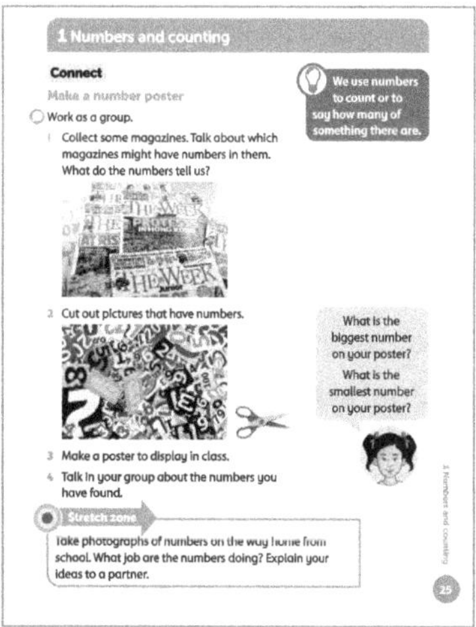

Differentiation

Supporting: Students listen to their peers pronouncing numbers and repeat the numbers correctly.

Consolidating: Students find numbers and can say what they are being used for. They recognise numbers up to 20.

Extending: Students recognise numbers larger than 20.

Stretch zone: *Take photographs of numbers on the way home from school. What job are the numbers doing? Explain your ideas to a partner.*

Students use digital cameras (or draw pictures) to create a presentation of numbers that they see in or around school, or on their way home from school.

 ## Reflection time

Ask one member of each group to present the poster to the rest of the class. Use the 'two stars and a wish' peer assessment technique (each group gives feedback on two things that they like and one thing they would change in each poster).

Differentiated outcomes	
All students	should recognise and write most of the numbers from 1 to 20.
Most students	will recognise and read and write numbers from 1 to 20.
Some students	may recognise and read and write numbers from 1 to 100.

1 Numbers and counting

Global skills

- **Creative skills:** problem solving
- **Real-world skills:** presenting information
- **Interpersonal skills:** communication
- **Self-development skills:** reflecting on learning

Student Book

With young children, assessment activities are most effective when carried out as an everyday classroom activity. Students should have number lines available with both the digits and the numbers in words so they can refer to this to support them.

Watch as children count the beads. Listen as they count to check that they say the number words in order and do not omit any numbers. Listen to check that they understand the last number they say is the total. If there are errors, count the beads for students so that they can hear the correct sequence.

It may help to create flash cards with the numerals and words that are in the review. You can use these to support some students.

Encourage children to use a number line and a 100-square for question 2. Extend by asking a range of similar questions. For example, *Using a number line find '1 more' than [a given number]* or *Using the 100-square find '10 more' than [a given number]*.

Answers

Student Book page 26

1 Check that the students have drawn the correct number of beads and filled in the correct numbers and words.

2 20, 30

Practice Book

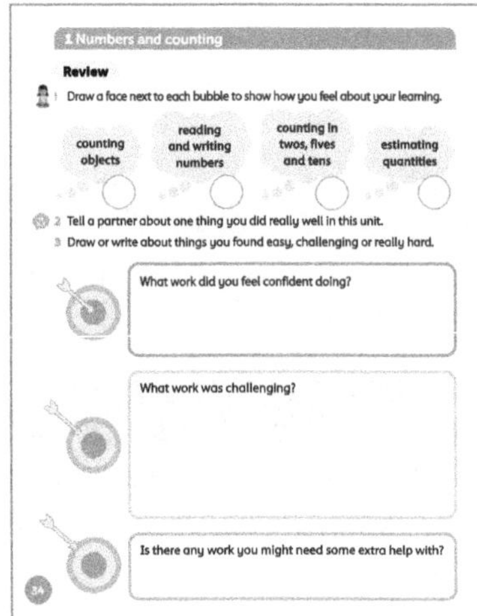

With young children it is appropriate to complete this as a whole-class discussion. You may choose to keep a record of the class discussion or a copy of the review page for your own records. Use the Student Book to briefly remind the students of the areas of mathematics

Give out the Student Book to support pairs of students as they discuss and answer the questions in the Practice Book.

Allow students plenty of time for discussion before asking them to share their responses with the rest of the class. If students complete this assessment at home, encourage them to discuss this with adults.

Make a note of areas that students still feel unsure about. As counting is at the heart of many of the other units you can revisit these areas regularly. You can also build counting into everyday practice: for example, counting how many students are in the class each day, counting objects as you give them out, and so on.

Additional material

There are additional end-of-unit assessments available on the *Oxford Owl* website.

2 Number bonds

Big Idea

As part of everyday life, students need to be able to identify numbers, understand the number system and solve problems. Whole numbers are used to tell the size of a set of objects or people. Number operations are introduced using concrete (solid) objects by joining together and splitting sets to develop the concepts of addition and subtraction (+ and –).

Counting everyday objects helps students to understand that the last number of a count tells how many are in the set or group. However, when counting, there is no requirement to count the objects in a particular order, only that each object is counted once. Conservation of number (the understanding that a set of objects can be rearranged but the amount stays the same) is important for understanding bonds or pairs of numbers that have the same total.

Within the operations, students will begin to understand that there is a variation of contexts For example, when adding, the student might count both sets to be combined, or learn to count on from the number in the first set. For subtraction, the notions of 'take away' and 'difference' need to be explored, since the first involves an operation on a single set, whereas the second involves comparison of two sets.

Look out for

- **Students who are confused about the sequence of the numbers.** Saying number sequences out loud and as a group can help students to learn them, but practical, hands-on experience of using, comparing

and calculating with numbers and quantities, and the development of mental methods, is of crucial importance.

- **Students who have poor understanding and development of language and vocabulary, which can stop some mathematics from 'happening'.** Give plenty of practical and real-life opportunities for developing mathematical language so that students learn to express their thinking using the correct vocabulary. Use charts and posters with numbers and matching pictures to link the two and enable students to have rich and varied opportunities to discuss, explain and question the contexts in which they find numbers.

- **Students who have difficulties in developing mental methods.** Support this by using jottings and visual images, such as number lines linked directly to the physical experience of counting objects or manipulatives.

Possible misconceptions

- **Students may misuse one-to-one correspondence when counting by missing out numbers as they count.**
- **Students may think of the '=' sign as meaning 'the answer is',** instead of it showing that two things have the same value.

Key vocabulary

- +, add, make, total, altogether, most, fewer, –, take away, one less, how many have gone?, how many are left?, number sentence, addition sentence, subtraction sentence

Coverage in lessons

Learning focus	Learning outcomes (the ENC objectives)
Number bonds for 6, 7, 8, 9	Read, write and interpret mathematical statements involving addition (+), subtraction (–) and equals (=) signs. Represent and use number bonds and related subtraction facts within 20.
Number bonds for 10	Read, write and interpret mathematical statements involving addition (+), subtraction (–) and equals (=) signs. Represent and use number bonds and related subtraction facts within 20.
Missing numbers	Solve one-step problems that involve addition and subtraction, using concrete objects and pictorial representations, and missing number problems such as $7 = \square - 9$.

2 Number bonds

Engage   Student Book page 27

Big question

- How many different ways are there to make 6, 7, 8 and 9?

Global skills

- **Creative skills:** problem solving
- **Interpersonal skills:** communication / teamwork

Key vocabulary

- total, altogether

Resources

- cubes or other objects to help with counting
- 0–10 number line

Language support

Using practical resources such as counters or cubes will help students to see the quantity of ten rather than it being an abstract idea. Ask students to arrange their ten counters in different ways but agree that there will always be ten. Suggest they draw the bugs, then sort the ten counters onto the bugs. This will show the number of legs to draw.

 Introductory activity

Ask students to look at page 27 in the Student Book in pairs. If you have access to an IWB, you could display the Student Book Engage page and discuss the Big idea of the unit with students. Ask students to talk to their partner about what they notice in the images. Give time for discussion. Ask pairs:

Which bug has the most legs?

Which bug has the fewest legs?

Ask pairs to look at two of the bugs. *How many legs do these two bugs have **altogether**? How many different **totals** can you find for two bugs?* Tell students to count only the legs they can see (so the grasshopper has 5 legs).

Ask pairs to find two bugs that have eight legs altogether.

 Main activity

Tell students that they are having a party for bugs. They can invite any bugs they like, but the bugs must each come with a friend, and each pair of bugs must have a total of ten legs. Ask students to draw pairs of bugs that can come to the party. They should try to find as many different pairs as they can.

Allow enough time for all or most students to complete their drawings.

Differentiation

Supporting: Use a 0–10 number line to help students with counting. Encourage students to count each set of legs. Record the number sentence for students.

Consolidating: Ask students to find a different way of making the total.

Extending: Ask students to check that they have all the different ways of making totals for two bugs.

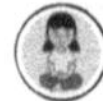 Reflection time

Ask students to share the pictures of their bug party.

How many pairs of bugs did you draw?

There are 11 possible pairs of bugs, with numbers of legs as follows:

(0, 10), (1, 9), (2, 8), (3, 7), (4, 6), (5, 5), (6, 4), (7, 3), (8, 2), (9, 1), (10, 0)

In the feedback, ask: *Who found all 11 pairs?*

2A Number bonds for 6, 7, 8, 9

Discover
Student Book page 28 • Practice Book page 35

Specific learning focus
- Begin to know number bonds to 6, 7, 8, 9 and 10.
- Understand addition as counting on and combining two sets.
- Use the = sign to represent equality.

Global skills
- **Creative skills:** investigating
- **Interpersonal skills:** communication / teamwork

Key vocabulary
- number bonds, addition, altogether, total, add, make, equals, take away

Resources
- counters, paper plates

Language support
Model the vocabulary so that students can hear the correct pronunciation. Help students who are struggling with '=' and '+' to know how to read them when they see them in number sentences.

 ## Introductory activity

Take 6 counters in your hand and show the class. Ask students to close their eyes and imagine the counters. *I take one away. How many do I have now?* (5) *How many do I put back to make 6 again?* (1)

Write on the board '6 **take away** 1 is 5; 5 **add** 1 **makes** 6'.

Repeat the activity with 6 take away 2. *How many are left?* (4). Write on the board '6 take away 2 is 4; 4 add 2 makes 6'. Repeat with '6 take away 3' and again record the number sentence on the board.

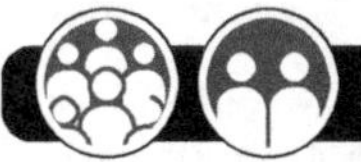 ## Main activity

Introduce this activity as being from page 28 of the Student Book, which looks at ways of adding to make 6 and 8. Show six counters. *How many different ways can we make 6?*

Show two paper plates. *I can put six counters here. How many will go on the other plate?* (0) Move one of the counters onto the other plate. *How many on this plate?* (5) *How many on that plate?* (1) *How many altogether?* (6)

Repeat, taking another one from the first plate so that there are four on one plate and two on the other. *How many are there altogether?*

Ask students to work with a partner to find all the different ways to make 6. They should draw pictures

in the Student Book to show the different ways they find.

Show students how to record the different ways by drawing them, for example,

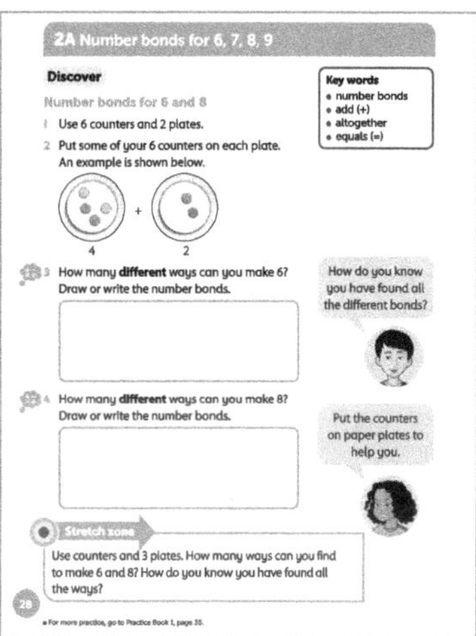

Instead of writing 'makes' write '='. Explain that this is an **equals** sign. It means that what is on one side of the sign is equal to what is on the other side of the sign.

Pairs should check that they have all the different arrangements. Explain that these are called **number bonds** to 6.

Some pairs should then move on to find pairs that total 8.

Differentiation
Supporting: Some students may have difficulty with mental imagery. Use a strategy where they can hear counters being added or taken out, such as putting counters into a tin. Record the number sentences for these students.

Consolidating: Ask students to make sure that they have all the different ways of making 6 or 8.

Extending: Ask students to explain how they know they have all the different arrangements.

Stretch zone: *Use counters and 3 plates. How many ways can you find to make 6 and 8? How do you know you have found all the ways?*

This activity is designed to challenge students to find more partitions of a number, beyond finding the two-number bond for a given total. Here, students will be adding three single digits to find a total, for example, $1 + 2 + 3 = 6$. Explain to students that some number sentences represent the same partitioning, since for example, $6 = 2 + 1 + 3 = 3 + 2 + 1 = 1 + 2 + 3$, and these should simply be counted as one way. Students need to be encouraged to record their thinking in a way that is meaningful to them. They should be encouraged to begin to think systematically to look for all possibilities:

$0 + 0 + 6 = 6$	$1 + 1 + 4 = 6$
$0 + 1 + 5 = 6$	$1 + 2 + 3 = 6$
$0 + 2 + 4 = 6$	$2 + 2 + 2 = 6$
$0 + 3 + 3 = 6$	

Share the different ways students found of making 6 and 8. *Do we have all the ways? How can we find out?* Show students how to order the information.

Look at the plates that have 6 + 0. What is one fewer than 6? (5) Show the plate with five. *How many does the other plate have?* (1) *What is one fewer than 5?* (4) Show the plate with 4. Continue with this pattern, looking for plates with 3, 2, 1, 0. *Look at the pattern of the numbers. What do you see?* (The number sentences before 3 + 3 are the same as the ones after 3 + 3 but the numbers are the opposite way around.) As each way of making 6 is found, write the number sentence on the board.

Practice Book: Students can complete Practice Book page 35. This can be done directly after the main activity, as homework, or as the focus of a separate mathematics session to help students consolidate their learning and build fluency. Look at the example together, explaining that the '6' is the whole and matches the total number of circles in the shape to the right and that the '2' and '4' are the parts; students will need to shade these numbers of circles in the shape.

Differentiated outcomes	
All students	should find two ways to make 6.
Most students	will find all the different ways to make 6.
Some students	may record systematically and explain how they know they have all the different ways of making 6 and 8.

Student Book page 28

3 $0 + 6 = 6$ $4 + 2 = 6$

   $1 + 5 = 6$ $5 + 1 = 6$

   $2 + 4 = 6$ $6 + 0 = 6$

   $3 + 3 = 6$

4 $0 + 8 = 8$ $5 + 3 = 8$

   $1 + 7 = 8$ $6 + 2 = 8$

   $2 + 6 = 8$ $7 + 1 = 8$

   $3 + 5 = 8$ $8 + 0 = 8$

   $4 + 4 = 8$

Practice Book page 35

1 5 circles shaded one way and 1 another way

2 3 circles shaded one way and 3 another way

3 6 circles shaded

4 6 circles shaded one way and 2 another way

5 7 circles shaded one way and 2 another way

6 8 circles shaded

Stretch zone: Number bonds to 8

0	+	8	8
1	+	7	8
2	+	6	8
3	+	5	8
4	+	4	8
5	+	3	8
6	+	2	8
7	+	1	8
8	+	0	8

2A Number bonds for 6, 7, 8, 9

Explore   Student Book page 29 • Practice Book page 36

- Begin to know number bonds to 6, 7, 8, 9.
- Understand addition as counting on and combining two sets.

- **Interpersonal skills:** communication / teamwork

- number bonds, addition, altogether, total, add (+), equals (=)

- sets of number cards 0–9, cubes or counters

The words 'add', 'take away', 'total' and the number names need to be clearly modelled throughout and connected to the correct symbols '+', '−' and '='.

Ask students to hold up five fingers, and then three more. *How many fingers are you showing?* (8) *Show me a different way of making 8.* (e.g. 3 + 5, 4 + 4) Repeat, asking students to show you five fingers again, and then two more. *How many are you showing me?* (7) *Show me a different way.* (3 + 4)

Discuss the different ways of making the same number. *How many ways can you make 6, using your fingers?* Ask students to talk to their partner. Give time for the activity, then share ideas and write them on the board: (1 + 5) (2 + 4) (3 + 3). *Can we have six add something? Why not?* (Because there are not six fingers on one hand.)

Main activity

Tell students they are going find as many different ways of making numbers as they can. Give each student a number card. Tell them they need to make a total of 5.

Look at your number. What number do you need to add to your number to make 5? Use your fingers to help you.

Give an example, such as: if they have the number card 4, they need the number card 1 to make 5. Ask each student to find another student with the card that they need. They then sit in that pair.

Allow time for students to complete the activity. Ask pairs to tell you the numbers on their cards. Check that they total 5.

Why have some of you not got a partner? (Because their number is more than 5.)

Repeat the activity, asking all students to find the number they need to make 7. Allow enough time for students to complete the activity. *Why do some of you not have a partner?*

Repeat, making a total of 9, so that everyone can find a partner.

Ask students to complete page 29 in the Student Book individually. Explain that they need to make a number bond for each of the numbers in the rectangles. Point out that in each model one of the parts has been given to them already. Work through the example together, making the number bond using cubes or counters. Say, *Two and four make six.*

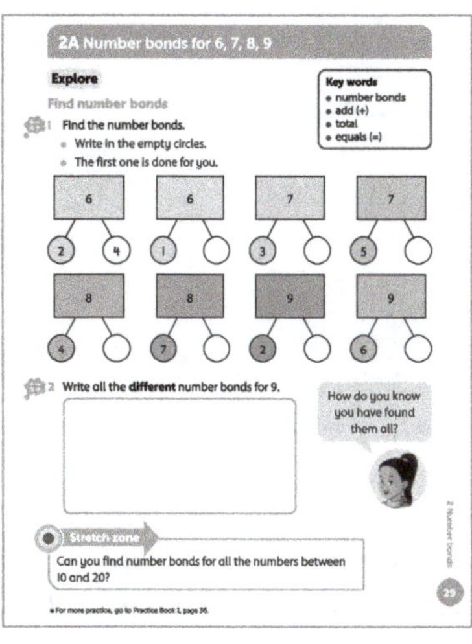

Differentiation

Supporting: Once students realise that a hand of fingers is 5, they don't have to count them all. They can start at 5 then count on to 6, 7, 8, … (If students prefer to count knuckles, this can be used instead of fingers. This will help to get rid of counting errors.)

Consolidating: Continue to ask for alternative answers for totals to 6, 7, 8 and 9.

Extending: Ask students to partition 6, 7, 8 and 9 using 3 numbers.

Stretch zone: *Can you find number bonds for all the numbers between 10 and 20?*

By finding all the number bonds to 20, students will begin to discover patterns when they try to order their number bonds. They will also discover that both numbers in a bond can be the same, in this instance, 10 + 10 = 20. Including '0' as a number provides the trivial examples of 0 + 20 and 20 + 0, but this brings up the notion that 0 is the additive identity, not changing any number it is added to.

Reflection time

Ask students for all the different ways they found to make 9. *Did you find them all? Are any number bonds more difficult than others? Who has any good ways of remembering them?*

Practice Book: Students can complete Practice Book page 36. This can be done directly after the main activity, as homework, or as the focus of a separate mathematics session to help students consolidate their learning and build fluency. Complete the first two questions together for bonds to 6, using cubes or counters, to model systematic working. For each question say, e.g. *One and five make six.*

Differentiated outcomes	
All students	should find partners in the main activity by matching the correct numbers to make the totals.
Most students	will record number bonds with numerals.
Some students	may work systematically to find all number bonds for 9.

Answers

Student Book page 29

1   6: 2, **4**   6: 1, **5**   7: 3, **4**   7: 5, **2**

     8: 4, **4**   8: 7, **1**   9: 2, **7**   9: 6, **3**

2   9 + 0   8 + 1   7 + 2

     6 + 3   5 + 4   4 + 5

     3 + 6   2 + 7   1 + 8

Practice Book page 36

0 + 6 = 6	0 + 7 = 7	0 + 8 = 8
1 + 5 = 6	1 + 6 = 7	1 + 7 = 8
2 + 4 = 6	2 + 5 = 7	2 + 6 = 8
3 + 3 = 6	3 + 4 = 7	3 + 5 = 8
4 + 2 = 6	4 + 3 = 7	4 + 4 = 8
5 + 1 = 6	5 + 2 = 7	5 + 3 = 8
6 + 0 = 6	6 + 1 = 7	6 + 2 = 8
	7 + 0 = 7	7 + 1 = 8
		8 + 0 = 8

Stretch zone:

0 + 9 = 9	4 + 5 = 9	7 + 2 = 9
1 + 8 = 9	5 + 4 = 9	8 + 1 = 9
2 + 7 = 9	6 + 3 = 9	9 + 0 = 9
3 + 6 = 9		

2B Number bonds for 10

Specific learning focus

- Begin to know number bonds to 10.
- Understand addition as counting on and combining two sets.
- Use the = sign to represent equality.

Global skills

- **Creative skills:** investigating
- **Interpersonal skills:** communication / teamwork

Key vocabulary

- number bonds, addition, altogether, total, add (+), equals (=)

Resources

- ten-bead string (5 red/5 white)
- number cards 0–9: a standard set for each student and a large set for teacher use; ones-cubes
- coat hanger with ten pegs

Language support

Make number and spot cards so that there is a visual image of the meaning of each numeral. Also support with numbers written in words, for example

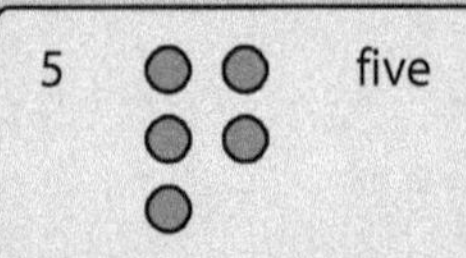

Introductory activity

Show the class the bead string. Count how many beads are on the string. Move the beads along the string as you count. Split the beads into two separate groups of 5 red and 5 white.

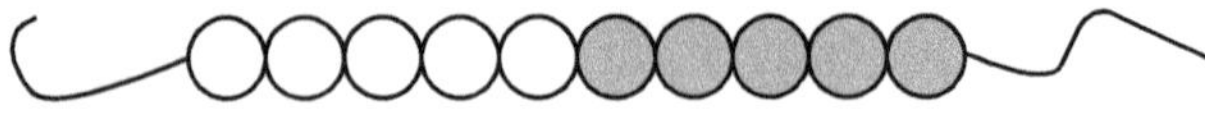

How many are red? (5)

How many are white? (5)

How many altogether? (10)

Split the beads in a different way.

On this side, I have 5 red and 3 white. How many is that? (8)

How many are on the other side? (2)

How many beads altogether? (10)

Repeat the activity, splitting the beads in a different way. Put the beads back together. Cover one bead with your fingers. *How many beads are hiding?* (1)

Demonstrate what you mean by hiding your face behind your hands and say, *I am hiding my face!*

Show the hiding bead. Repeat with other numbers. Always go back to counting the complete string of 10.

Main activity

Give each student a set of number cards 0–9 and ask them to place the cards in front of them so you can see them all. Explain that you are going to hold up a large number card and they will hold up the card that makes a total of 10. Hold up 9.

What number do I add to 9 to make 10? (1)

Write on the board: 9 + 1 = 10.

Hold up 1. *What number do I add to 1 to make 10?* (9)

Give time for students to choose their number. Write on the board: 1 + 9 = 10.

Continue playing the game until all of the numbers have been used.

Ask students to complete page 30 of the Student Book individually. Give students 10 cubes each to help them build number bonds to 10.

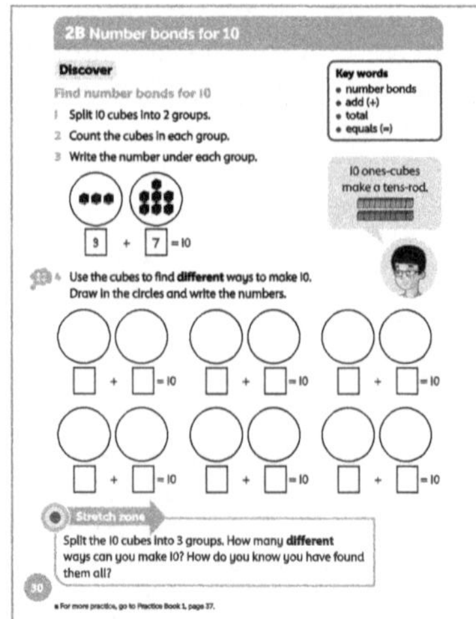

Differentiation

Supporting: Ask students to verbally give all the bonds that make 10. Write the bonds in the Student Book for them.

Consolidating: Ask students to show you the different bonds using cubes.

Extending: Ask students to look for number bonds to 20, using what they know about bonds to 10 to help them.

Stretch zone: *Split the 10 cubes into 3 groups. How many different ways can you make 10? How do you know you have found them all?*

This activity extends the notion from 2A Discover, where a given total (in this case, 10) is not simply partitioned into two numbers as number bonds, but partitioned into three sets in a range of ways. For example, $10 = 6 + 3 + 1$, or $10 = 5 + 3 + 2$ and so on. Explain to students that some number sentences represent the same partitioning, since for example, $10 = 7 + 2 + 1 = 2 + 7 + 1 = 1 + 7 + 2$ and these should simply be counted as one way. Finding all the combinations, even without repeats (ignoring order), will be challenging for students unless they have a systematic way of recording the different partitions:

$1 + 1 + 8$	$2 + 2 + 6$	$3 + 3 + 4$
$1 + 2 + 7$	$2 + 3 + 5$	
$1 + 3 + 6$	$2 + 4 + 4$	
$1 + 4 + 5$		
$1 + 5 + 4$		

Reflection time

Clip ten pegs onto a coat hanger. *Count the pegs with me: 1, 2, 3, 4, …, 10.*

Ask a student to push one of the pegs to one side.

How many pegs are there altogether? How many pegs at this end? How many pegs at that end?

Read the number sentence to the class (e.g. $1 + 9$ makes 10) and ask them to say it back. Ask a student to split the pegs to make a different number sentence. Record this by using '='.

Practice Book: Students can complete Practice Book page 37. This can be done directly after the main activity, as homework, or as the focus of a separate mathematics session to help students consolidate their learning and build fluency. Relate the bar model representation to the part-whole models on page 29 of the Student Book, emphasising the missing parts and wholes. Model the example using cubes and encourage students to use them throughout.

Differentiated outcomes	
All students	should list the number bonds to 10 in the Student Book.
Most students	will quickly find the bond to 10 in the Introductory activity.
Some students	may explore number bonds to 20.

Student Book page 30

Find different ways to make 10. Check that students have drawn the correct numbers of cubes to match the number sentence.

$10 + 0 = 10$	$9 + 1 = 10$
$8 + 2 = 10$	$7 + 3 = 10$
$6 + 4 = 10$	$5 + 5 = 10$
$4 + 6 = 10$	$3 + 7 = 10$
$2 + 8 = 10$	$1 + 9 = 10$
$0 + 10 = 10$	

Practice Book page 37

1 $7 + 3 = 10$

2 $1 + 9 = 10$

3 $2 + 8 = 10$

4 $5 + 5 = 10$

5 $8 + 2 = 10$

6 $6 + 4 = 10$

7 $3 + 7 = 10$

Stretch zone: Check that students have written the number sentence that matches their colouring. For example, if a student has coloured 2 yellow circles and 8 red circles, then they should have written $2 + 8 = 10$ or $8 + 2 = 10$.

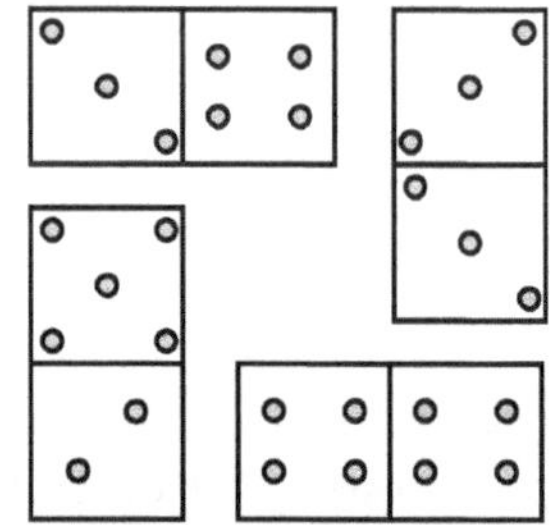

Explore Student Book page 31 • Practice Book page 38

Specific learning focus

- Begin to know number bonds to 10.
- Understand addition as counting on and combining two sets.

Global skills

- **Creative skills:** problem solving
- **Interpersonal skills:** communication / teamwork

Key vocabulary

- number bonds, addition, altogether, total, add (+), equals (=)

Resources

- set of dominoes for each group, or domino cards cut out from Resource Sheet 1 (available on the Oxford Owl website), copied onto card
- number cards 0–10
- counters
- ten small objects (e.g. plastic figures, pebbles, cubes), feely bag

Language support

Joining in with class counting allows students to develop instant recall.

 **Introductory activity**

Explain what table tennis is, if necessary. Then say that you are going to play a game of table tennis, but without a ball! Explain you are going to use numbers instead of a ball. You will 'hit' a number to them and they must 'return' the number that makes 10. For example, 9 (1), 6 (4), and so on. Increase speed as students become more confident.

 **Main activity**

Demonstrate a dominoes game to the whole class by playing the game with three students. Put the dominoes (or domino cards from Resource Sheet 1) face down on the table. Students choose four each, then turn them over to see the spots. They should match up dominos to make a total of 10 at the join. This is possible with 6 and 4, and 5 and 5.

Now they should arrange the four dominoes as a square. They must try to make all the sides total 10 by adding the total spots on one domino with one side of spots on another domino.

Students can move the dominoes as much as they want to. If the sides don't total 10, they can choose another domino from the pile. They must replace a domino for each one they take so that they always have four dominoes. They can swap a domino with another player.

Ask students to play the game in groups of four. As they work, ask *Can you find another way to add three of the numbers to make 10?*

Ask students to complete page 31 in the Student Book. Check that they are clear on how scales balance before they begin. If you have maths balance scales, model the activity first using these. Explain the activity, asking questions such as *What number is on one of the sides of each balance scale? How many numbers do you need to put on the other side of each scale to balance?* (2) *What will the total of these numbers equal?* (10)

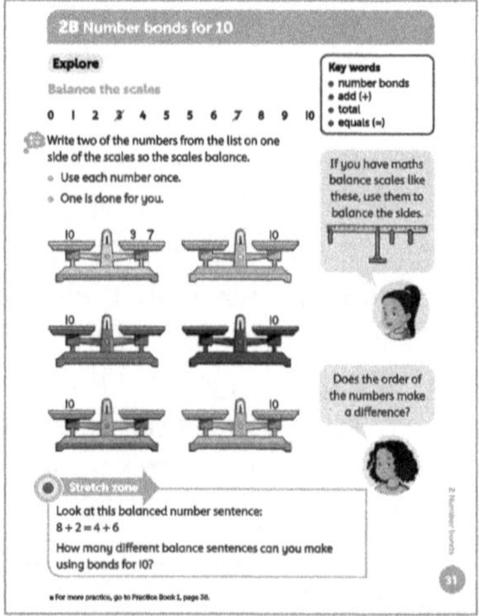

Differentiation

Supporting: Give students number cards or counters and ask them to count on a number line to help them with the Student Book activity.

Consolidating: Ask for alternative ways of making 10 with three numbers that include a number greater than 6. (e.g., 10 = 7 + 2 + 1)

Extending: Ask students if they can make a larger domino square so that the totals are always 10.

Stretch zone: *Look at this balanced number sentence: 8 + 2 = 4 + 6*

How many different balance sentences can you make using bonds for 10?

This activity is designed to show students that pairs of number bonds to 10 are all equal to each other. Ask students to work out the total on each side of the equals sign (10) and explain that because the totals are the same this is a balanced number sentence. By choosing a pair of number bonds for 10, students can write balanced number sentences, which also helps them understand the notion of '='.

Keeping a recorded list of the different possibilities should help students to omit any repeats, and they may begin to notice ways or ordering the sentences and spotting patterns.

 **Reflection time**

Put ten small objects (e.g. cubes) into a feely bag.

Ask a student to choose a number from 1 to 10. Ask a different student to take that number of objects out of the bag and say how many are left. Check together. Write it as a number sentence on the board. Repeat several times.

Practice Book: Students can complete Practice Book page 38. This can be done directly after the main activity, as homework, or as the focus of a separate mathematics session to help students consolidate their learning and build fluency.

Differentiated outcomes	
All students	should find number bonds to 10.
Most students	will find several groups of three numbers that total 10.
Some students	may investigate using all the dominos to make totals of 10.

Answers

Student Book page 31

Check that the numbers that students have written on the empty side of the scales total 10. The order of the numbers in each pair doesn't matter.

Practice Book page 38

Check that the number sentence written beneath each domino is correct and that it matches the numbers of spots drawn on the domino. For example, students could have written $8 + 2 = 10$, $3 + 7 = 10$.

2C Missing numbers

Discover Student Book page 32 • Practice Book page 39

Specific learning focus

- Find a number 2 more or 2 less than 20, recording the jumps on a number line.
- Within the range 0 to 30, say the number that is 1 or 10 more or less than any given number.

Global skills

- **Creative skills:** problem solving
- **Interpersonal skills:** communication / teamwork

Key vocabulary

- count on, count back, missing number, total, add

Resources

- counting stick, number cards 0–20, sticky tack

Language support

Make speech bubbles such as, 'How many more?' and 'How many less?' with arrows to show counting on and counting back, to remind students that 'how many more' means counting on (forwards arrow) and 'how many less' means counting back (backwards arrow).

 **Introductory activity**

Stick number cards 1–10 on a counting stick held horizontally to form a number line. **Count on** (forwards) and **count back** to 10. Remove some numbers and count again. Ask students to count all the numbers from 1 to 10 including those that have been removed from the stick.

*How did you know what the **missing numbers** were?* Change the numbers to 10–20, leaving some missing, and count on and back. Ask some students to fill the gaps on the stick with the missing numbers by counting along and adding the number cards where they are missing.

 **Main activity**

Put number cards 2 and 6 on the counting stick. Jump from 2 to 6 in ones, pointing to each number in turn. Count the jumps as you point. Write on the board $2 + 4 = 6$. Repeat with two different numbers. Ask each pair to draw their own 0–20 number line. They choose two numbers, then find the number of jumps from one number to the other. Each pair then writes what they do – first writing the missing number sentence and then the completed number sentence, e.g. $7 + [\ ] = 9$ then $7 + 2 = 9$.

Ask students to complete page 32 in the Student Book. They can use the number line at the top of the page to help and should work in pairs to support each other.

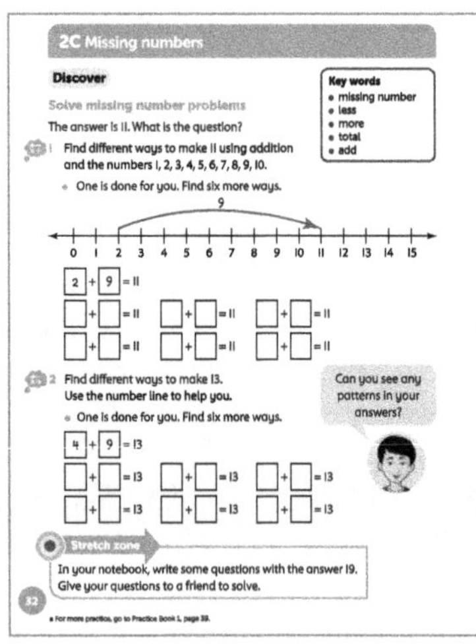

Differentiation

Supporting: Use extra number cards to support the missing numbers activity so that students can move the numbers around.

Consolidating: Ask students to explain their strategies.

Extending: Ask students to make up their own missing number activity like those on page 32 of the Student Book.

Stretch zone: *In your notebook, write some questions with the answer 19.*

Give your questions to a friend to solve.

In this activity, students are asked to find some number bonds to 19. They should be encouraged to use what they have learned to systematically record various possibilities from 0 + 19 to 19 + 0. Some students might use their knowledge of number bonds to 20 to help them.

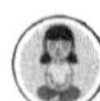 Reflection time

Check that students have completed the boxes in the Student Book correctly. Call out two numbers and ask, *How many more to make …?* or *How many jumps to make …?* Record the answers on the board.

Encourage students to look for patterns in their answers in the number bonds to 11 and 13. *What do you notice about the two number used?* For example, they may notice that one number is odd and the other is even.

Show students that, since each pair adds to 11 (or 13), the pairs are equal to each other, for example, If 2 + 9 = 11 and 5 + 6 = 11, then 2 + 9 = 5 + 6. *Can you recall what we call this?* (Balanced number sentence)

Practice Book: Students can complete Practice Book page 39. This can be done directly after the main activity, as homework, or as the focus of a separate mathematics session to help students consolidate their learning and build fluency.

Differentiated outcomes	
All students	should understand how a number line can help in calculation.
Most students	will be able to use the number line to find missing numbers.
Some students	may support others in understanding the use of the number line.

Answers

Student Book page 32

Check that students' additions are correct and total 11 each time for question 1, and 13 each time for question 2.

Practice Book page 39

1 15

2 38

3 17

4 19

5 4

6 24

7 22

8 9

9 6 + 4 = 10

10 3 + 6 = 9

11 16 − 7 = 9

12 1 + 4 = 5

2C Missing numbers

Explore 1   Student Book page 33 • Practice Book page 40

Specific learning focus

- Complete addition and subtraction number sentences.

Global skills

- **Creative skills:** problem solving
- **Interpersonal skills:** communication / teamwork

Key vocabulary

- missing number, how many more? how many less? number sentence, addition sentence, subtraction sentence

Resources

- number cards 0–20
- counting stick, sticky tack
- paper cup and 10 cubes per pair
- recording sheet of cups and cubes (as described below)

Language support

Display addition and subtraction stories with concrete resources (e.g. 3 red cubes and 5 blue cubes interlocked), words (e.g. '5 blue cubes and 3 red cubes make 8 cubes') and symbols +, – and = (e.g. 5 + 3 = 8) to reinforce the connection between the different representations.

 **Introductory activity**

Show the class the counting stick. Hold the stick vertically and stick on the number cards 6–15. Ask students to count forwards and backwards.

Say, *Close your eyes. I am going to take some numbers away.* Remove five or six of the numbers. *Open your eyes. Count with me up and back down.* Students should count all the numbers in sequence, including those where the number card is missing from the stick. Repeat with different start and end numbers, changing the orientation of the stick.

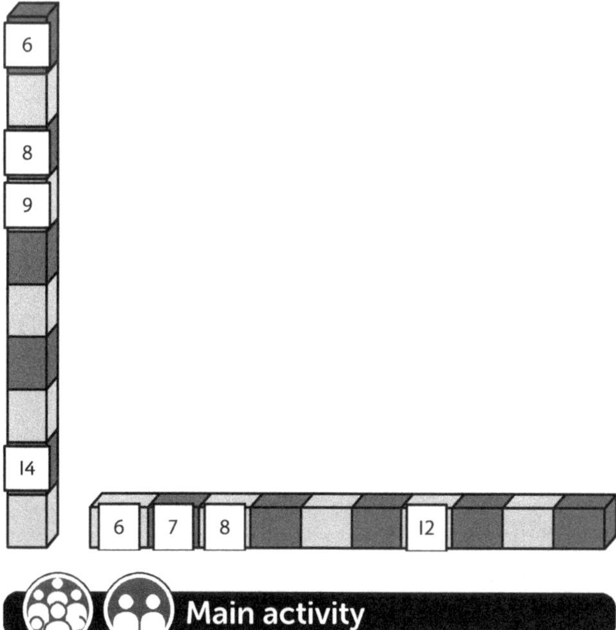

 **Main activity**

Give each pair of students a paper cup and 10 cubes. Demonstrate the activity by showing the class an empty cup. Count 6 cubes out loud then hide 2 of them in the cup. *How many cubes do I have left?* (4) *How many are inside the cup?* Record this on the board as shown below.

Total   6

Outside the cup 4

Inside the cup ?

Record this as a **number addition sentence:** 4 + [   ] = 6 and say, *4 outside and some inside makes 6. How many are inside?* (2)

Next, write the **subtraction sentence** 6 – [   ] = 4 and say, *I have 6, I hide some in the cup and have 4 left. How many are in the cup?* (2)

Repeat with a different start number, and change the numbers of cubes. Include examples with more than one missing number box.

When students are confident with the task, ask them to work in pairs to count out cubes and hide some under the cup for their partner to work out. They should record the sentence using a box for the missing number, e.g. 3 + [   ] = 7; 7 – [   ] = 3

Ask students to complete page 33 in the Student Book. They can use the number line at the top of the page to help and should work in pairs to support each other. They may also choose to continue to use cubes alongside the number line or to check their answers.

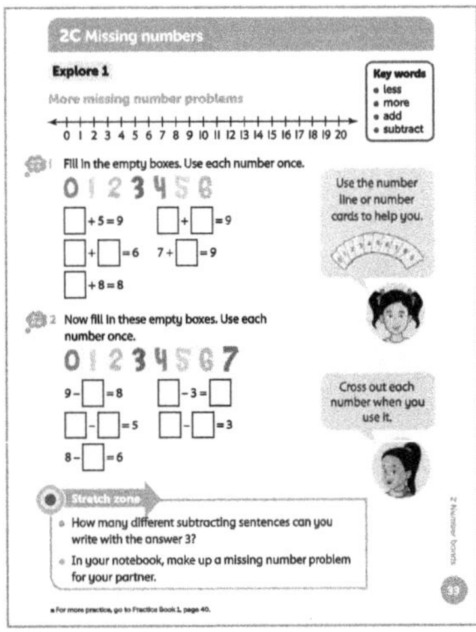

Differentiation

Supporting: As students record number sentences, you can help them write the answers by, for example, saying 'add' and recording '+'.

Consolidating: Ask students to read number sentences aloud as they write them.

Extending: Ask students to construct their own missing-number problems with pictures and number sentences.

Stretch zone: *How many different subtracting sentences can you write with the answer 3?*

In your notebook, make up a missing number problem for your partner.

Students explore number sentences that contain two numbers that have a difference of 3. If they work systematically, they might start with $3 - 0 = 3$, and then add 1 to each number in the calculation to get $4 - 1 = 3$, $5 - 2 = 3$, and so on. You could encourage them to use the biggest numbers they can to get an answer of 3, e.g. $20 - 17 = 3$.

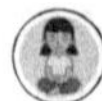 **Reflection time**

Invite two students to come to the front of the class. One student takes a handful of cubes (e.g. 8 in total) and counts them out loud with the class. With their back to the class, the second student takes the cubes and puts some in one hand and some in the other. The students then hides one hand behind their back and faces the class with the other hand palm up, showing the cubes.

Ask the class, *How many cubes are hiding in his/her other hand?*

How do you know? How can we write this as a number sentence?

Repeat with other numbers of cubes and pairs of students.

Practice Book: Students can complete Practice Book page 40. This can be done directly after the main activity, as homework, or as the focus of a separate mathematics session to help students consolidate their learning and build fluency. Remind students of what the + and – signs represent and provide them with cubes so that they can 'build' each number sentence.

Differentiated outcomes	
All students	should be able to count the total number of cubes and the part that is not hidden, and be able to work out the other hidden part.
Most students	will be able to record number sentences.
Some students	may construct their own missing-number problems and work out the answers.

Answers

Student Book page 33

1 $4 + 5 = 9$

$1 + 5 = 6$

$0 + 8 = 8$

$3 + 6 = 9$

$7 + 2 = 9$

2 $9 - 1 = 8$

$5 - 0 = 5$

$8 - 2 = 6$

$6 - 3 = 3$

$7 - 4 = 3$

Practice Book page 40

Check that students have used the addition sign (+) and subtraction sign (–) correctly in their number sentences and that the number sentences are correct.

2C Missing numbers

Explore 2   Student Book page 34 • Practice Book page 41

Specific learning focus

- Complete missing number sentences.

Global skills

- **Real-world skills:** presenting information
- **Interpersonal skills:** communication

Key vocabulary

- missing number, how many more? how many less/fewer?

Resources

- number cards 0–20
- counting stick
- paper plates, counters, cubes

Language support

Use speech bubbles and arrows to show the meanings of 'count on' and 'count back'. Model the use of language for the number sentences – add, equals, total, altogether – by pointing to each part of a number sentence as you read it aloud, reinforcing the words for each numeral and symbol.

 Introductory activity

Hold up a number card drawn from a set 0–10. Ask students to say the number. Now ask them which other number it needs to be paired with to make 10. How do they know? Ask students to count on together from the card number to 10.

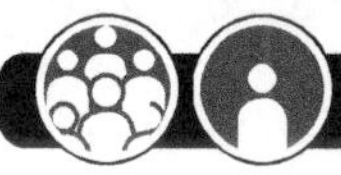 Main activity

Put 9 cubes on a plate. Tell students the cubes represent cupcakes. *How many cakes have I got on the plate? What is the total?* (9)

I would like to put some on another plate to give to my friend. How could I do that? Ask a student to come up and separate the cubes, so there are some on each plate (e.g. 6 and 3).

Count together how many are on each plate and how many there are altogether. Write the number sentence to represent the partition, e.g. 3 + 6 = 9.

Put all the cubes back onto one plate and ask another student to come and put some on the empty plate. Count the cubes on each plate and write the number sentence to show how many.

Students can then work on the Student Book page 34 individually. Give each student 12 cubes to help them work out how many balls to draw for each.

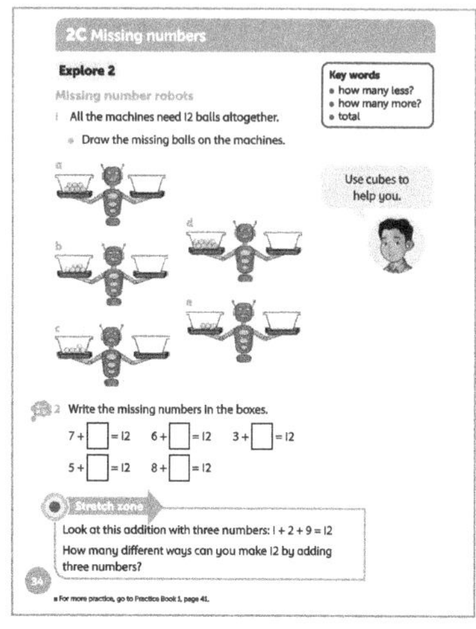

Differentiation

Supporting: Give students cubes to help them represent each question.

Consolidating: Ask students to read number sentences aloud as they write them.

Extending: Ask students to construct their own missing-number problems with pictures and number sentences.

Stretch zone: *Look at this addition with three numbers: 1 + 2 + 9 = 12. How many different ways can you make 12 by adding three numbers?*

Check that students understand how to add three numbers and write the number sentence. There are many different ways to do this.

 Reflection time

Ask students to describe some of the ways they shared the 9 cubes between the plates. *Did you count by counting on or by counting back? How many different ways did you find to share the cubes between two plates?*

Use 11 cubes and share them out on 3 plates. Cover one of the plates and ask students how many are covered. How did they work it out?

Practice Book: Students can complete Practice Book page 41. This can be done directly after the main activity, as homework, or as the focus of a separate mathematics session to help students consolidate their learning and build fluency. Students could record each combination they make as a number sentence as well.

Differentiated outcomes	
All students	should be able to count the number of cubes and work out how many on one plate if they know how many in total.
Most students	will count how many on each plate and write the number sentence.
Some students	may make their own missing cubes problem using 9 cubes split over 3 plates in different ways.

Student Book page 34

1 a 5 **b** 6 **c** 7 **d** 4 **e** 9

2 7 + 5 = 12 6 + 6 = 12 3 + 9 = 12

5 + 7 = 12 8 + 4 = 12

Practice Book page 41

1–4 Check that students have put a number on each side of the scales so that the total is 15 each time.

5 7 + 8 = 15

6 5 + 10 = 15

7 6 + 9 = 15

8 8 + 7 = 15

9 3 + 12 = 15

Stretch zone: Check that students have completed the number sentence to show three numbers that add to make 15. There are many different ways to do this.

2 Number bonds

Connect Student Book page 35

Big idea

- There are lots of different ways to make a number. We can use number bonds to make 6, 7, 8, 9 and 10.

Global skills

- **Creative skills:** investigating
- **Real-world skills:** presenting information / financial literacy
- **Interpersonal skills:** communication / teamwork

Key vocabulary

- number bonds, addition, altogether, total, add (+), equals (=)

Resources

- a selection of real or plastic American 1-, 5-, 10- and 25-cent coins (or another currency that does not involve making amounts over 50)

Language support

Write any name and value equivalences on the board for reference, e.g. 10c = 10 cents = 1 dime.

Introductory activity

Ask students how many ways there are to add two numbers to make 4. (three: 0 + 4, 1 + 3, 2 + 2) There are five ways if 3 + 1 and 4 + 0 are included. Write them on the board. *Will there be more ways to make 5? Tell me a way to make 5.*

Write each one on the board as students say them: 0 + 5, 1 + 4, 2 + 3, 3 + 2, 4 + 1, 5 + 0 (6 ways).

What do you notice about 3 add 2 and 2 add 3? (Both use the same numbers to make the answer 5.) *Does it matter which order the numbers are in? What about 5 add 3 and 3 add 5?*

Main activity

Tell students they are going to imagine using a sweet dispenser that takes coins to buy different sweets. Show them an image of dispenser if they are unfamiliar with them. The machine will only take the exact amount for each purchase – they can only use coins that total the price of the sweets. So they have to decide which coins to use to make each amount.

The machine will take the following coins: 1¢, 5¢, 10¢, 25¢. Sweets cost 30¢. Which coins could you put in the machine to pay for the sweets? (e.g. 25¢ + 5¢) Could you use a different set of coins for the same amount? (e.g. 10¢ + 10¢ + 10¢) Is there another way? (e.g. thirty 1¢ coins)

Ask students to work in pairs to see what other amounts they can make using exactly two different coins.

Students can then work on the Student Book page 35 in pairs to make 10 cents. If students have had limited experience with coins, take some time to discuss that coins have different values, and that amounts can often be represented using different combinations of coins. You may also choose to do this activity working as a whole class.

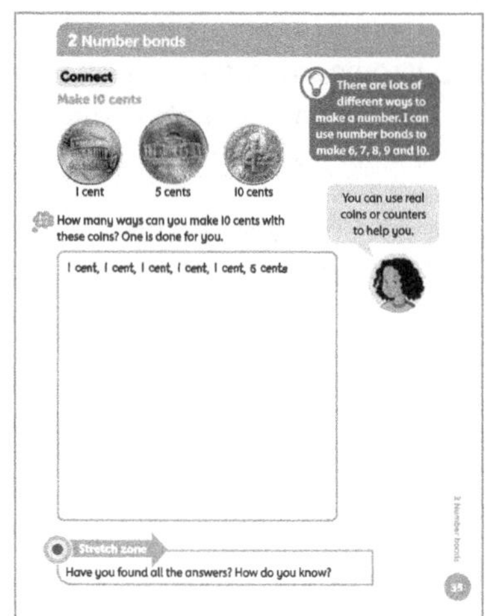

Differentiation

Supporting: Ask students to work in mixed-attainment pairs for support, and give them real coins or counters to help them.

Consolidating: Ask students for alternative ways of finding totals.

Extending: Ask more confident students to support less confident students and to explain how to work out a total of two coins.

Stretch zone: *Have you found all the answers? How do you know?*

Students will need to record possible ways of making the amount in different coins, and then check for any repeats. By starting with each coin in turn from the largest value, they will find there are only four different ways of doing this.

10 cents

5 cents, 5 cents

5 cents, 1 cent, 1 cent, 1 cent, 1 cent, 1 cent

1 cent, 1 cent, 1 cent, 1 cent, 1 cent, 1 cent, 1 cent, 1 cent, 1 cent, 1 cent

 **Reflection time**

Ask students to share some of their ways of making an amount with exactly two different coins. List their responses until they have shared all the possible combinations.

Can they see any patterns in the coins chosen each time? Is there a systematic way to list them to check that all combinations have been found?

Differentiated outcomes	
All students	should work out how to make the given amounts in at least one way.
Most students	will work out more than one way to make an amount.
Some students	may work out all possible ways of making a given amount.

2 Number bonds

Review Student Book page 36 • Practice Book page 42

Global skills

- **Creative skills:** problem solving / exploring
- **Real-world skills:** presenting information
- **Interpersonal skills:** communication
- **Self-development skills:** reflecting on learning

Student Book

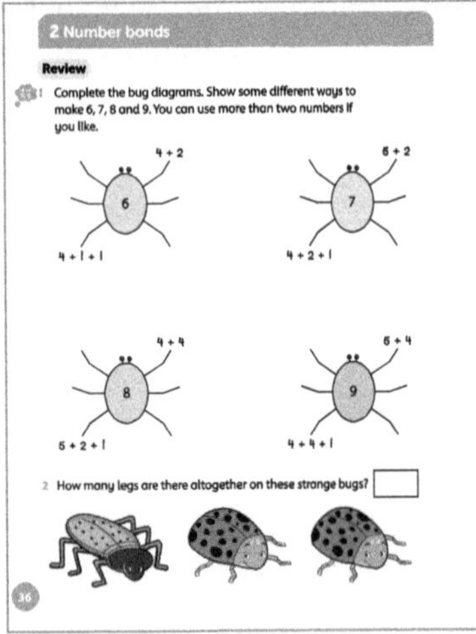

With young children, assessment activities are most effective when carried out as an everyday classroom activity. Students should have number lines available with both the digits and the numbers in words so they can refer to these to support them. Some students may also need cubes to help them.

Watch as students complete the bugs with different number bonds. Listen to them as they count, to make sure that they say the number words in order and do not omit any numbers. Also listen to check that they understand that the number in the middle of the bug is the total. If there are errors, add the numbers for the students and read them out so they can hear the correct sequence.

It may help to create flash cards with the numbers and words that are in the review. You can use these to support for some students.

Encourage students to use a number line for question 2. Extend by asking a range of similar questions or asking the students to make their own bug leg problems.

Answers

Student Book page 36

1 Students record their answers in different ways, writing out additions. These are possible answers for 6, using up to three numbers:

$3 + 3 = 6$ $2 + 2 + 2 = 6$ $3 + 2 + 1 = 6$

$4 + 1 + 1 = 6$ $5 + 1 = 6$ $0 + 6 = 6$

The possible answers for 7, 8, 9 are more numerous. Check that students' additions total the correct number.

2 14 legs

Practice Book

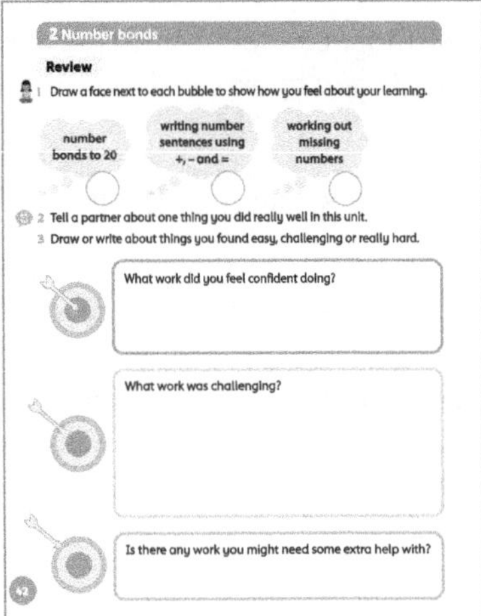

With young children it is appropriate to complete this as a whole-class discussion. You may choose to keep a record of the class discussion or a copy of the review page for your own records. Use the Student Book to briefly remind students of the areas of mathematics.

Give out the Student Book to support pairs of students as they discuss and answer the questions in the Practice Book.

Allow students plenty of time for discussion before asking them to share their responses with the rest of the class. If students complete this assessment at home, encourage them to discuss this with adults. Make a note of areas that students still feel unsure about.

As counting is at the heart of many of the other units you can revisit these areas regularly. You can also build counting into everyday practice: for example, counting how many students are in the class each day; counting objects as you give them out, and so on.

Additional material

There are additional end-of-unit assessments available on the *Oxford Owl* website.

Overview

Big Idea

Exploring numbers and developing number sense is important. Some students come to school with very little experience of how the number system works.

To promote this development, use these strategies:

- Speaking – encourage students to count out loud and explain their thinking. Model correct language and vocabulary.
- Listening – model more efficient counting strategies and count on from another number. Listen carefully to students.
- Reading – look at a collection of objects and say how many without counting. Show students how to read numbers such as tens and ones.
- Writing – explore informal writing. Delay formal recording until students have a good understanding of numbers and the number system.

Also focus on the following key ideas:

- Early number work involves more than counting.
- Place value understanding can lead to number sense and efficient strategies for calculating.
- There is a number word and symbol that tells us how many are in a group.
- One-to-one correspondence and counting can be used to compare groups and find which has more, which has fewer, or whether the groups are the same.
- Objects can be ordered in a specific ordinal sequence. Ordinal words, such as first, second, third, fourth and fifth, describe ordinal positions.
- The use of practical equipment is crucial in developing number sense. Materials that can be grouped (e.g. straws, lolly sticks, interlocking cubes) enable students to build a sense of tens by grouping.

Look out for

- **Students who recite numbers in order, but are not counting.** Practise one-to-one correspondence – matching one object to another object, or a number name to that number of objects. For example, the student should be able to say '1, 2, 3, 4, …' while counting or moving one object for each number said aloud. If an object is counted twice or one of the objects is missed, one-to-one correspondence has not been mastered.
- **Students who have difficulty with place value because it is abstract**. Young students are concrete learners, meaning they need sensory experiences to develop their learning. Using manipulatives, such as base-ten equipment, can help students connect the abstract with more concrete representations.
- **Students who have an inadequate part-whole knowledge for numbers 0–10 (understanding that numbers can be split into two parts that make up the whole) or an inability to trust the count.**

Possible misconceptions

- **Students may think counting can only be done in 'ones'.** They may not be able to recognise 2, 5 and 10 as countable units, affecting their ability to count large groups efficiently as they only count in units of 1.

Key vocabulary

- more, less, fewer, more than, less than, fewer than, how many
- lowest, highest
- ones, tens, between, digits
- ordinal numbers, first, second, third, …, tenth, 1st, 2nd, 3rd, …, 20th, order
- place-value table, 2-digit number
- equal parts, odd, even, pattern, repeating pattern
- 10 more, 10 less, above, below, diagonally, count on, count back

Learning focus	Learning outcomes (the ENC objectives)
More and less	Given a number, identify 1 more and 1 less.
Between	Identify and represent numbers using objects and pictorial representations including the number line, and use the language of: equal to, more than, less than (fewer), most, least.
Tens and ones	Count, read and write numbers to 100 in numerals; count in multiples of twos, fives and tens.
Partitioning	Represent and use number bonds and related subtraction facts within 20.
Ordering numbers	Count to and across 100, forwards and backwards, beginning with 0 or 1, or from any given number.
Even and odd	Count, read and write numbers to 100 in numerals; count in multiples of twos, fives and tens.

3 Exploring numbers

Engage   Student Book page 37

Big Question

- What special things can we say about numbers?

Global skills

- **Real-world skills:** research
- **Interpersonal skills:** communication / teamwork

Key vocabulary

- more than, less than, first, second, third, ….., tenth, order

Resources

- large 0–20 number line for the front of the class
- set of large number cards 0–20
- sales brochures or catalogues

Language support

Listen to the vocabulary being used between pairs of students during the activity. Demonstrate 'more' and 'less' on the number line. Label the number line using arrows to show that numbers to the left are smaller and numbers to the right are larger.

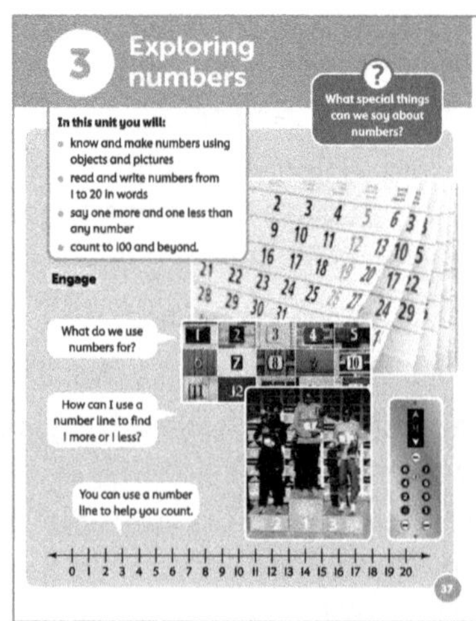

Introductory activity

Sit the class on the floor. Show them the number cards 0–20 in order. Ask students to say the numbers with you as you look at the cards.

Shuffle the cards and place them face down on a table. Pick up the top card and show it to the class. *What number is this? Do you think the next number will be more or less? If you think it will be more, put your hands in the air* (demonstrate as you say this). *If you think it will be less, put your hands on the floor* (demonstrate). Turn the next card over. *What number is this? Is it more or less? Who was right?*

Put the card at the bottom of the pile and pick the top card again. Repeat the game until you get back to your first number.

 **Main activity**

If you have access to an IWB, you could display page 37 of the Student Book. Use it as a starting point for discussing with students where they see numbers in their life at home, out walking or at school. Say that sometimes numbers are joined together to make bigger numbers (e.g. on a bus), sometimes they are in order. *Count with me to 10: 1, 2, 3, 4, …, 10.* Hand out brochures and catalogues and ask students to work in pairs to find a picture that has the numbers 1–10 in order (e.g. phone, calculator, calendar).

Draw a row of five runners on the board with blank race numbers on their vests or choose five students to be 'runners'. *These runners are about to start the race in order.*

Choose a number card from 1 to 10 and write the number on the **first** runner on the board (or give the number card to the first student 'runner'). Ask students to talk to a partner to decide what the next number will be. Ask a student to come to the front and write the next four numbers on the board (or give a card to the next four 'runners').

Ask a student to come to the front and select a number card at random. They hold it up. Ask pairs to write down a number that is **less than** this number. Wait five seconds and say, *Show me.* Repeat for a number that is **more than** the card number. Again, wait five seconds and say, *Show me.*

Use a number line to consolidate counting to 20.

Differentiation

Supporting: Count using the number line so that students connect the numeral with the words.

Consolidating: Focus on 1 more than and 1 less than. *What is 1 more than …? What is 1 less than …?*

Extending: *What is 2 more than …? What is 2 less than …?* See how far students can count beyond 20.

 **Reflection time**

Sit students in a circle on the floor. Give them number cards in a random order. Count round the circle 1, 2, 3, …

Are the numbers in the right order? (No.) *We need to change them. We need 2 after the 1.* Ask the student after 1 to change places with the student holding 2, so that 1 and 2 are in the correct order. *Let's count again: 1, 2, … Is that right?* Repeat until they are all in the correct order.

3A More and less

Discover Student Book page 38 • Practice Book page 43

Specific learning focus

- Use 'more' or 'fewer' to compare two quantities of cubes.

Global skills

- **Creative skills:** investigating
- **Interpersonal skills:** communication / teamwork

Key vocabulary

- more than, less than, fewer than, how many

Resources

- interlocking cubes
- two dishes (one red, one blue) per group
- number cards 0–20
- 0–20 number track

Language support

Put labels on the number track to show 'more than' and 'fewer than'. Count with students so that they hear the correct vocabulary modelled.

Introductory activity

Show the class the interlocking cubes.

Hold up number card '7'. *What is this number?* **How many cubes do I need to get?**

Choose a student to come to the front of the class and collect that number of cubes.

Show number 3. *What is this number? How many cubes do I need to get?* Choose a student to come to the front of the class and collect that number of cubes.

Do I have more cubes for the number 7? Or more cubes for the number 3? How can we find out? Allow students to offer some suggestions. We can count them or we can match them. If there are more cubes for the number 7, then we say there are **fewer** cubes for number 3.

Have one student stand up and hold up 7 fingers and another stand up and hold up 3 fingers. *Which student is holding up more fingers? Which student is holding up fewer fingers? We can say that 7 is more than 3, and 3 is less than 7.*

Put 7 and 3 on the number track. *7 is further along the track. That means that 7 is more than 3. What number is more than 7? Look at the track to help you.* Accept any answer between 8 and 20. *Who can tell me a number that is less than 3?* Accept 0, 1, 2.

Write on the board:

3 is less than 7.

3 cubes are fewer than 7 cubes.

When we count numbers, we say 'less than', but when we count cubes we say 'fewer than'.

Main activity

Ask students to work in pairs; give each pair two dishes and some interlocking cubes. Tell them they are going to work in pairs to find out as much as they can about numbers, which numbers are more or less than other numbers, and record their work on page 38 of the Student Book. If you do not have enough red and blue dishes for pairs to have one of each, use different descriptions for the dishes as appropriate, e.g. *Liam and Sasha, you have a spotty dish and a red dish. Each time you read blue, think about your spotty dish.*

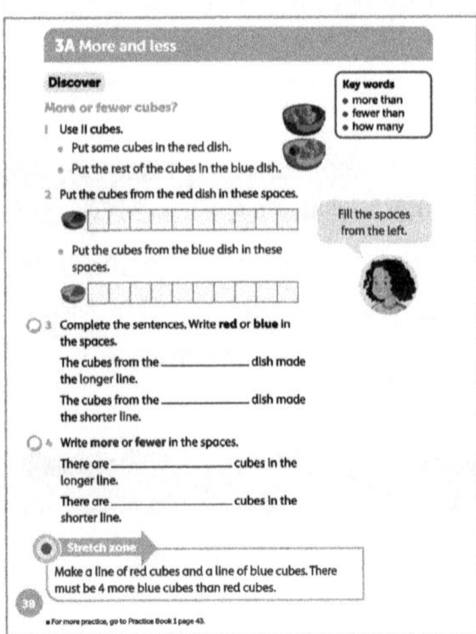

Tell me what you know about numbers. How did you find out which of two numbers was more or less? Did you have to count every time? What else could you do?

Differentiation

Supporting: Use the number track to support counting. Label the blank number tracks in the Student Book.

Consolidating: Ask for other numbers that are 'more' or 'less' than the numbers of cubes in the dishes.

Extending: Compare three numbers by making a track which is in between the two tracks they have made and providing a third dish.

Stretch zone: *Make a line of red cubes and a line of blue cubes. There must be 4 more blue cubes than red cubes.*

Students need to be aware of the number of cubes in each line and be able to compare them. They should notice that one line is longer than the other, but they need to apply their ability to count to make sure there are four more blue cubes than red cubes.

 **Reflection time**

Say you are thinking of a number of cubes. Choose different students to ask you questions, e.g. Is it more or fewer than another number of cubes? After ten questions, choose a student to tell you your number. (As students play this game, they get better at asking questions and may not need ten questions.) The student who guesses the number correctly can choose a number for the rest of the class to find.

Practice Book: Students can complete Practice Book page 43. This can be done directly after the main activity, as homework, or as the focus of a separate mathematics session to help students consolidate their learning and build fluency. Look at the worked example and ask students to count how many in the first column. (5) How many in the second? (9) Explain that, using the pictures of cubes for support, they need to find out how much more or less the first number is than the second number. Read out the answer: 5 is 4 less than 9.

Differentiated outcomes	
All students	should compare two groups accurately.
Most students	will be able to say numbers that are more or less than a given number.
Some students	may work with numbers of cubes beyond 20.

Answers

Student Book page 38

Observe students as they work through the activity and check that they make the connection between comparing the lengths of the cube lines and which line has more cubes.

Practice Book page 43

1   7 is 6 more than 1.

2   13 is 5 more than 8.

3   11 is 2 more than 9.

4   3 is 3 less than 6.

5   9 is 3 less than 12.

3A More and less

Explore Student Book page 39 • Practice Book page 44

Specific learning focus
- Use 'more' or 'less' to compare two numbers.

Global skills
- **Creative skills:** investigating
- **Interpersonal skills:** communication / teamwork

Key vocabulary
- more than, less than, between

Resources
- 'more' and 'less' cards (several of each), number cards 0–20
- large 0–20 number line for display
- dice, counters

Language support

Put labels on the number line to show 'more than' and 'less than'. Count with students so that they hear the correct vocabulary modelled. Help them to formulate questions to compare numbers using 'more than' and 'less than'.

Introductory activity

Divide the class into two teams. Place the 'more' and 'less' cards face down in a pile on a table (shuffled in a random mix). Place two sets of number cards face down in two piles on the table. Choose one student from each team to take one of the number cards from their pile. Tell them to face the class and show them the number on their card. Choose a card from the 'more and less' pile. Show it to the class. If the card says 'less', the person holding the lower number must say, for example, *I win because 5 is less than 8*. They win a point for their team. If the card says 'more', the person holding the higher number must say, for example, *I win because 8 is more than 5*. They win a point for their team. Keep track of the points on the board. Keep playing until all the students have had a turn. The team with the most points wins.

Main activity

Display a 0–20 number line. Send three students out of the room so that they cannot see what you are doing. Point silently to one of the numbers on the line, e.g. 8. Say to the rest of the class, *This is our secret chosen number.*

Ask the three students to come back in. *You need to guess which number we have chosen. You may ask one question each.* Encourage students to ask questions such as, *Is it more than 5?, Is it less than 10?*, rather than, *Is it 7?*

After asking one question each, they talk together to guess the secret number. Repeat the game with another three students, choosing a different number for them to guess.

Ask students to complete questions 1 and 2 on page 39 in the Student Book individually, using a number line. They answer question 3 individually and then discuss their answers for question 3 in pairs. Can they explain and show their partner how they know they are correct? Ask students to complete question 4 in pairs.

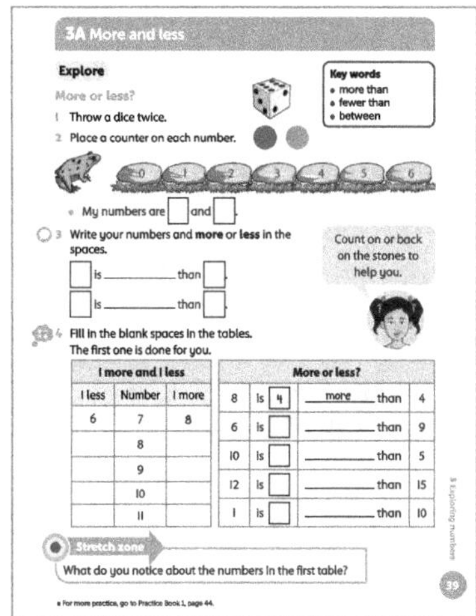

Differentiation

Supporting: Continue to count on a number line to model counting. Just use 'more' and 'less' in the second part of question 4 in the Student Book (ignore the second column).

Consolidating: Ask for 2 more or 2 less than a range of numbers on the number line. Just use 'more' and 'less' in the second part of question 4 in the Student Book.

Extending: Ask students to find the difference in the final second part of question 4 in the Student Book.

Stretch zone: Display page 39 of the Student Book on the IWB. *What do you notice about the numbers in the first table?*

Students may respond by describing how the numbers show '1 more than' and '1 less than', but they could also notice that the numbers in each row of the table form a counting sequence in ones.

Reflection time

Draw a ladder on the board with 20 rungs. Write 0 at the bottom rung.

Shuffle the number cards and place them face down. Choose students to take a card and write the number at the correct position on the ladder on the board. As numbers are positioned on the ladder, students can use these numbers as reference points to write their own number.

Is that number more than 0? Is it less than 5?

When the ladder is filled, count up and down.

While students are playing, listen to their language and vocabulary. Do they understand 'more' and 'less'? Allow students to use the number line as support.

Practice Book: Students can complete Practice Book page 44. This can be done directly after the main activity, as homework, or as the focus of a separate mathematics session to help students consolidate their learning and build fluency. Students may find it useful to use a counter to mark numbers on the number track and then physically move it left and right one space to find 1 more and 1 less.

Differentiated outcomes	
All students	should understand 'more than' and 'less than'.
Most students	will find 2 more and 2 less using a number line for support.
Some students	may begin to find the difference between two numbers less than 20 without using a number line.

Answers

Student Book page 39

4 7	8	9
8	9	10
9	10	11
10	11	12

More or less?

8	is 4 more than	4
6	is 3 less than	9
10	is 5 more than	5
12	is 3 less than	15
1	is 9 less than	10

Practice Book page 44

1 10	11	12
2 11	12	13
3 12	13	14
4 13	14	15
5 14	15	16
6 15	16	17
7 16	17	18
8 17	18	19

Stretch zone:

11 is 3 less than 14

4 is 3 more than 1

17 is 6 more than 11

4 is 3 more than 1

2 is 7 less than 9

3B Between

Discover Student Book page 40 • Practice Book page 45

Specific learning focus

- Use 'more' or 'less' to compare two numbers and give a number that lies between them.

Global skills

- **Creative skills:** exploring
- **Interpersonal skills:** communication / teamwork

Key vocabulary

- before, after, highest, lowest, between

Resources

- set of large number cards 0–20 for class use, set of number cards 0–20 per pair
- 'more' and 'less' cards, five of each, shuffled together
- two counters per student for table work
- 0–20 number line

Language support

Use classroom routines to model the key vocabulary, such as when lining up for playtime: *Yusuf, stand between Anna and Olga.*

Introductory activity

Mark out a number line on the floor and label each mark using large number cards 0–20. Choose two students to stand anywhere they like on the number line. They may stand where they want to, but there must be at least one mark **between** them.

I am going to stand between these two students on the number line. Stand on any number between them. *What does 'between' mean? Look where I am standing. I am between them.*

Repeat with other students in different positions on the line.

Using a full set of small number cards 0–20, give three students a card each (e.g. 4, 7, 9).

Who has a number between 2 and 5? Can you stand on your number on the line? Ask the other two students to tell you their numbers and then to stand on their places on the line. *Who is standing between two students? What is*

the number of the person between the others? What are the three numbers? Which numbers come between the other two students?

Repeat with other students until all cards have been chosen.

Main activity

Remove labels from the class number line. Give out one large number card 0–20 per pair of students (make sure all cards are used – some pairs can have more than one card if necessary). Choose the pair with number 11. Ask them to put it on the line. *What number is this? Who has the number that comes **before** 11?* Ask that pair to put their number on the line. *Who has the number that comes **after** 11?* That pair puts their number on the line. Read all three numbers in the sequence – 10, 11, 12. *Which is the **highest** number? Which is the **lowest**?*

Next, place 8 and 14 and so on until you have built up the whole number line.

Give each pair a set of number cards 0–9. Explain that they are going to play a game. Then read aloud the instructions on page 40 in the Student Book.

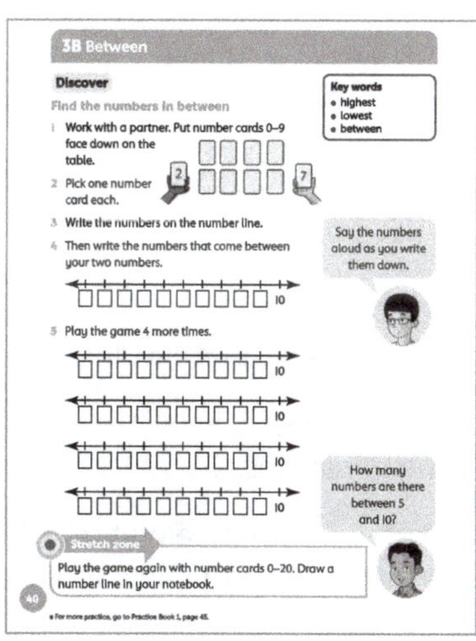

Repeat the game several times.

Differentiation

Supporting: Continue counting along the number lines and ask what the numbers on the cards are.

Consolidating: Use the language '1 more', '1 less' and 'between' and ask what is '1 more' or '1 less' than numbers on the cards, or the number between the cards.

Extending: Ask students to score the game by adding the numbers on the cards together. Model with cubes, if necessary.

Stretch zone: *Play the game again with number cards 0–20. Draw a number line in your notebook.*

Students can deepen their understanding of how to place numbers between given numbers by repeating this game.

 Reflection time

Assign eight students a number between 1 and 8. The rest of the class take it in turns to give instructions to each of the eight students to get them into the correct order. They must use '1 more', '1 less' and 'between' each time.

Practice Book: Students can complete Practice Book page 45. This can be done directly after the main activity, as homework, or as the focus of a separate mathematics session to help students consolidate their learning and build fluency. Students may find it useful to use two counters to mark the first and second number on the number track to support them in seeing the numbers in between.

Differentiated outcomes	
All students	should understand before, after, more, less and number in between.
Most students	will be able to name a number in between using a number line and say 1 more and 1 less.
Some students	may be able to find the totals of the cards in the game.

Answers

Student Book page 40

Observe students as they work through the activity and check that they correctly identify the highest and lowest numbers from each game.

When marking, check that the number lines have been completed correctly.

Practice Book page 45

Students should have written one of the numbers shown in the column 'Number between'.

First number	Number between	Second number
8	9 or 10	11
12	13 or 14 or 15 or 16	17
15	16 or 17 or 18	19
9	10	11
1	2 or 3 or 4 or 5 or 6 or 7	8
16	17 or 18 or 19	20
5	6 or 7 or 8 or 9 or 10 or 11	12
11	12 or 13	14

Stretch zone:
5 (22, 23, 24, 25, 26)
5 (32, 33, 34, 35, 36)
5 (33, 34, 35, 36, 37)

3B Between

Explore Student Book page 41 • Practice Book page 46

Specific learning focus

- Use 'more' or 'less' to compare two numbers and give a number that lies between them.

Global skills

- **Creative skills:** problem solving
- **Interpersonal skills:** communication / teamwork

Key vocabulary

- more, less, between, before, after

Resources

- number cards 1–20, number line (string or a washing line), pegs

Language support

Look out for students who have difficulty with the key vocabulary. Give them just one option such as, *I am between 6 and 8. What am I?* Allow them to use a small number line or ruler with numbers as a visual aid.

 **Introductory activity**

Display number cards pegged to a washing line in the wrong order. *What is wrong with this number line? We need to put it in the right order!*

Remove the number cards.

Ask a student to choose a number card. Ask them to peg that number on the line and, as they do, they should say, *'I have …. It goes between … and …'.* Continue until all of the numbers are placed.

 **Main activity**

Remove every second number from the 1–20 string number line leaving only the odd numbers, as below:

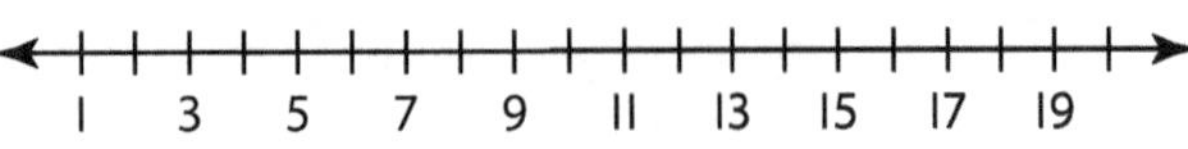

Ask students to count from 1 to 20, taking turns to say a number. If a student says a number which is missing from the number line, they stand up. When the count is over, give each standing student their 'missing number'. Ask them to peg it in the right place. *Are all the numbers in the right place?* Count together from the beginning to check. Repeat, removing every third number.

Repeat the activity, this time removing the numbers 3, 6, 8, 10, 14, 15, 17. As each student replaces their missing number, ask them to tell you what he/she is doing.

Write sentences on the board for students to use as they replace the numbers on the line:

I have ___________ .

It goes after ___________ and before ___________ .

It goes between ___________ and ___________ .

Together, look at page 41 in the Student Book. Say that there are groups of numbers, arranged in order in a line, and some of the numbers are missing. Look at the caterpillar and ask *What is the first number?* (0) *What is 1 more?* (1) *What is the last number?* (7) *What is 1 less?* (6) Ask students to record these numbers and continue to ask themselves questions in this way to find each missing number in each number sequence.

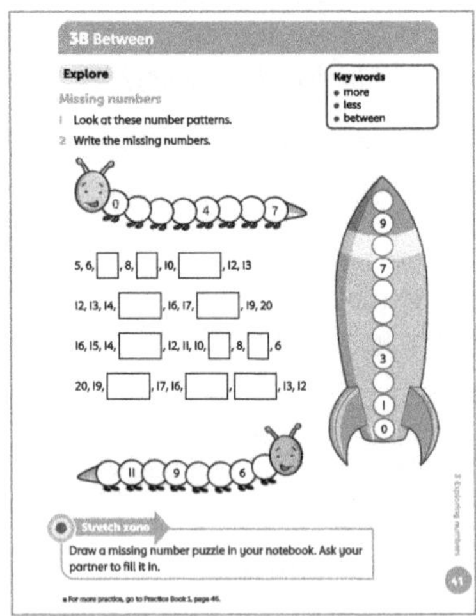

Differentiation

Supporting: Give students number lines for support.

Consolidating: Ask students to tell you what numbers come next in the sequences.

Extending: Ask students to develop their own sequences – counting forwards and backwards in twos.

Stretch zone: *Draw a missing number puzzle in your notebook. Ask your partner to fill it in.*

Following on from trying to work out what the missing numbers in a sequence are, students construct their own number line puzzle for a partner to solve, so they make decisions about which numbers to remove in a pattern.

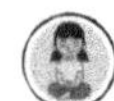 **Reflection time**

Use the washing line and pegs. *I am more than 10 but less than 14. What could I be?* Students who say 11, 12 or 13 can peg these number cards on the line. *I am between 6 and 9. What could I be?* Students who say 7 or 8 can peg them on the line. Continue using the words 'between', 'more' and 'less' until all of the numbers are on the line.

Practice Book: Students can complete Practice Book page 46. This can be done directly after the main activity, as homework, or as the focus of a separate mathematics session to help students consolidate their learning and build fluency. Remind students to check whether each number sequence is going forwards or backwards to help them decide what the missing numbers are.

Differentiated outcomes	
All students	should count confidently to 20, identifying the missing numbers.
Most students	will complete the questions in the Student Book without using a number line.
Some students	may develop their own number sequences both forwards and backwards.

Student Book page 41

Observe students as they work through the activity and check their answers to the 7 questions in the Student book.

Practice Book page 46

Students should complete each number pattern, counting forwards or backwards as required.

3C Tens and ones

Discover Student Book page 42 • Practice Book page 47

Specific learning focus

- Begin partitioning 2-digit numbers into tens and ones and vice versa.

Global skills

- **Creative skills:** problem solving / exploring / investigating
- **Real-world skills:** research / presenting information / interpreting information / financial literacy

Key vocabulary

- tens, ones, partition, place-value table

Resources

- place-value table to display, sticky tack
- straws, elastic bands
- interlocking cubes, small box or container (per pair)
- digit cards 0–9
- number cards 0–20
- base-ten equipment

Language support

Some students may have difficulty with counting to 20. Ensure you clearly model the pronunciation of each number for them and give them lots of practice at counting as a class.

Introductory activity

Show the class a bundle of straws (27). Drop them on the table. *How can we count these?* Say 1, 2, 3, 4, 6, 8 as you pick them up – lose your count on purpose as you pick them up.

Is it easy to count all of these? There are so many I keep making a mistake!

Suggest putting them in piles of 10, so they will be easier to count. Choose two students to count 10 each. Put a band round each bundle of 10. *We have two lots of 10 but there are still some on the table. Count with me to find how many are left.*

1, 2, 3, …, 7. We have two lots of ten and 7 more.

Draw or display a **place-value table** on the board and write 2 in the **tens** column and 7 in the **ones** column. *Look at the number I have written.* Write 27. *That tells us there are 2 tens* (point to the 2) *and 7 ones* (point to the 7).

Tens	Ones
2	7

 **Main activity**

Ask students to take 15 cubes from the interlocking cubes on their table. Write 15 on the board. *How many tens has 15 got?* (1) *How many ones?* (5)

Make 15 using one lot of ten by joining 10 cubes together to make a tower, and five single cubes (ones). Ask students to copy you.

Hold the 10 tower in the air. *How many ones do you have on the table?* (5) *How many tens?* (1) Repeat for 23 cubes.

Give individual students or pairs a small box and ask them to fill it with the cubes from their tables. Draw students' attention back to the place-value table from the introductory activity and ask them to draw one in their notebooks; it should take up roughly half of a page. Ask students to complete the rest of the activity on page 42 in the Student Book individually.

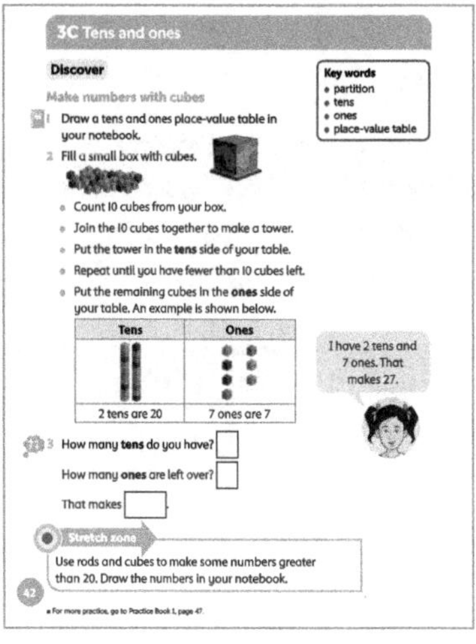

Differentiation

Supporting: Ask students to use number lines and the number cards to practise counting to 20.

Consolidating: Ask students to make all the teen numbers using the interlocking cubes, and to label them.

Extending: Make a 2-digit number using digit cards and ask students how many tens and ones are in the number.

Stretch zone: *Use rods and cubes to make some numbers greater than 20. Draw the numbers in your notebook (both as a picture of rods and cubes and write the number they show).*

This activity will extend some students to making and recording numbers larger than 20. They focus on counting the tens-rods first to see how many tens, and then the cubes for ones.

 **Reflection time**

Stick numbers from a set of digit cards 0–9 onto a place-value table on the board and ask students to tell you the number it represents, e.g. put a 3 in the ones column and a 1 in the tens column to make 13. Repeat several times and then ask students to take your role, with the other students saying the number. Initially limit this to numbers to 20. Ask if any students know numbers bigger than 20.

Practice Book: Students can complete Practice Book page 47. This can be done directly after the main activity, as homework, or as the focus of a separate mathematics session to help students consolidate their learning and build fluency. Provide students with base-ten equipment, interlocking cubes or similar to support them to partition numbers into tens and ones.

Differentiated outcomes	
All students	should count up to 20 seeing 'teen' numbers as a 'ten' and some 'ones'.
Most students	will understand partitioning for numbers from 20 to 30.
Some students	may understand partitioning into tens and ones beyond 20.

Answers

Student Book page 42

Observe students as they work through the activity and look at their answers to the questions in the Student book, checking that they have written the number for the total correctly.

Practice Book page 47

Number	Tens	Ones
35	30	5
24	20	4
52	50	2
29	20	9
41	40	1
19	10	9
33	30	3
11	10	1
46	40	6
7	0	7
60	60	0

3C Tens and ones

Explore Student Book page 43 • Practice Book page 48

Specific learning focus

- Begin partitioning 2-digit numbers into tens and ones and vice versa.

Global skills

- **Creative skills:** exploring
- **Interpersonal skills:** communication / teamwork

Key vocabulary

- partition, tens, ones, place value, digit

Resources

- digit cards 1–9: a large set for the whole class, and small sets for pairs
- base-ten equipment
- large 100-square for the front of the class, and small 100-squares for each table

Language support

Model the words you want students to use and learn. Do the activity alongside them, using the correct vocabulary and language structure as you work, writing the words on the board for the students to refer to as they complete the activity.

 **Introductory activity**

Count to 100 as a whole class. Use the large 100-square to support the counting. Focus on the 'tens' as you count, emphasising 'twenty', 'thirty', 'forty' and so on.

Place the large digit cards face down. Choose two students to take two cards each. Ask them to make a number with the **digits** on their cards, for example, make 26 or 62 with digits 2 and 6.

Ask them to choose two more cards each. This time the first card is the tens number and the second card is the ones number.

What is the number you have just made? How many tens? How many ones? Is it more than your first number? Is it less than your first number? How do you know? Did you count?

Main activity

Ask students, in pairs, to repeat the Introductory activity with the set of digit cards 1–9 on their table. Explain that the student in each pair who has the larger number each time wins one point. For each number, students make it with base-ten equipment, say how many tens and ones, and explain why their number is larger or smaller.

Where is the best place to put the larger digit? (In the tens place) Why?

Students stop when one of them has 10 points. Repeat the game, but this time the student with the smaller number wins a point.

Ask students to complete page 43 in the Student Book in pairs. When they have completed the questions, they can find 10 more than each number and represent it using base-ten equipment.

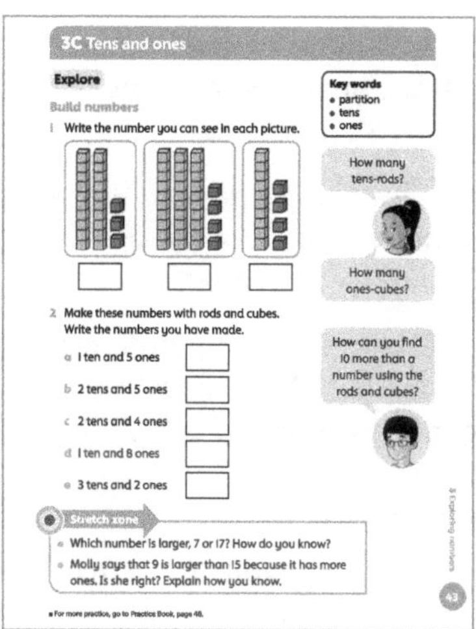

Differentiation

Supporting: Count rods and cubes with students, using the 100-square as support.

Consolidating: Ask students to tell you the value of the digits in their numbers. *Which number has more ones? Can you explain why this does not necessarily mean the number is larger?*

Extending: Ask students to make numbers 10 more and 10 less than the numbers they made with rods and cubes.

Stretch zone: Which number is larger, 7 or 17? How do you know?

Molly says that 9 is larger than 15 because it has more ones. Is she right? Explain how you know.

This activity will enable students to show that they understand that the position of the digits indicates the value of the numeral in that position: its **place value**.

Reflection time

Tell students that you want to make the largest number you can with the digits 4 and 5. *What is my number?* (54) Repeat with 6 and 2. (62)

Ask students to work with a partner. On the count of 3, each of them holds up 1, 2, 3, 4 or 5 fingers. If one of them shows 3 fingers and the other shows 5 fingers, they could choose to make 35 or 53.

Each time, ask students to write the two numbers they make and circle the larger number.

Ask pairs to share some of their numbers.

Has anyone made a number larger than 60? Has anyone made a number smaller than 12? What is the largest and smallest number you can make? (55 and 11) Why?

Practice Book: Students can complete Practice Book page 48. This can be done directly after the main activity, as homework, or as the focus of a separate mathematics session to help students consolidate their learning and build fluency. Students can continue to use base-ten equipment to make each number and to complete the addition sentences.

Differentiated outcomes	
All students	should make a number using rods and cubes.
Most students	will partition the rods and cubes and say what each is worth.
Some students	may record the number sentence in tens and ones.

Answers

Student Book page 43

1   23 (2 tens and 3 ones)

    34 (3 tens and 4 ones)

    14 (1 ten and 4 ones)

2   **a** 15   **b** 25   **c** 24   **d** 18   **e** 32

Practice Book page 48

1   27   **2**   44   **3**   18

4   21   **5**   35   **6**   18

7   22   **8**   36   **9**   26

10 16

3D Partitioning

Discover Student Book page 44 • Practice Book page 49

Specific learning focus

- Partitioning 2-digit numbers into tens and ones and reverse.

Global skills

- **Real-world skills:** presenting information
- **Interpersonal skills:** communication / teamwork

Key vocabulary

- partition, tens, ones, place value, 2-digit number

Resources

- large 100-square for the front of the class, and small 100-squares for each table
- base-ten equipment

Language support

Help students relate the vocabulary of rods and cubes to tens and ones. They count the rods and cubes and then make a number comprising tens and ones.

 **Introductory activity**

Write the number 26 and say it aloud, then ask a student to come to the front of the class and make the number using tens-rods and ones-cubes. Ask them to explain how they know they have made 26.

Now separate the rods from the cubes and count them separately.

How many rods are there? (2) *How many 10s? What number is this?* (20)

How many cubes are there? (6) *How many ones? What number is this?* (6)

Write on the board 26 = 20 + 6

 **Main activity**

Ask a student to choose a **2-digit number** from the 100-square. Ask another student to make that number using rods and cubes and explain how they know they have made the right number. Then ask the first student to separate the rods from the cubes and say how many of each there are and what the value of each is. Explain that separating the tens and the ones (rods from the cubes) is called **partitioning**.

Ask students to work in pairs. One will choose a 2-digit number from the 100-square, the other will make it with rods and cubes, then the first will partition and record the number sentence.

Ask students to complete page 44 in the Student Book in pairs. They will need to draw a place-value table in their notebooks large enough to represent two numbers with base-ten equipment. As pairs complete questions 2 and 3, listen to their explanations of how they know their base-ten representations are correct. They should be saying, for example, that 4 in 45 is 4 tens and the 5 is five ones so they must have four tens-rods and five ones-cubes.

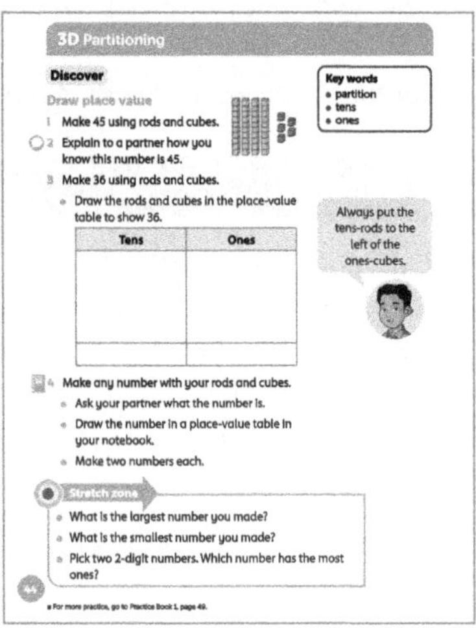

Differentiation

Supporting: Count rods and cubes with students, using the 100-square as support.

Consolidating: Ask students to tell you the value of the digits in their numbers. *Which number has more ones? Can you explain why this does not necessarily mean the number is larger?*

Extending: Ask students to make numbers 10 more and 1 more and 10 less and 1 less than the numbers they made with the rods and cubes.

Stretch zone: *What is the largest number you made? What is the smallest number you made? Pick two 2-digit numbers. Which number has the most ones?*

This activity will help students partition a 2-digit number and then identify the value of each digit. Check that they correctly identify the tens and ones.

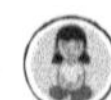 **Reflection time**

Ask pairs of students to share one of their numbers and explain how they partitioned it. Did they always make the correct number? Did they partition correctly? Could they say what each part was worth?

Practice Book: Students can complete Practice Book page 49. This can be done directly after the main activity, as homework, or as the focus of a separate mathematics session to help students consolidate their learning and build fluency. Students can continue to use base-ten equipment to make each number before drawing rods and cubes.

Differentiated outcomes	
All students	should make a number using rods and cubes.
Most students	will partition the rods and cubes and say what each is worth.
Some students	may record the number sentence in tens and ones.

Student Book page 44

1 and **2** Listen to make sure students can explain how they know they have made 45 using rods and cubes.

3 and **4** Check that students have formed 36 correctly with rods and cubes and recorded the number correctly in the table, as well as for the other numbers they make.

Practice Book page 49

	Number	Tens	Ones	Rods/cubes
1	28	20	8	2 rods, 8 cubes
2	18	10	8	1 rod, 8 cubes
3	17	10	7	1 rod, 7 cubes
4	7	0	7	0 rods, 7 cubes

3D Partitioning

Explore 1 Student Book page 45 • Practice page 50

Specific learning focus

- Partitioning 2-digit numbers into tens and ones and vice versa.

Global skills

- **Creative skills:** exploring
- **Real-world skills:** presenting information
- **Interpersonal skills:** communication / teamwork

Key vocabulary

- partition, tens, ones, place value, 2-digit number

Resources

- digit cards 1–9: a large set to use with the whole class and small sets for students
- large 100-square for the front of the class, and small 100-squares for each table
- base-ten equipment

Language support

Listen to how students use the vocabulary of rods and cubes, tens and ones and check that they understand that separating the rods and cubes is called partitioning.

 Introductory activity

Ask a student to choose two number cards from a set of digit cards 1–9 and use them to form a 2-digit number. For example, if they choose cards 3 and 7, they could make 37 or 73.

Now ask another student to make that number using rods and cubes, explaining how they know they have made the right number. For example, 30 is 3 tens so I use 3 rods, and I show 7 ones with 7 cubes. Ask them to partition the rods and cubes and say how much each is worth (30 and 7, or 70 and 3).

Write on the board 37 = 30 + 7 (or 73 = 70 + 3).

 Main activity

Students work in pairs using a set of digit cards 1–9 and rods and cubes. One student makes a 2-digit number by picking two of the digit cards. The other student makes that number using rods and cubes and says how they know they have made the right number. Then the first student separates the rods from the cubes and says how many of each and what their values are.

The students take turns until they have used all the digit cards.

Ask students to make a new pair to complete page 45 in the Student Book. Take students through the steps of the activity, choosing a student to be your 'partner'. Explain that your partner should not see the number you are making until you have both had a chance to make it, draw it and write it in numerals.

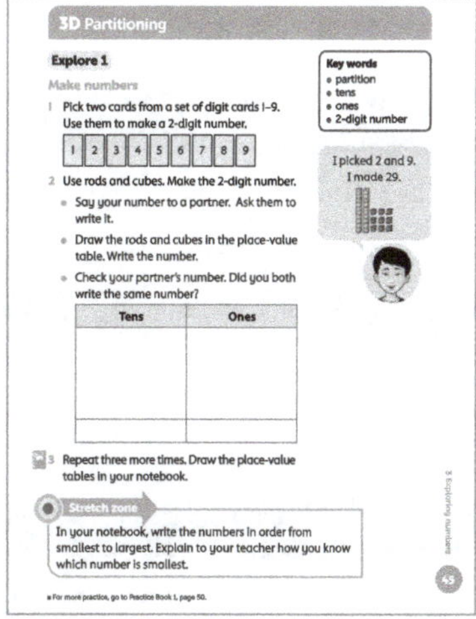

Differentiation

Supporting: Count rods and cubes with students, highlighting the value of the tens and ones in each number, e.g. *24, two tens and four ones.*

Consolidating: Ask students to find their numbers on the 100-square.

Extending: Ask students to make numbers 10 more and 10 less than the numbers they made with the rods and cubes.

Stretch zone: *In your notebook, write the numbers in order from smallest to largest. Explain to your teacher how you know which number is smallest.*

This activity combines partitioning with ordering 2-digit numbers. Check that students have ordered the numbers correctly.

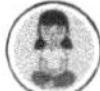 ## Reflection time

Ask students to look at the numbers they made. *Who made the largest number? Who made the smallest number?* Ask if they can put their numbers in order of size by looking at which number had the most rods (tens)?

Ask further questions such as: *Is [ ] the largest number you can make with one set of digit cards? Is [ ] the smallest? How do you know?* (The largest number of 10s (rods) you can have is 9 and the next largest digit is 8 so the largest number would be 98.)

Practice Book: Students can complete Practice Book page 50. This can be done directly after the main activity, as homework, or as the focus of a separate mathematics session to help students consolidate their learning and build fluency. Each student will need a set of digit cards 1–9 to complete this activity. Work through the example together. Read each number sentence as *Forty-three is equal to forty and three. Forty-three is four tens and three ones.* Ask students to read 34 = 30 + 4 in the same way. Provide students with base-ten equipment to complete the activity.

Differentiated outcomes	
All students	should make a number using rods and cubes.
Most students	will partition the rods and cubes and say what each is worth.
Some students	may record the number sentence in tens and ones.

Answers

Student Book page 45

For each number students make, check they have formed it correctly using rods and cubes, then recorded it correctly in the table as tens and ones.

Practice Book page 50

For each pair of numbers students make, check they have recorded them correctly in the table as 2-digit numbers, number sentences and written which is the larger number.

3D Partitioning

Explore 2   Student Book page 46 • Practice page 51

Specific learning focus

- Partitioning 2-digit numbers into tens and ones and vice versa.

Global skills

- **Creative skills:** exploring
- **Interpersonal skills:** communication / teamwork

Key vocabulary

- partition, tens, ones, place value, 2-digit number

Resources

- digit cards 1–9: a large set to use with the whole class and small sets for pairs
- large 100-square for the front of the class, and small 100-squares for each table
- base-ten equipment

Language support

Listen to how students use the vocabulary of rods and cubes, tens and ones and check that they understand that separating the rods and cubes is called partitioning. Check that they are saying the numbers correctly.

 ## Introductory activity

Show students a set of digit cards 1–9. Ask a student to choose three of the cards (e.g. 2, 5, 7). Ask them to make three different 2-digit numbers using two of the three cards. The possible numbers are: 25, 27, 52, 57, 72, 75.

 ## Main activity

Ask a student to pick three cards (e.g. 1, 4, 8) from a set of digit cards 1–9 and make four of the possible 2-digit numbers (14, 18, 41, 48, 81, 84). Ask them to choose one of the numbers and make it using rods, thinking first about how many tens (rods) they need and then how many ones (cubes) they need. Ask students to share their representations, saying which number they made and describing it. ('This is 18. It has one ten and eight ones. I made it with one tens-rod and eight ones-cubes.')

Explain that students are going to work in pairs with a set of digit cards 1–9. One of them chooses three cards, and together they make a set of four 2-digit numbers using the chosen digits. Then they make the numbers using rods and cubes and record the numbers by writing them in figures and drawing the rods and cubes on page 46 of the Student Book.

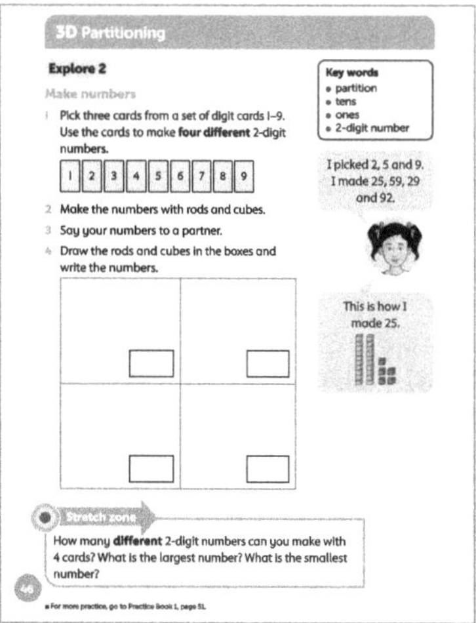

Differentiation

Supporting: Arrange digit cards to help students form different 2-digit numbers.

Consolidating: Ask students to find their 2-digit numbers on the 100-square.

Extending: Ask students to find all six possible 2-digit numbers from three digit cards.

Stretch zone: *How many different 2-digit numbers can you make with four cards. What is the largest number? What is the smallest number?*

Check that students have made all possible 2-digit numbers with the cards they have picked (12 in total). For example, if they choose 2, 3, 5 and 8, they can make the following:

23 and 58, 23 and 85, 32 and 58, 32 and 85, 25 and 38, 25 and 83, 52 and 38, 52 and 83, 28 and 35, 28 and 53, 82 and 35, 82 and 53.

The largest number is 85 and the smallest number is 23.

Reflection time

Ask students to describe any patterns they could see in their sets of numbers. How did they know they had found different numbers?

Practice Book: Students can complete Practice Book page 51. This can be done directly after the main activity, as homework, or as the focus of a separate mathematics session to help students consolidate their learning and build fluency.

Differentiated outcomes	
All students	should make a number using digit cards and rods and cubes.
Most students	will make more than one different 2-digit number using three digit cards.
Some students	may find all combinations of 2-digit numbers from three cards.

Answers

Student Book page 46

Check that students have made a set of four different 2-digit numbers using three cards. Check that they can say the numbers correctly, then form them with rods and cubes and record the numbers correctly.

Practice Book page 51

Check that students have made a set of four different 2-digit numbers using three cards. Check that they can partition the numbers correctly, then form them with rods and cubes and record the numbers correctly.

3E Ordering numbers

Discover   Student Book page 47 · Practice page 55

Specific learning focus

- Order numbers to at least 20, positioning on a number track, and describe their position using ordinal numbers.

Global skills

- **Creative skills:** investigating
- **Interpersonal skills:** communication / teamwork

Key vocabulary

- first, second, third, fourth, …, tenth, ordinal numbers

Resources

- large number cards 1–10, large ordinal number cards (1st, 2nd, 3rd, …, 10th)

Language support

Use ordinal numbers throughout the school day as often as possible.

 Introductory activity

Ask students to listen carefully while you count, *1, 2, 3, 5, 6.*

What did you notice? That's right. I left out number 4.

Repeat with 6, 7, 8, 10, 12. *Which numbers did I leave out this time?* (9, 11)

Ask students to put their hand up when they realise there is a number missing, *10, 9, 8, 7, 5, 4. What was the missing number?* (6)

Give ten students each a number card 1–10 and ask them to line up in order. *Which number should come **first**?* (1) *Which number comes after 1?* (2) *Two is the **second** number.* Continue until you reach 10.

Show the class the ordinal number cards. Say that when we talk about the order in which numbers come, we call them **ordinal numbers**.

Place the 1–10 and 1st–10th cards around the room. Each student should find one card. Ask them to try to find their partner. For example, the student with '3' number card finds the student with the '3rd' card. When they have found their partner, they make a line at the front of the classroom. *Are you in the right place?* They must not talk to anyone apart from their partner. They look at the pair in front of them and the pair behind. Once they are all happy with their order, they read the ordinal cards out loud, in order.

Repeat the game with students who did not find a matching card last time starting the game.

Ask students to complete page 47 of the Student Book in pairs. Discuss what a traffic jam is before they begin. Consider pairing students who have stronger writing skills with those who have weaker writing skills.

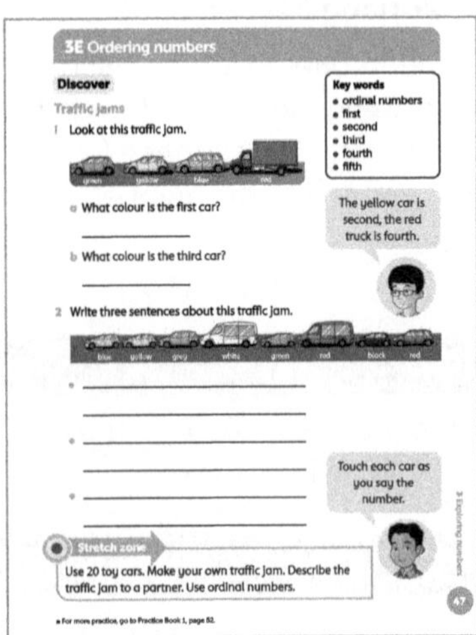

Differentiation

Supporting: Play a matching game using small sets of digit and ordinal number cards. Model the language for students.

Consolidating: Ask students additional questions relating to the Student Book questions, e.g. *The yellow car leaves the line of cars. In what position is the black car now?*

Extending: Ask students to predict what the ordinal numbers will be between 10 and 20.

Stretch zone: *Use 20 toy cars. Make your own traffic jam. Describe the traffic jam to a partner. Use ordinal numbers.*

Check that students can describe the positions of the vehicles in their toy traffic jam using the correct ordinal numbers. Help them to say 'the 5th car' rather than 'car number 5', for example.

Reflection time

Students sit with a 'talk partner'. They should think of any examples of using ordinal numbers in everyday life. Share all the different ideas with the class. When you dismiss the class, dismiss them one by one using ordinal numbers.

Practice Book: Students can complete Practice Book page 52. This can be done directly after the main activity, as homework, or as the focus of a separate mathematics session to help students consolidate their learning and build fluency.

Differentiated outcomes	
All students	should match ordinal and cardinal (counting) numbers and place ordinal numbers in the correct order.
Most students	will continue the pattern of ordinal numbers up to 20.
Some students	may continue the pattern of ordinal numbers beyond 20.

Answers

Student Book page 47

1 a Green **b** Blue

2 Check that students can use ordinal numbers to describe the traffic jam, for example 'the yellow car is 2nd'.

Practice Book page 52

1 5th   fourth unmatched pattern

2 11th   second unmatched pattern

3 1st   sixth unmatched pattern

4 18   first unmatched pattern

5 9th   eighth unmatched pattern

6 14th   fifth unmatched pattern

7 20th   third unmatched pattern

8 3rd   seventh unmatched pattern

3E Ordering numbers

Explore 1 Student Book page 48 • Practice page 53

Specific learning focus

- Order numbers to at least 20, positioning on a number track; use ordinal numbers.

Global skills

- **Creative skills:** problem solving
- **Interpersonal skills:** communication / teamwork

Key vocabulary

- first, second, third, fourth, …, tenth, ordinal numbers

Resources

- ordinal 1st–20th number line per student
- large ordinal number cards 1st–20th
- A4 pieces of paper, sticky tack, counters
- 20-bead string

Language support

For the Main activity, ensure that the ordinal numbers are repeated throughout by you and the students. Repeating each number as often as possible will help students to pronounce the words correctly (these are particularly difficult sounds to pronounce).

 **Introductory activity**

Give each student an ordinal 1st–20th number line. Ask them to circle ten numbers on the line.

Say you are going to show them some ordinal number cards. If they have that number circled on their line, they put a line through it. When they have crossed through all their numbers, they should hold their paper in the air.

Read and show the ordinal numbers one at a time. Display them so that the class can see. Leave time for students to find and cross though the number. At the end, check that all ten numbers have been called.

 **Main activity**

As a class, make up a funny story. Each part of the story will be illustrated with a simple drawing.

Start by asking students what happened first and then draw a simple illustration of this on a piece of paper.

What happened next? Allow one student to make up the next part of the story and again make a simple drawing of what happened on a piece of paper. *This is the second part of our story.*

Repeat six times until the story is finished. Each time say what part of the story it is, for example, *This is the fifth part of our story.*

Stick the pictures you have drawn onto the board in a random order.

Show the large ordinal number cards 1st to 6th. Ask one student to stick the first number card next to the first part of the story. *What happened first?* Repeat, asking, *What was the second part of the story? What was the third part? What happens before the third part? What happens after?* and so on until the story is correctly labelled.

Ask students to complete page 48 in the Student Book individually. Ask students further questions about the counters, e.g. *In what position is the brown counter?*

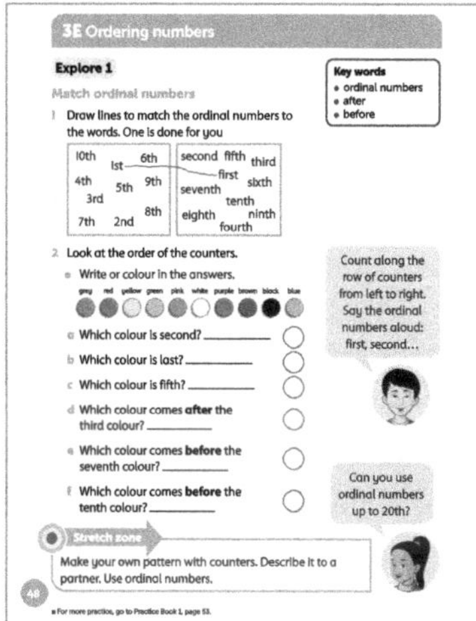

Differentiation

Supporting: Repeat the matching game from the previous lesson to support students with the first question in the Student Book.

Consolidating: Pairs should ask each other their own questions about the coloured counters in the Student Book, e.g. *What is the position of the green counter? Which colour is 8th?*

Extending: Ask students to create their own pattern with counters and create questions for their peers.

Stretch zone: *Make your own pattern with counters. Describe it to a partner. Use ordinal numbers.*

Check that students can make a pattern that they can describe it with ordinal numbers.

 **Reflection time**

Ask students to work with their partner and think of a question about ordinal numbers to ask the class, e.g. *What comes before the 10th number?* Give time for discussion between pairs. Choose one pair to ask their question. That pair can ask the next pair and so on until all the students have asked their question.

Practice Book: Students can complete Practice Book page 53. This can be done directly after the main activity, as homework, or as the focus of a separate mathematics session to help students consolidate their learning and build fluency. Extend to looking at ordinal numbers up to 20 using a bead string.

Differentiated outcomes	
All students	should match ordinal and cardinal (counting) numbers and place ordinal numbers in the correct order.
Most students	will continue the pattern of ordinal numbers up to 20.
Some students	may continue the pattern of ordinal numbers beyond 20.

Answers

Student Book page 48

1 Make sure that students are correctly drawing a line to match the ordinal numbers to the cardinal (counting) numbers.

2 a Red

  b Blue

  c Pink

  d Green

  e White

  f Black

Practice Book page 53

1 3-wheeled vehicle

2 caravan

3 lorry

4 scooter

5 digger

6 The caravan is the ninth vehicle.

7 The motorbike is the seventh vehicle.

8 The car is the eighth vehicle.

9 The coach is the fifth vehicle.

10 The digger is the sixth vehicle.

3E Ordering numbers

Explore 2 Student Book page 49 • Practice Book page 54

Specific learning focus

- Order numbers to at least 20, positioning on a number line; use ordinal numbers.

Global skills

- **Creative skills:** problem solving / exploring / investigating
- **Real-world skills:** research / presenting information / interpreting information / financial literacy

Key vocabulary

- first, second, third, fourth, …, tenth, ordinal numbers, pattern, repeating pattern

Resources

- ordinal 1st–20th number line per student
- ordinal number cards 1st–20th
- string and coloured beads

Language support

Ensure that the ordinal numbers are repeated throughout by both you and the students. Repeating each number as often as possible will help students to pronounce the words correctly. Help students with the unusual ordinal numbers that do not end in 'th'– 1st, 2nd, 3rd. At the start of the introductory activity, talk through the colour names of the beads with students.

 **Introductory activity**

Show students a bead string threaded with a mixture of colours, up to 20 beads. Point to the first bead and ask, *What colour is the 1st bead?* Point to the next bead and ask, *What colour is the 2nd bead?* Repeat this for the first 10 beads.

 **Main activity**

Show students the bead string and point to a bead at random, for example, the 7th bead. *How many beads is this one from the start?* Count together as you point to each bead in turn. When you get to the one you have chosen, you will have counted to 7, so tell students this is the 7th bead. Repeat using a different bead.

Students work in pairs. They take turns to thread a coloured bead onto a string until there are 20 beads. Then one student points to a bead, the other counts from the start and says what position that bead is in. They do this for 5 beads each. Then they complete the activities on page 49 of the Student Book.

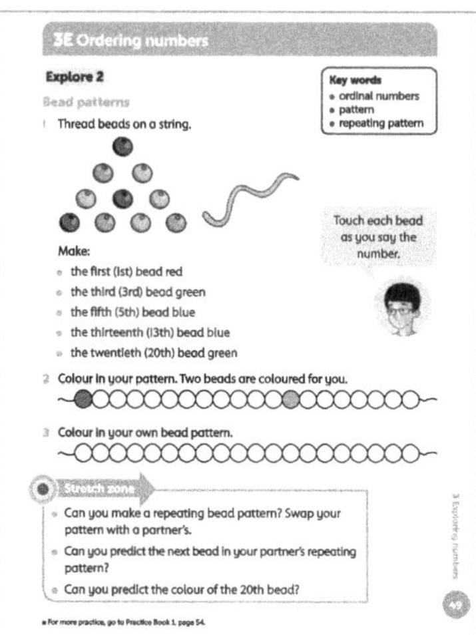

Differentiation

Supporting: Count with students to find how many beads.

Consolidating: Pairs should ask each other their own questions about the beads on their string, e.g. What is the position of the green bead? Which colour is the 8th bead?

Extending: Ask students to create their own **pattern** with beads and create a set of questions for their peers.

Stretch zone: *Can you make a repeating bead pattern? Swap your pattern with a partner's.*

Can you predict the next bead in your partner's repeating pattern?

Can you predict the colour of the 20th bead?

This activity offers students the opportunity to be creative in designing a pattern, then to use their critical thinking to determine someone else's **repeating pattern** and predict the colour of a bead somewhere along the pattern.

Ask different pairs of students to show their bead pattern and describe some of the colours in it by their positions. Ask them questions such as, *Which position is this blue bead?*, or *What colour is the 16th bead?* Get students to think of their own question to ask about someone else's bead pattern.

Practice Book: Students can complete Practice Book page 54. This can be done directly after the main activity, as homework, or as the focus of a separate mathematics session to help students consolidate their learning and build fluency. Students will need coloured pencils to complete the activity. You may also ask students to draw a line from their description to the corresponding bead, particularly if their pattern includes several beads of the same colour.

Differentiated outcomes	
All students	should count the beads up to a given position.
Most students	will say which ordinal number matches a given position.
Some students	may extend beyond the 20th position.

Answers

Student Book page 49

Check that students have threaded the beads according to the instructions, and then coloured the pattern to match for question 2.

Practice Book page 54

Check that students have matched the description of the pattern using ordinal numbers to the pattern they created.

3F Even and odd

Discover Student Book page 50 • Practice Book page 55

Specific learning focus

- Count on in twos, beginning to recognise odd/even numbers to 20 as 'every other number'.

Global skills

- **Creative skills:** exploring
- **Interpersonal skills:** communication / teamwork

Key vocabulary

- odd, even, equal parts

Resources

- interlocking cubes
- strips of paper for zigzag books, large pre-made zigzag book

Language support

Help students understand the concept of 'even' and 'odd' in numerical terms. Explain that 'even' does not mean the same as 'equal'.

 **Introductory activity**

Ask students to work in pairs and use cubes to build towers of 11, 12, 13, 15, 18 and 20 cubes. Ask them to break each tower so that there is the same number of cubes in each part. Give time to complete the activity, then ask, *Did all the towers break into two **equal parts**?*

Ask students to now take a set of four cubes and a set of nine cubes. Ask students to pair up cubes in each set to find if the number is **odd** or **even**. Tell them that if every cube has a partner, it is an even number. If they do not, it is an odd number. They should find that 4 is even and 9 is odd. Refer back to the towers in the Introductory activity. Agree that all numbers are either odd or even. Even numbers can be split into two equal parts, odd numbers cannot.

Students work on the activities on Student Book page 50 individually. For activity 2, students scoop (demonstrate a scoop by cupping your hand and making a scooping action) cubes and try to arrange them in two equal parts to find if they have 'scooped' an odd or even number. Students count and record the number in their Student Books. Discuss what they have discovered, and any patterns they see.

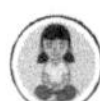

Differentiation

Supporting: Allow students to use cubes to test for even and odd numbers.

Consolidating: Ask students to describe what makes a number even or odd.

Extending: Ask students to predict whether a number is even or odd. Use numbers over 20.

Stretch zone: *Can you explain to a partner the difference between an odd and an even number?*

Listen to students' explanations to be sure that they understand even and odd in terms of being able to make a set of pairs with an even number.

Reflection time

Give each student a strip of paper folded to make a zigzag book (unnumbered) with 12 'pages'. Show them how to write one number (1–12) on each 'page'.

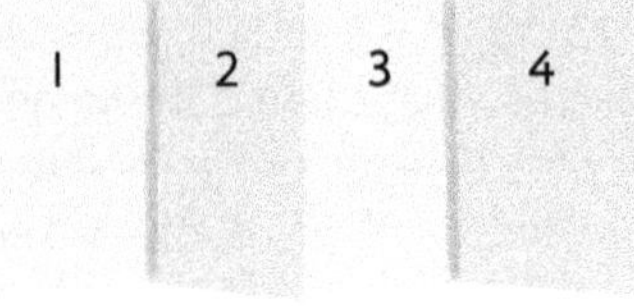

Show an example of a zigzag book you have made and turn it slightly so that the odd numbers can be seen, but not the even numbers. Students recite the sequence of odd numbers to 11. Make sure you use the word 'odd'.

Turn the book the other way so that the even numbers are showing and repeat the activity saying the word 'even'.

Cover a number on one page and challenge students to tell you the number that is hidden and whether it is odd or even.

Practice Book: Students can complete Practice Book page 55. This can be done directly after the main activity, as homework, or as the focus of a separate mathematics session to help students consolidate their learning and build fluency. Ask students to remind you what odd and even numbers are, and to give you an example of each.

Differentiated outcomes	
All students	should be able to build towers and match cubes to find out if a number is odd or even.
Most students	will be able to describe what is meant by even and odd.
Some students	may predict which numbers are odd and which are even.

Student Book page 50

1 8 gloves – even, 1 shoe – odd, 10 socks – even.

2 Answers will vary because students are scooping cubes to find numbers to work with.

Practice Book page 55

1, 2 Check that students have coloured alternate numbers in the grid different colours.

3, 4 Students should have written eight even numbers and eight odd numbers.

3F Even and odd

Explore 1
Student Book page 51 · Practice Book page 56

Specific learning focus

- Count on in twos, beginning to recognise odd/even numbers to 20 as 'every other number'.

Global skills

- **Creative skills:** problem solving / investigating
- **Interpersonal skills:** communication / teamwork

Key vocabulary

- odd, even

Resources

- 0–100 number line, one per pair
- cardboard strips
- large 0–100 number line
- interlocking cubes, ten-frames and two colours of counters

Language support

While students are working on the activity in the Student Book, ask them to explain how they know that a number is odd or even. If necessary, correct the language of the explanations and ensure that they are using 'odd' and 'even' correctly.

 **Introductory activity**

Use the large number line. Ask a student to choose a number between 20 and 30. Circle that number.

Tell the class you are going to draw jumps of 2 from 0. Ask them if they think you will land on the chosen number. Use thumbs up for yes, thumbs down for no. Draw the jumps while the class say out loud the numbers you land on. Repeat with other end (target) numbers.

 **Main activity**

Write target numbers on the board – include both odd numbers and even numbers. Students work in pairs to choose one of the target numbers to circle on their number line. They start at 0 and use the jump size of 2 (to give even numbers) to investigate whether they can reach their target in jumps of that size. They then use a different target number. Encourage students to predict which numbers will and won't work.

Students can record their work in two lists – those numbers that can or cannot be reached in jumps of 2.

Repeat the activity, but using jumps of 2 starting from 1 (to give odd numbers). *How can we tell if a large number is odd or even? Look at the last digit. If that is even, the whole number is even. If it is odd, the whole number is odd.*

Ask students to complete page 51 in their Student Book individually or in pairs. Before they begin, explain how they can make a path, using the speech bubble example for reference. Provide cubes or ten-frames and counters so that they can build numbers to check if they are odd or even.

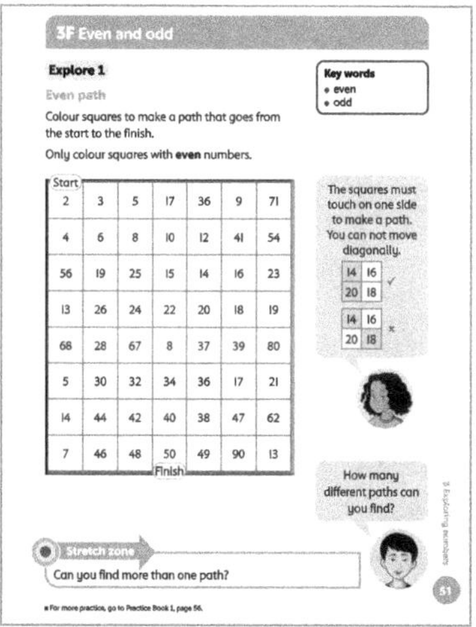

Differentiation

Supporting: Students who have difficulty with the physical act of drawing equal-size jumps, or have trouble understanding the language, might find using a 'size jumper' (a cardboard strip the same length as the jump size they are working on) helpful.

Consolidating: Ask students to describe what makes a number even or odd.

Extending: Ask students to predict whether a number is even or odd. Use numbers over 20.

Stretch zone: *Can you find more than one path?*

Students can look for different paths that are linked by even numbers. The path they take does not have to be in sequence.

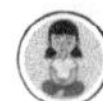 **Reflection time**

Students sit in a circle and count round the circle in ones. If their number is odd, they stand up; if it is even, they put their hands on their head. Repeat the activity, changing the actions and/or giving students different numbers, for example, if their number is even, wave their hands, if it is odd, sit down.

Practice Book: Students can complete Practice Book page 56. This can be done directly after the main activity, as homework, or as the focus of a separate mathematics session to help students consolidate their learning and build fluency. Provide two colours of counters and ten-frames so that students can build number sentences and check if the total is odd or even. Work through the example together before they begin.

Differentiated outcomes	
All students	should be able to make jumps of 2 from one even (or odd) number to get to another even (or odd) number.
Most students	will be able to describe what is meant by even and odd.
Some students	may predict which numbers are odd and which are even.

3F Even and odd

Explore 2 Student Book page 52 • Practice Book page 57

Specific learning focus

- Relate counting on and back in tens to finding 10 more/less than a number (< 100).

Global skills

- **Creative skills:** problem solving / exploring
- **Interpersonal skills:** communication / teamwork

Key vocabulary

- 10 more, 10 less, counting on, counting back, above, below, diagonally

Resources

- large 100-square for front of class, standard 100-squares, counters, 2 dice per pair
- interlocking cubes, ten-frames

Language support

Make sure students understand **above** and **below** as positions on the 100-square. Use arrows to help. These can be angled to show **diagonally**.

 **Introductory activity**

Review addition and subtraction of 10, counting down and up the 100-square. Start on different numbers. For example, start with 32. *What is **10 more**?* (42) *What is **10 less**?* (22) *Which digit changes?* (the 3/tens) *Which stays the same?* (the 2/ones) Repeat with other numbers,

Student Book page 51

Check that students have only coloured even numbers to get from the start to the finish. One possible route is: 2, 4, 6, 8, 10, 12, 14, 20, 22, 8, 34, 40, 50

Practice Book page 56

Check that the students have written correct additions and that they have correctly identified the answers as odd or even. For example:

6 + 1 = 7   7 is an odd number

Stretch zone: True. Students could explain by saying that an odd number is an even number with 1 added to it. When you add two odd numbers the two 1s add to make 2, which added to two even numbers gives another even number.

including numbers which allow students to practise saying numbers they may find harder to pronounce, such as 19 and 90.

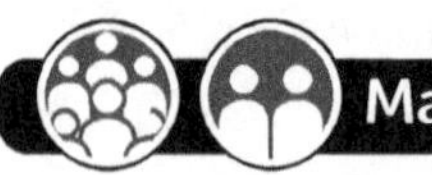 **Main activity**

Show students how to add 11 by adding 10 and then 1 more. Choose a number on the 100-square, for example, 42. Explain that moving down one place on the square is the same as adding 10. *What number does that make?* (52) *Is 52 odd or even? How do you know? How can we check?* Make 52 with cubes and split into two equal groups or represent 52 with counters on ten-frames. Then move one place to the right, which is the same as adding 1 more. *What number does this give?* (53). *How much has been added on in total?* (11) Repeat using different start numbers. Ask if they can see the pattern of the position of the new number (diagonally right from the start number). *Is 53 odd or even?* (odd) *How do you know?* (Because 52 is even, and 53 is only 1 more than 52, so it is odd.)

Show students how to add on 9 by adding 10 and moving back 1. Repeat using different start numbers. Ask if they can see the pattern of the position of the new number (diagonally left from the start number).

Show the 100-square with only 45, 46, 55, 56 showing. Ask students questions to find and uncover the other numbers such as, *What is 10 less than 45?* (Also ask for 9 more or less, 10 more, and 11 more or less.) If a student gives a correct response, they uncover the number. Continue until all of the numbers are showing.

Ask students to play the game on page 52 in the Student Book in pairs. Encourage students to explain to their partner how they know if a number is odd or even. Students should check their partner's moves after each turn and build numbers using counters and ten-frames or cubes to check if they are odd or even.

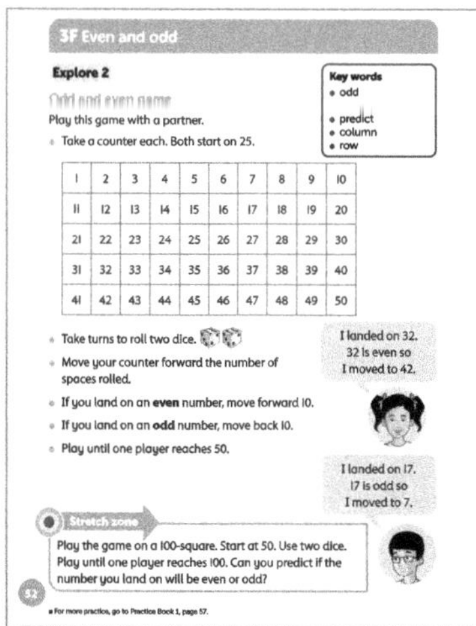

Differentiation

Supporting: Support students in placing counters accurately on the 100-square.

Consolidating: Ask students to explain how they know if a number is odd or even.

Extending: Say, *If you add 10 to an even number you still have an even number. If you add 10 to an odd number you still have an odd number. True or false? Can you explain your thinking?*

Stretch zone: *Play the game on a 100-square. Start at 50. Use two dice. Play until one player reaches 100. Can you predict if the number you land on will be even or odd?*

This activity gives students practice at **counting on** and **counting back** with numbers above 50, and asks them whether they can predict the odd-ness or even-ness of the numbers they land on.

 Reflection time

Using the 100-square, give instructions for counting on 9, 10 and 11. Use a mixture of all three. Give time for all students to find the number. For students who enjoy a challenge, speed up the questions. Continue to ask if numbers are odd or even. Encourage students to begin to notice patterns. *Do you notice a pattern when, for example, you add 9 to an odd number? What about when you add 9 to an even number?*

Practice Book: Students can complete Practice Book page 57. This can be done directly after the main activity, as homework, or as the focus of a separate mathematics session to help students consolidate their learning and build fluency. Give students a 100-square to support them in counting in twos and fives or have them use the 100-square on page 39 of the Practice Book for reference. They may also find it useful to use a counter to move along the 100-square as they count up.

Differentiated outcomes	
All students	should be able to use the 100-square to support their calculations.
Most students	will notice the patterns in the 100-square.
Some students	may be able to explain the patterns in the 100-square.

Answers

Student Book page 52

Listen to students as they play the game and check that they are adding 10 or subtracting 10 correctly each time. Have they remembered how to identify odd and even numbers?

Practice Book page 57

The start number is odd.

7: 9, 11, 13, 15, 17, 19, 21, 23

7: 12, 17, 22, 27, 32, 37, 42, 47

The start number is even.

6: 8, 10, 12, 14, 16, 18, 20, 22

6: 11, 16, 21, 26, 31, 36, 41, 46

3 Exploring numbers

Connect Student Book page 53

Big Idea

- We can say how a number is made up of tens and ones. We can say if a number is even or odd. We can use numbers to order things.

Global skills

- **Creative skills:** problem solving / exploring / investigating
- **Real-world skills:** research / presenting information / interpreting information / financial literacy
- **Interpersonal skills:** communication / teamwork / leadership
- **Self-development skills:** reflecting on learning

Key vocabulary

- first, second, third, fourth, …., tenth, ordinal numbers, more, less

Resources

- number line, coloured cubes or counters

Language support

Make a number display in the classroom. Show ordinal and cardinal numbers. Show how to make tens and ones.

 **Introductory activity**

Sort students in the class into two groups – boys and girls. Count how many of each. *Are there more boys or more girls? How do you know? Are there fewer boys or fewer girls? How do you know?*

 **Main activity**

Choose a group of ten students to make a line at the front of the class. Ask a range of questions about the positions of students in the line, making clear which end of the line is the start. For example, ask, *Is the 3rd student a boy or a girl? What colour top is the 7th student wearing?*

Give each pair of students a group of about ten coloured cubes (or counters). Ask each pair to arrange their cubes into groups of different colours. Then ask them to make a statement about their cubes and tell their partner, who can check if they are correct. For example, 'There are more red cubes than yellow cubes.' Ask each student to make up two statements for their partner to check.

Now ask students to arrange their cubes in a line and choose one end to be the front of the line. Ask them to make a statement about their line of cubes, for example, *The 5th cube in the line is blue.* Ask each student to make up two statements for their partner to check.

Look at the pictures on page 53 of the Student Book together. As a class describe the pictures and compare the objects. Use questions to help focus students' descriptions on odd and even numbers, ordinal numbers and describing numbers using place-value language. *How many babies and adults altogether? Is that an odd or an even number? How many tens? How many ones?* Students can then complete questions 1 and 2 individually and then share their asnwers with a partner.

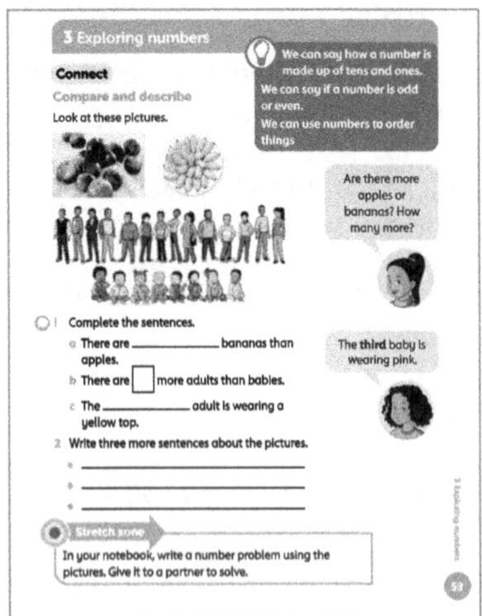

Differentiation

Supporting: Help students to compare the numbers in two groups by finding them on the number line to see which is larger.

Consolidating: Encourage students to use precise language when making their statements, using vocabulary of more/fewer and ordinal numbers.

Extending: Ask students to sort other objects by their own criteria before making statements about the number and position, for example a set of toys sorted by size or colour or number of wheels.

Stretch zone: *In your notebook, write a number problem using the pictures. Give it to a partner to solve.*

Check that students have written a suitable number problem related to the pictures.

 **Reflection time**

Ask students to discuss what they noticed about the statements they were making in the first part of the main activity. Were they just saying which group had more or fewer objects than another, or were they also saying how many more or fewer? For example, *There are 3 fewer red cubes than blue cubes.*

Model the vocabulary of positional language to include more detailed statements such as, *There are blue cubes in the 4th and 7th positions. There are 2 green cubes between the 5th and 8th positions.*

Differentiated outcomes	
All students	should count two groups of objects to compare them.
Most students	will compare two groups or find a position in a line and make a descriptive statement.
Some students	may make more detailed comparisons, for example, 'there are 6 more girls than boys'.

Student Book page 53

1 a There are more bananas that apples.

   b There are 6 more adults than babies.

   c The 12th adult is wearing a yellow top.

2 Students write three sentences about the pictures. Encourage them to include comparing words, such as 'more' and 'fewer' as well as to include amounts in their comparisons, e.g. There are 5 more apples than babies.

Stretch zone: Encourage students to draw simple sketches with clearly defined groups. Ask them to think of the question they want their partner to answer before they begin drawing.

3 Exploring numbers

Review Student Book page 54 • Practice Book page 58

Global skills

- **Creative skills:** problem solving
- **Real-world skills:** presenting information
- **Interpersonal skills:** communication
- **Self-development skills:** reflecting on learning

Student Book

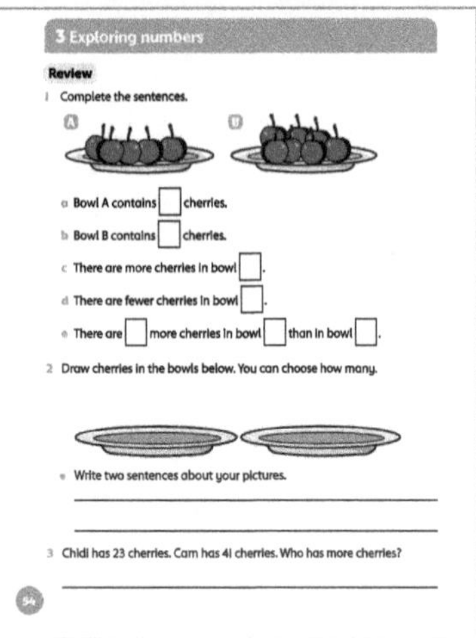

With young children, assessment activities are most effective when carried out as an everyday classroom activity. Students should have interlocking cubes, number lines and 100-squares available with both the digits and the numbers in words so that they can refer to these to support them.

Watch as students compare the number of cherries in the bowls. Listen to them as they count, to make sure that they say the number words in order and do not omit any numbers. Listen to check that they understand the last number they say is the total. If there are errors, add the numbers for the students so that they can hear the correct sequence.

It may help to create flash cards with the numbers and words that are in the review. You can use these as support for some students.

Encourage students to use a number line for questions 2 and 3. Extend by asking a range of similar questions or asking the students to make their own cherries and bowls problems.

Answers

Student Book page 54

1 **a** Bowl A contains 5 cherries.

   b Bowl B contains 7 cherries.

   c There are more cherries in bowl B.

   d There are fewer cherries in bowl A.

   e There are 2 more cherries in bowl B than in bowl A.

2 Check that students' sentences match their cherry pictures.

3 Cam has more cherries.

Practice Book

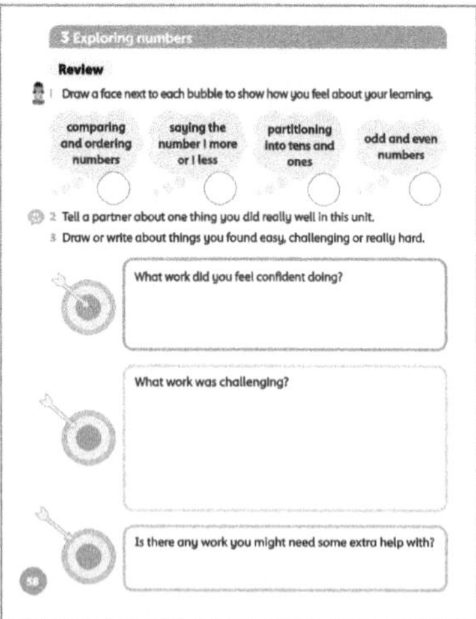

With young children it is appropriate to complete this as a whole-class discussion. You may choose to keep a record of the class discussion or a copy of the review page for your own records. Use the Student Book to briefly remind the students of the areas of mathematics

Give out the Student Book to support pairs of students as they discuss and answer the questions in the Practice Book.

Allow students plenty of time for discussion before asking them to share their responses with the rest of the class. If students complete this assessment at home, encourage them to discuss this with adults. Make a note of areas that students still feel unsure about.

As counting is at the heart of many of the other units you can revisit these areas regularly. You can also build counting into everyday practice: for example, counting how many students are in the class each day; counting objects as you give them out, and so on.

Additional material

There are additional end-of-unit assessments available on the *Oxford Owl* website.

4 Addition

Overview

Big Idea

Students will be beginning to understand that adding is what happens when the objects in two or more smaller groups come together to make one big group.

- Addition of whole numbers is based on sequential counting with whole numbers (1, 2, 3, 4, 5 and so on).
- Many different problem situations can be represented by part-part-whole relationships and addition and this relationship can be represented using a part-whole diagram.

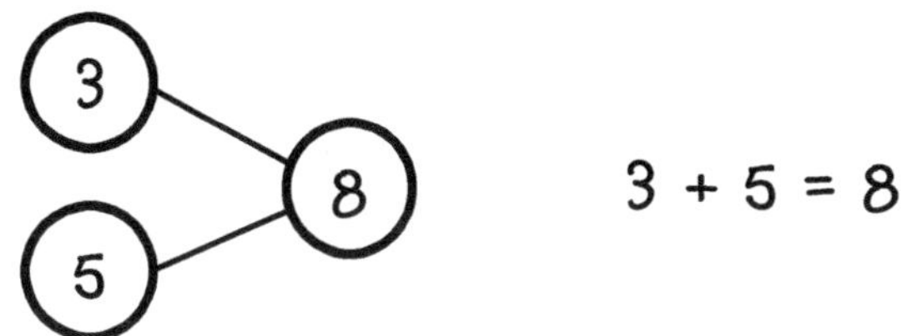

$$3 + 5 = 8$$

- The context of a problem situation such as 'I have 3 apples and my sister has 2. How many are there altogether?' and its interpretation can lead to different representations.

When assessing students, ask questions where students can apply their knowledge to other arithmetic or problem-solving situations. For example, instead of asking *What is 13 + 4?* say *If you know 3 + 4 = 7, can you use that to find 13 + 4?*

Look out for

- **Students who depend too heavily on their fingers when counting on or adding.** To help them make totals greater than 10, introduce other resources, such as number lines, as early as possible.

- **Students who have difficulty recalling facts.** Don't teach several facts at one time. Move slowly through them and give plenty of time for understanding. Approach addition in different ways, such as physical (jumps along a number track), songs or chants, drawings or pictures, or games. Encourage students to use what they know about the commutative property of addition, so that when asked to solve, for example, 4 + 9, they could change it to 9 + 4 to make it easier and make fewer errors.

Possible misconceptions

- **Students may include the first number in a sum when counting on**, e.g. for 5 + 3, they may say *five, six, seven*, instead of *six, seven, eight*.

- **Students may think numbers must be added in the order they are written.** Students do not yet understand that numbers can be added in any order and the total will not change. Therefore, they do not see that it is easier to start with the larger number when adding, e.g. they may count on 8 from 3 rather than counting on 3 from 8 when solving 3 + 8.

Key vocabulary

- add, addition, order, plus (+), altogether, equals (=), make, total

- count on, number sentence, jump, bridging 10, part, whole

Coverage in lessons

Learning focus	Learning outcomes (the ENC objectives)
Combining sets	Read, write and interpret mathematical statements involving addition (+) and equals (=) signs.
Adding on a number line	Add single-digit and 2-digit numbers to 20, including 0.
Counting on	Add single-digit and 2-digit numbers to 20, including 0.
Bridging 10	Represent and use number bonds within 20.
Addition word problems	Solve one-step problems that involve addition, using concrete objects and pictorial representations.

4 Addition

Engage
Student Book page 55

Big question

- How can I use objects or number lines to help me add?

Global skills

- **Creative skills:** problem solving
- **Interpersonal skills:** communication / teamwork / leadership

Key vocabulary

- add, addition, altogether, total, +, =

Resources

- cubes, a large beach ball with numbers 0–10 written on it, number lines, counters

Language support

Singing and saying number songs and rhymes as a whole class will help pronunciation and number name recall. Teachers can model the words, emphasising any difficult parts.

Students will also need support with simple number sentences, so always model these clearly on the board.

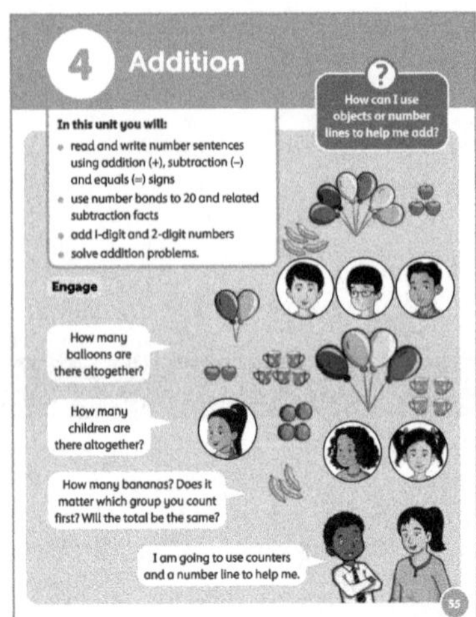

Introductory activity

Use songs, rhymes and stories, such as '5 little ducks', '5 speckled frogs', '10 in the bed' to help students become confident with numbers when counting forwards or backwards.

Main activity

If you have access to an IWB, you could display the Student Book Engage page 55 and discuss the Big idea of the unit with students. Ask students questions such as *How can we find how many students there are **altogether**? Where would you start counting? Perhaps you might start with the group of 3 and **add** on the two pairs underneath and the single student?* Establish that there are 8 students.

Now count the number of balloons. *How many balloons are there? Perhaps you could start by counting the largest group first?* Agree that there are 12 balloons. Repeat for the other items on the picture. Ask students to share their strategies for counting.

Differentiation

Supporting: Ask students to find the **total** of different sets of objects on page 55 of the Student Book using a number line and counters as support.

Consolidating: Ask students to select three groups of objects on page 55 of the Student Book, and to find the total number of objects in them.

Extending: Ask students to find the total of four or more groups of objects on page 55 of the Student Book.

Reflection time

Use a beach ball with numbers written on it. Ask students to stand in a circle (or face to face in pairs). Throw or roll the beach ball backwards and forwards between students.

Tell students that, whenever they catch the ball, they should say aloud the numbers that are under (or nearest to) their thumbs. They should add the two numbers that are under their thumbs before they pass the ball to the next person.

Keep playing the game so that all students have more than one turn each. If students are playing in pairs, limit it to five turns each.

This can be extended to adding three numbers. This time choose a start number, e.g. 3, then throw the ball and add to 3 the two numbers under the thumbs. Check the addition by adding the numbers in a different order.

Discover 1   Student Book page 56 • Practice Book page 59

Specific learning focus
- Understand addition as combining two sets; record related addition sentences.

Global skills
- **Creative skills:** investigating
- **Interpersonal skills:** communication

Key vocabulary
- add, addition, altogether, total, +, =, plus, equals

Resources
- paper plates, see-through jar or pot
- 10 cut-out gingerbread men, 10 counting bears or counters per student, digit cards 0–9

Language support

Ask key questions such as, *How can we make 10?*

Help students with the language used to describe the different ways of separating and combining to make 10. Reinforce the vocabulary of **addition** and **total**, and the words associated with '+' (**plus**) and '=' (**equals**).

 **Introductory activity**

Show the class a cookie jar with five gingerbread men in it. *This is my cookie jar and these are my 5 gingerbread men cookies.* Remove a cookie and put it on a plate. *How many are left in the jar?* (4)

Demonstrate this as you count the number that are left off the plate. Put them all together again and count five.

Repeat, putting different numbers on the plate and asking how many are left. Always ask: *How many?* when you put them back together.

 **Main activity**

Exchange the gingerbread men for 10 counting bears or counters and have a jar and two plates next to you.

Put two bears in the jar for the moment and count out the remaining bears. *We have 8 bears. How many should go on each plate?* (Any pair of numbers that totals 8.)

Wait for a student response and place that number on the plate.

Write the different combinations for the plates on the board: 0 + 8, 1 + 7, 2 + 6 and so on.

Ask individual students to sort the bears on the plates. Repeat with different numbers in the jar and on the plates. Ask students if, however they are arranged, there could ever be more than 10 bears in total. *Why not?* (Because the total at the start and at the end will always be the same, however they are grouped.)

Ask all students to complete page 56 of the Student Book using counting bears, imagining they are splitting them between two plates. Show students how to draw a simple stick bear before they begin so that they can focus on the addition rather than the drawing.

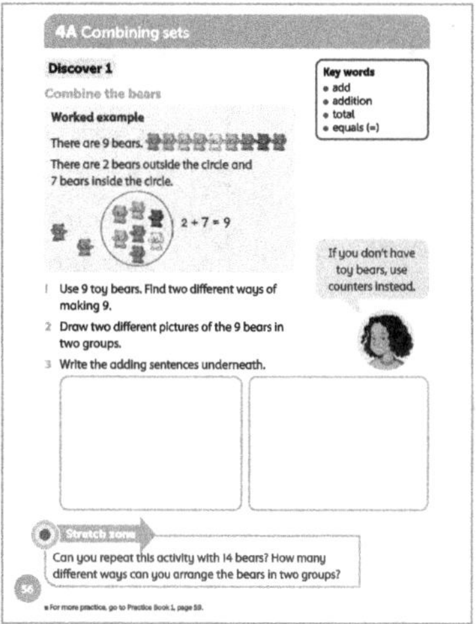

Differentiation

Supporting: You can help write the answers for the students as they tell you how they have split the bears into two groups.

Consolidating: Encourage students to be systematic in finding totals to 9.

Extending: Ask students how many different ways they think are possible.

Stretch zone: *Can you repeat this activity with 14 bears? How many different ways can you arrange the bears in two groups?*

Check that students can find different partitions of 14 and record the addition sentences correctly. For example::

$1 + 13 = 14$

$2 + 12 = 14$

Students could write $0 + 14 = 14$ or $14 + 0 = 14$, but this would mean no bears in one group, which isn't a group of bears!

 **Reflection time**

Use 10 bears, a jar and a plate. Challenge students to find all the different ways of separating the bears into the jar and onto the plate, and record each way as a number sentence, e.g. $4 + 6 = 10$. Repeat until all the ways of making 10 are recorded.

Practice Book: Students can complete Practice Book page 59. This can be done directly after the main activity, as homework, or as the focus of a separate mathematics session to help students consolidate their learning and build fluency. Students will need digit cards 0–9 to complete the activity. They may use counters or draw diagrams to model their addition sentences.

Differentiated outcomes	
All students	should model ways of making totals of 10 or less using two groups of counting bears with an adult recording for them.
Most students	will find a range of ways of making totals of 10 or less with two numbers.
Some students	may work systematically to find as many ways as they can of making totals of 10 or less with two numbers, recording using number sentences.

Student Book page 56

Students are finding ways of making 9 bears in total. The following answers are possible:

1 1 and 8, 2 and 7, 3 and 6, 4 and 5.

2 & 3 Check that students have made the right partitions and recorded them correctly as addition sentences:

$1 + 8 = 9, 2 + 7 = 9, 3 + 6 = 9, 4 + 5 = 9$

Stretch zone: $1 + 13 = 14, 2 + 12 = 14, 3 + 11 = 14,$
$4 + 10 = 14, 5 + 9 = 14, 6 + 8 = 14, 7 + 7 = 14$

Practice Book page 59

Check that the students' additions are correct. The totals may be greater than 10.

Stretch zone: There are several ways of making 9 by adding two or three digits.

4A Combining sets

Discover 2 Student Book page 57 · Practice Book page 60

Specific learning focus

- Understand addition as combining sets; record related addition sentences.

Global skills

- **Creative skills:** investigating
- **Interpersonal skills:** communication

Key vocabulary

- add, addition, altogether, equals, total, +, =

Resources

- cubes, paper plates, paper cups, see-through jar or pot, 10 cut-out gingerbread men
- large 0–20 number line for the front of the classroom

Language support

Ask key questions such as, *How can we make 10 by adding 3 numbers?*

Using practical resources to explore the different ways of combining 3 amounts to make 10 will develop understanding of bonds to 10.

 **Introductory activity**

Show the class ten gingerbread men and 3 empty plates. Put one gingerbread man on a plate. *How many are left?* (9). Now put 2 on a second plate. How many are on plates altogether? (3) How many are left? (7). Put the remaining 7 on the third plate.

Write on the baord $1 + 2 + 7 = 10$. Put the gingerbread men back in the jar, counting to 10.

Repeat, putting different numbers on each plate to show another way of making 10 by adding three numbers.

 **Main activity**

Exchange the gingerbread men for 10 cubes and have the jar and three plates next to you.

Partition the 10 cubes into 3 groups and put one group on each plate. Do this for different partitions. Write the different combinations for the plates on the board: $1 + 3 + 6; 1 + 1 + 8; 1 + 2 + 7$ and so on.

Ask individual students to sort the cubes on the plates. Repeat with different numbers on the plates. Ask students, *However they are arranged, could there ever be more than 10? Why not?* (Because the total at the start and at the end will always be the same, however they are grouped.)

Model the activity on page 57 of the Student Book. Place three cups so that all students can see them. Ask one student to count out 10 cubes and put two cubes in the first cup. Count together to see how many are left. Then place three cubes in the next cup. Count out how many are left. Place five cubes in the last cup. Write $2 + 3 + 5 = 10$.

Ask students to copy the example on page 57 of the Student Book and then complete the questions using cups and cubes.

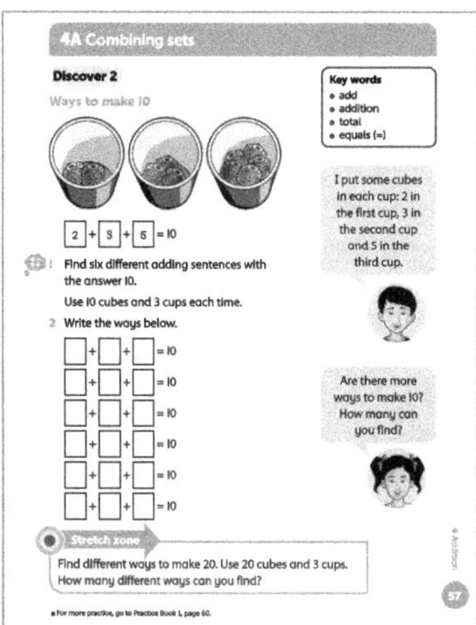

Differentiation

Supporting: You can help write the answers for the students as they tell you how they have split the cubes into the cups.

Consolidating: Encourage students to be systematic in finding totals to 10.

Extending: Ask students how many different ways they think are possible.

Stretch zone: *Find different ways to make 20. Use 20 cubes and 3 cups. How many different ways can you find?*

This activity reinforces the idea of partitioning a number into three in different ways. Check that students have found as many ways as possible without repetition and without groups of 0, and have recorded them systematically.

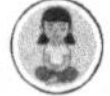 **Reflection time**

Challenge students to find all the different ways of separating the cubes into three cups, and record each way as a number sentence, e.g. $1 + 3 + 6 = 10$. Repeat until all the ways of making 10 are recorded, without repetition and without groups of 0 (9 in total)

Practice Book: Students can complete Practice Book page 60. This can be done directly after the main activity, as homework, or as the focus of a separate mathematics session to help students consolidate their learning and build fluency. Model how to use a number line to help, for example for $1 + 3 + [\,] = 10$, mark 1 and 10 on the number line, count on 3 from 1 to 4 and then work out how many jumps of 1 to get to 10 (6). Show students that they can rearrange the order of the numbers (even the missing ones) to make it easier to find the missing number, e.g. change $1 + [\,] + 3 = 10$ to $1 + 3 + [\,] = 10$.

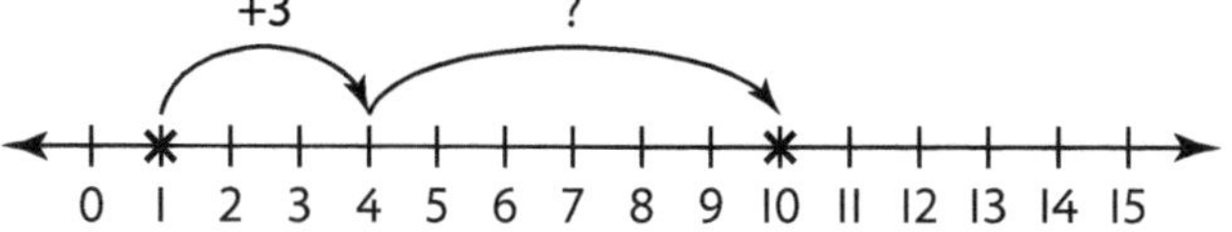

Differentiated outcomes	
All students	should model ways of making 10 using cubes and cups with an adult recording for them.
Most students	will find a range of ways of making 10 using cups and cubes and record them.
Some students	may work systematically to find as many ways as they can of making 10 with three numbers, recording using number sentences.

Answers

Student Book page 57

There will be different responses to this, for example $6 + 3 + 1 = 10$. Check that students have found six different ways.

Practice Book page 60

1 $1 + 6 + 3 = 10$

2 $6 + 3 + 1 = 10$

3 $5 + 4 + 1 = 10$

4 $5 + 1 + 4 = 10$

5 $4 + 4 + 2 = 10$

6 $2 + 4 + 4 = 10$

There are several ways of making 15 in three groups, e.g. $9 + 4 + 2 = 15$, $11 + 3 + 1 = 15$. Check that the partitions made by the students are correct.

Stretch zone: There are many different ways to add three numbers to make 15, so students will only find some of them. Check that students have formed correct additions, without repetition, and look for a systematic approach to recording them.

4A Combining sets

Explore 1
Student Book page 58 • Practice Book page 61

Specific learning focus
- Understand that changing the order of addition does not change the total.

Global skills
- **Creative skills:** investigating
- **Interpersonal skills:** communication / teamwork

Key vocabulary
- addition, add, total, +, =, order

Resources
- cubes in three different colours
- number cards 0–10
- table-top 0–20 number lines
- strips of paper

Language support
Use visual images to enable students to understand addition, focusing on the part-whole model and the vocabulary that supports it, for example, *7 is the whole, 4 is a part, what is the other part?*

 **Introductory activity**

Choose six students to come to the front of the class. Give each student a number card from a set of 0–10.

Say a number from 1 to 6 (e.g. 4). If any student has the number on their card that, when added to the number you say, totals 10, they give their card to a sitting student and leave the game. Repeat several times until all of the number cards have been used. The winner is the last student left standing with a number card.

 **Main activity**

Give each pair or group a table-top number line and two strips of paper of different lengths to fit along the number line, e.g. 5 and 2. Ask students to place the two paper strips end to end along the number line so that the numbers match.

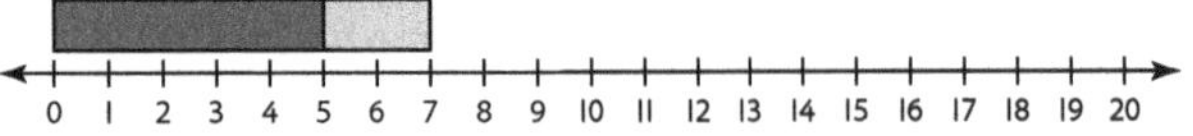

What is the last number on the line? (7) Write on the board $5 + 2 = 7$.

Repeat three or four times with different lengths of paper. Ask students to change the **order** of the strips

of paper. *Does it matter which way round the strips are placed?* (No) *Is the total always the same?* (Yes) Write the new addition sentence to reinforce that order does not matter and the total stays the same.

Ask students to work through page 58 in the Student Book. They will need some cubes in three different colours. After they have completed the questions, ask *What do you notice about your answers? What do you notice the numbers you are adding?* Agree that it does not matter in which order you add. The answer is the same.

Check their understanding by asking questions such as, *I know that $3 + 9 = 12$. What other addition do I know from this?*

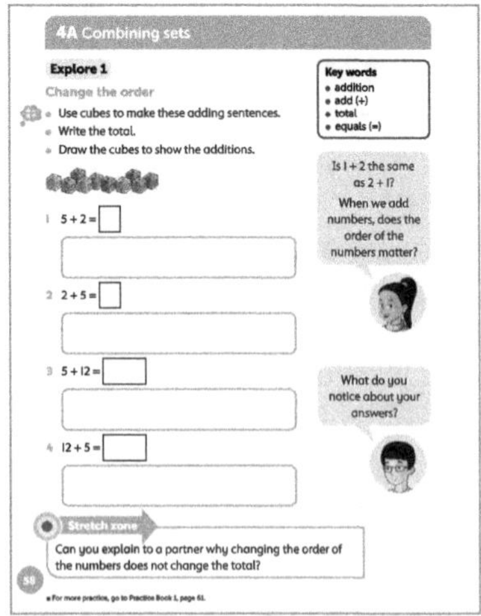

Differentiation
Supporting: Students should place the numbers strips, then count the first number and mark the number line, and then count on the second number.

Consolidating: Encourage students to look for number bonds that make 10 to help find totals greater than 10.

Extending: Ask students if they can use number bonds to help them find the total. Check by counting on.

Stretch zone: *Can you explain to a partner why changing the order of the numbers does not change the total?*

Check that students can see that the order of addition does not change the total. They should be able to check with different number pairs.

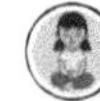 **Reflection time**

Make a giant number line in the class or outdoors using chalk. Start at 0. Ask two students to choose two number cards from a 1–10 set. *If we add these two numbers together, what number will we land on?* Ask them to talk to their partner, using the language of addition: add, more, equals.

Practice Book: Students can complete Practice Book page 61. This can be done directly after the main activity, as homework, or as the focus of a separate mathematics session to help students consolidate their learning and build fluency. Work through the example together, making sure that students understand that they can use the first number sentence to fill in the missing number or numbers of the second sentence.

Differentiated outcomes	
All students	should be able to add by counting both numbers on the number line.
Most students	will count on from the starting number to find the total.
Some students	may use number bonds to find the answer by recall.

Answers

Student Book page 58

1 $5 + 2 = 7$

2 $2 + 5 = 7$

3 $5 + 12 = 17$

4 $12 + 5 = 17$

Practice Book page 61

1 $5 + 7 = 12$ so ... $7 + 5 = 12$

2 $9 + 1 = 10$ so ... $1 + 9 = 10$

3 $8 + 6 = 14$ so ... $6 + 8 = 14$

4 $3 + 8 = 11$ so ... $8 + 3 = 11$

5 $5 + 9 = 14$ so ... $9 + 5 = 14$

6 $1 + 11 = 12$ so ... $11 + 1 = 12$

7 $7 + 8 = 15$ so ... $8 + 7 = 15$

8 $7 + 9 = 16$ so ... $9 + 7 = 16$

Stretch zone: As students choose their own numbers, answers will vary.

4A Combining sets

Explore 2   Student Book page 59 • Practice Book page 62

Specific learning focus

- Count objects up to 20, recognising conservation of number.

Global skills

- **Creative skills:** exploring
- **Interpersonal skills:** communication / teamwork

Key vocabulary

- number, count on, total, altogether, make

Resources

- fishing cards cut out from Resource Sheet 2 (available on the Oxford Owl website), copied onto card, one set for each pair
- cubes, beads or counters for counting
- two sets of number cards 1–10 for each pair

Language support

Place students in mixed-attainment pairs so they can hear peer modelling of the appropriate language. Count out the groups of cubes carefully, touching each one, and ask less confident language learners to repeat the process.

 Introductory activity

Teach the rhyme 'Once I caught a fish alive' to the class. Demonstrate the actions as you say it. Repeat it several times until the class can all say it with you.

1, 2, 3, 4, 5, once I caught a fish alive, 6, 7, 8, 9, 10, then I let it go again. (Hold up an extra finger each time.) *Why did you let it go? Because it bit my finger so. Which finger did it bite? This little finger on my right.* (Show the little finger on the right hand and wiggle it!)

Main activity

Show the class the fishing game cards (cut out from Resource Sheet 2). Explain how to play the game:

This is a game for two players. You each need a set of fish cards. Put all your fish together, face down, on a table. You will have 20 fish in total. Record your game on page 59 in the Student Book.

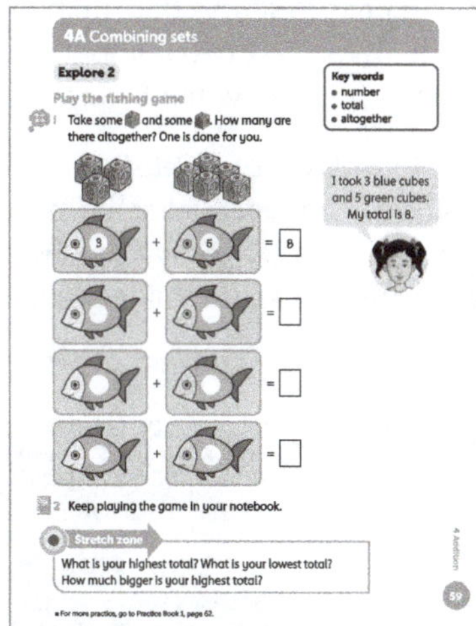

*Take turns to turn two of the numbered fish over. Look at the **numbers** on your fish. Get the same number of cubes to match them. Count all your cubes and write the total. Do not put the fish back. Keep fishing until all the fish have been caught.*

Look at the totals. Who has the highest total? Who has the lowest total? Do you have any totals that are the same?

Play the game three times. Ask students to record their results in their notebooks.

As the class are playing the game, walk around the classroom to make sure that the students understand the task. Look for students who find it difficult to **count on** from one number to another. Show them that they don't need to start from the beginning every time when they know the first number.

Differentiation

Supporting: Some students should complete the boxes showing the numbers of fish they 'caught'. Support them to count the totals and record them.

Consolidating: Ask students for a different way to make a total they have recorded, e.g., 8 fish is 3 fish and 5 fish but could also be 2 fish and 6 fish.

Extending: Some students could extend the game and turn three fish over each time.

Stretch zone: *What is your highest total? What is your lowest total? How much bigger is your highest total?*

Students can practise their addition skills by finding their highest and lowest totals. By counting on from the lowest to the highest they are beginning to experience calculating a difference for later work on subtraction as the inverse of addition.

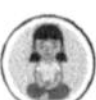 ## Reflection time

Each pair of students will need two sets of number cards 1–10. *Did anyone have the same totals in the game with different numbers?*

Demonstrate using 7. *The total was 7. Show me two numbers that **make** 7 (1 and 6, 2 and 5, 3 and 4). Does it matter which order the numbers are in? Give me an example. Is there a number that has only one way to make it using two other numbers?* (2 = 1 + 1)

Instead of asking pairs to simply hold up number cards, you could ask them to make two numbers using towers of cubes and hold these up with the number cards.

Practice Book: Students can complete Practice Book page 62. This can be done directly after the main activity, as homework, or as the focus of a separate mathematics session to help students consolidate their learning and build fluency.

Students may benefit from using counters or a similar concrete resource for support. Encourage them to compare representations to see if they can use them to complete other addition sentences: for example, question 5 has 1 more button than question 1 and the answer is 7 so the answer for question 5 will be will be 1 more than 7, which is 8.

Differentiated outcomes	
All students	should count out the correct number of cubes.
Most students	will record numbers accurately and add to find the total.
Some students	may begin to say the totals before they count the cubes.

Answers

Student Book page 59

Try to mark this activity as you walk around the classroom. As you look at their books, check that students have put out the correct numbers of cubes for each fish and that their totals are correct. You can write the correct numbers for those students who are not yet confident in writing numbers.

Practice Book page 62

1 6 + 2 = 8

2 5 + 6 = 11

3 4 + 7 = 11

4 2 + 7 = 9

5 7 + 2 = 9

6 8 + 5 = 13

7 8 + 4 = 12

Stretch zone: There are many ways to add three numbers to make 8, e.g. 1 + 2 + 5, 1 + 3 + 4, etc. Check that students have created three correct additions.

4A Combining sets

Explore 3   Student Book page 60 · Practice Book page 63

Specific learning focus

- Understand that changing the order of addition does not change the total.

Global skills

- **Creative skills:** problem solving
- **Interpersonal skills:** communication / teamwork

Key vocabulary

- addition, add, total, +, =

Resources

- interlocking cubes in two colours
- number cards 0–10

Language support

Use visual images to enable students to understand addition, focusing on the part-whole model and the vocabulary that supports it, for example, '7 is the whole, 4 is a part, what is the other part?'

Introductory activity

Choose 11 students to come to the front of the class. Give each student a number card from a set of 0–10.

Say a number from 0 to 10 (e.g. 4). The student with that number card steps forward. Say another number (e.g. 6). If a student with that number adds to the previous one to total 10, they both give their cards back and leave the game. Repeat several times until one student is left holding a number card.

 Main activity

Show students two rows of linked cubes, for example 3 and 4. *How many cubes are in each row?'* Put the two rows together to make a single row. Draw a 0–20 number line on the board and place the cubes along the number line so that the numbers on the line match the numbers of cubes.

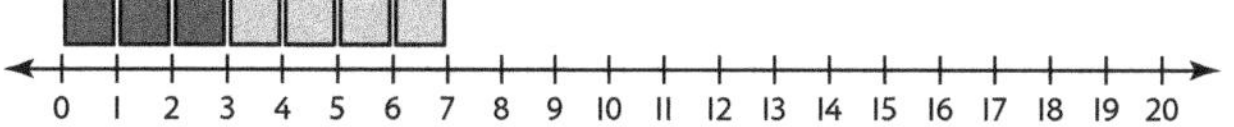

What is the last number on the line? (7) Write on the board: $3 + 4 = 7$.

Now swap the two groups of cubes around and show students that it now shows $4 + 3$. *Is the total number of cubes still the same?* (Yes) Write on the board: $4 + 3 = 7$.

Repeat three or four times with different numbers, each time changing the order of the lines of cubes. *Does it matter which way round the cubes are placed?* (No) *Is the total always the same?* (Yes)

Ask students to work through page 60 in the Student Book individually. They will need some cubes, in two different colours. After they have completed question 2, ask them to share their responses with a partner and discuss whether it matters in what order we add numbers.

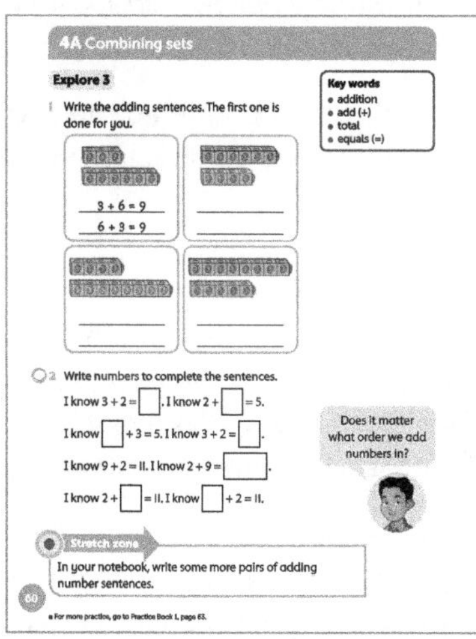

Differentiation

Supporting: Students should count the first number and mark the number line and then count on the second number.

Consolidating: Encourage students to count on from the first number of cubes they have.

Extending: Ask students if they can use number pairs to help them find the total. Check by counting on.

Stretch zone: In your notebook, write some more pairs of adding number sentences.

Check that students have written pairs of related addition sentences showing that reversing the numbers does not change the total.

Reflection time

Give students a number sentence, for example $6 + 2 = 8$. Ask them if they can think of another number sentence that totals 8 that is related to this one ($2 + 6 = 8$). Ask a student to model this with cubes and explain why the sentences are related.

Practice Book: Students can complete Practice Book page 63. This can be done directly after the main activity, as homework, or as the focus of a separate mathematics session to help students consolidate their learning and build fluency. Encourage students to look at previous questions to help them find the answers, e.g. If I know that $7 + 4 = 11$, and 8 is 1 more than 7, then $8 + 4$ will equal 1 more than 11, which is 12.

Differentiated outcomes	
All students	should be able to add by counting both numbers on the number line.
Most students	will see that swapping the line of cubes still makes the same total.
Some students	may recognise the total after turning the cubes without having to count or use the number line.

Answers

Student Book page 60

1 $6 + 4 = 10, 4 + 6 = 10$

  $4 + 8 = 12, 8 + 4 = 12$

  $8 + 5 = 13, 5 + 8 = 13$

2 I know $3 + 2 = 5$. I know $2 + 3 = 5$.

  I know $2 + 3 = 5$. I know $2 + 3 = 5$.

  I know $9 + 2 = 11$. I know $2 + 9 = 11$.

  I know $2 + 9 = 11$. I know $9 + 2 = 11$.

Practice Book page 63

1 $7 + 4 = 11$

2 $8 + 4 = 12$

3 $4 + 8 = 12$

4 $9 + 6 = 15$

5 $9 + 7 = 16$

Check that students have drawn the correct cubes next to each sentence.

Stretch zone: Make sure students' addition sentences are correct.

4B Adding on a number line

Discover
Student Book page 61 • Practice Book page 64

Specific learning focus
- Add two numbers by making a jump on the number line.

Global skills
- **Creative skills:** problem solving
- **Interpersonal skills:** communication / teamwork.

Key vocabulary
- add, addition (+), total, equals (=), jump

Resources
- large 0–20 number line for the front of the class
- number cards 1–10, interlocking cubes

Language support
Help students to use the language of addition and total to reinforce the procedure of adding using the number line.

 **Introductory activity**

Show students a 0–20 number line and a pile of 5 cubes. *I have 5 cubes. Attach the cubes one at a time, counting 1, 2, 3, 4, 5.* Explain that you are going to add some more cubes to the 5 already there. Show students 6 more cubes. Attach the 6 cubes one at a time to the 5, counting on as you join each one: 6, 7, 8, 9, 10, 11. Make sure the students know there are now 11 cubes in total.

Explain to students that you are going to show this addition on the number line. Start by pointing at 5 and say 1, 2, 3, 4, 5, 6 each time your finger moves on one place. When you get to 6, you will be pointing at 11. Mark the **jump** from 5 to 11 with an arrow and label it +6.

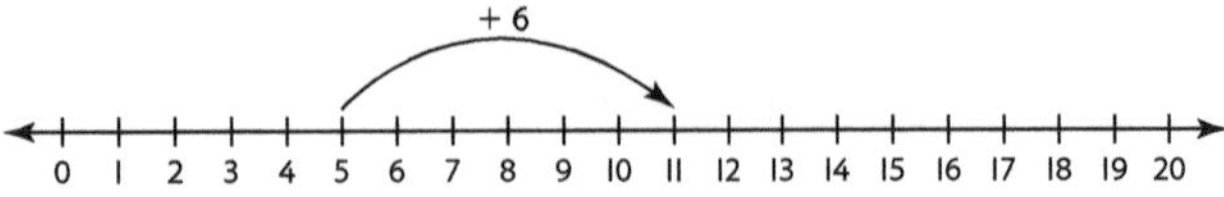

 **Main activity**

Ask two students to come out to the front of the class. One student picks a card at random from 1 to 10, and circles the number on the number line. The second student picks another card and counts on by the number shown on the card. For example, the first student picks 7 and the second student picks 5. The first student circles 7 on the number line, then the second student jumps on 5 places to reach 12, marking the jump with an arrow from 7 to 12.

Repeat with two more students and mark the new addition on the number line.

Ask students to work through page 61 in the Student Book in pairs, each picking a handful of cubes and finding the total. Support students to draw a number line in their notebooks as necessary, for example using a ruler to make a straight line and using centimetre markings to draw intervals. Alternatively, they could trace over one in the Student Book or Practice Book.

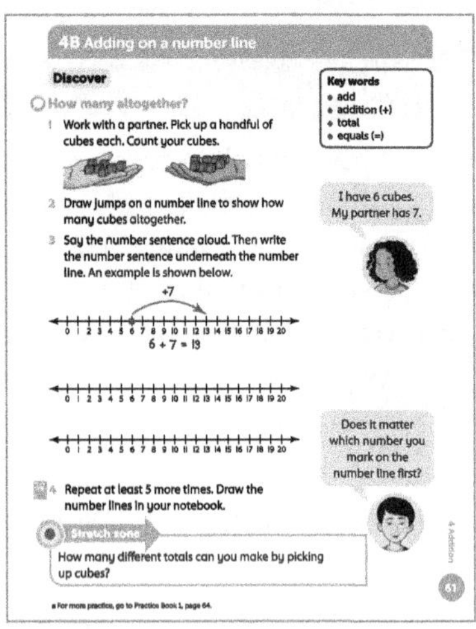

Differentiation
Supporting: Students should count the cubes, think about how to record the number and then match it to the number on the number line.

Consolidating: Encourage students to count on from the first number.

Extending: Ask students if they can use number bonds to help them find the total, for example, knowing that 7 + 5 = 12, mark it and check by counting on. They can then write the addition sentence.

Stretch zone: *How many different totals can you make by picking up cubes?*

Check that students have completed the addition number sentences and the number line jumps correctly.

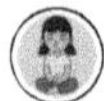 **Reflection time**

Show students the incomplete addition sentence 9 + 4 = [ ]. Ask, *Which number should we mark first on the number line?* (9). Then ask, *How many more need to be jumped?* (4) Complete the number sentence as 9 + 4 = 13.

Now show them 4 + 9 = [ ]. Some students may recall that this is the same as 9 + 4. Explain that starting on 4 and adding a jump of 9 gives the same total, but needs a bigger jump, so starting from the larger number is easier. Use the number line to demonstrate both methods.

Practice Book: Students can complete Practice Book page 64. This can be done directly after the main activity, as homework, or as the focus of a separate mathematics session to help students consolidate their learning and build fluency. Students will need two different-coloured pencils to complete the activity or alternatively they could shade the squares using different patterns, e.g. spots and diagonal lines.

Differentiated outcomes	
All students	should mark the first number on the number line.
Most students	will count on from the first number to find the total.
Some students	may write the addition sentence.

Answers

Student Book page 61

Check that the students have completed the number sentences and the jumps on the number line correctly.

Practice Book page 64

Check that students have correctly added the numbers, coloured in the squares and completed the number sentence correctly.

1 $1 + 8 = 9$

2 $8 + 2 = 10$

3 $3 + 5 = 8$

4 $6 + 6 = 12$

5 $7 + 4 = 11$

Stretch zone: Largest total is $9 + 9 = 18$. Smallest total is $0 + 0 = 0$.

4B Adding on a number line

Explore
Student Book page 62 • Practice Book page 65

Specific learning focus

- Add two numbers by making a jump on the number line.

Global skills

- **Creative skills:** problem solving
- **Real-world skills:** presenting information
- **Interpersonal skills:** communication / teamwork

Key vocabulary

- add, addition (+), total, equals (=)

Resources

- large 0–20 number line for the front of the class
- cubes

Language support

Help students to use the language of addition and total to reinforce the procedure of adding using the number line.

Introductory activity

Show students the number line. On it, mark a jump from one number to another, for example, from 8 to 15. *What was the starting number?* (8) Write this as the first number in the addition sentence. *How much was the jump?* (7) Write this as the second number in the sentence. Show students that we use the add **(+)** sign between the numbers to represent adding them together. *What number did the jump land on?* (15) Write this as the total in the sentence with the equals **(=)** sign. Also write the jump size over the arrow on the number line.

Main activity

Ask a student to come to the front of the class and mark a number on the number line, for example, 6. Ask another student to draw a jump from that number to another number, for example 13. Now ask students what the number sentence will look like for this jump. Help them to write it as $6 + 7 = 13$.

Repeat this with, for example, $9 + 3$ and then $5 + 8$. Explain that the second number in the sentence is always the same as the jump size.

Students can then complete Student Book page 62 in pairs. Some students may choose to use cubes to help them. Pairs take it in turns to describe how they found each of the missing numbers.

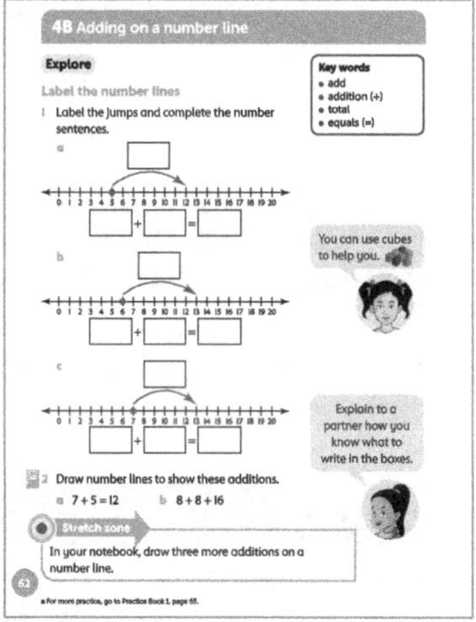

Differentiation

Supporting: Ask students to build the start and end numbers with cubes, and count on the 'difference' to find the size of the jump. Help students record this as a number sentence.

Consolidating: Ask students questions to help them explain their thinking, e.g. *What is the start number? How can you work out the size of the jump to your end number?*

Extending: Encourage students to see if they can use number bonds to help them find the total. They should check by counting on.

Stretch zone: *In your notebook, draw three more additions on a number line.*

Check that students have written addition number sentences correctly. Stretch them further by asking for three additions that have the same total.

 Reflection time

Show students a number line marked with a jump from 3 to 14. Ask how many the jump represents (11). *What number sentence represents this addition?* (3 + 11 = 14) Ask them to suggest two other additions with the same size jump as this, in other words, a jump of 11. Write the sentences on the board.

Practice Book: Students can complete Practice Book page 65. This can be done directly after the main activity, as homework, or as the focus of a separate mathematics session to help students consolidate their learning and build fluency. Ask students to look at the example and say what is the same and what is different from the previous page in the Practice Book. They should notice that they are adding three numbers rather than two. Using the example, show them how, for each number you add, you need to make another jump. Students may benefit from continuing to work with cubes to find totals.

Differentiated outcomes	
All students	should use cubes to represent the start and end numbers.
Most students	will count from the start number to the end number to complete the number sentences.
Some students	may be able to use their knowledge of number bonds to 20 to complete the number sentences.

Answers

Student Book page 62

1 a 5 + 7 = 12, jump size 7.

  b 6 + 7 = 13, jump size 7

  c 7 + 6 = 13, jump size 6.

2 Check that students have drawn the number line jumps correctly from the start number to the total.

Practice Book page 65

1 8 + 3 + 4 = 15

2 4 + 3 + 8 = 15

3 7 + 7 + 6 = 20

4 7 + 7 + 5 = 19

Stretch zone: Answers will vary as students choose their own three numbers. Check that they have written three correct additions of three numbers adding to make 20, e.g. 8 + 5 + 7 = 20.

4C Counting on

Discover Student Book page 63 • Practice Book page 66

Specific learning focus

- Add a single-digit number by counting on.
- Understand addition as counting on; record related addition sentences.

Global skills

- **Creative skills:** problem solving
- **Interpersonal skills:** communication / teamwork

Key vocabulary

- addition, add, total, altogether, +, =, make, number sentence

Resources

- set of large number cards 0–12 for class use, cubes in two different colours
- dice, counters

Language support

Use visual images to enable students to understand 'counting on'. Emphasise that, to count on, you don't count the number you are already on.

 Introductory activity

Choose six students to come to the front of the class. Give each of them a number card from 0 to 12. Ask them to stand in order from the smallest number to the largest number. Count as a whole class from 1 to 12. Some cards will be missing. When a number card is missing, make a tower of cubes to show the previous number, then add 1 cube in a different colour to make the tower into the missing number. If there is a gap of two numbers, add 2 cubes in the second colour, and so on.

Repeat with two number cards where there is a larger gap between them.

Main activity

Use the towers made in the Introductory activity. Hold one tower for students to see, for example, a tower made of three cubes of one colour and five of another.

What does this tower show you? (3 and 5 make 8) *How many do we have altogether?* (8) Turn the tower up the other way and repeat the questioning.

Agree that 3 and 5 make 8, and 5 and 3 make 8. Record this on the board.

Ask a student to pick a handful of cubes (all the same colour) and to join them to make a tower. Ask a different student to take some cubes of a different colour and to join them to make a tower. Join the two towers together. *Count with me, 1, 2, 3, 4, …*

We have five of this colour and two of this other colour. How many do we have altogether? (7) *Five and two more make seven.*

Show the tower the other way up. *Two and five more makes seven.* Write the two **number sentences** on the board. *Does it matter which way round the numbers are in this sentence?* (No)

Ask students to work through page 63 in the Student Book, playing the game twice to make at least two totals, and recording each total in two word sentences as well as a number sentence.

Ask them to read their sentences to a partner. Can their partner follow what they say to move along the number track?

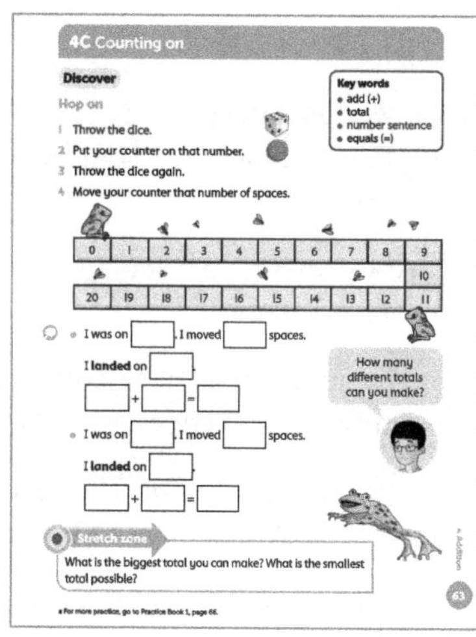

Differentiation

Supporting: Help students to write number sentences.

Consolidating: Ask students to check that reversing the two numbers does not change the total. Model with different towers.

Extending: Ask students to find three different numbers that also add together to make a given total.

Stretch zone: *What is the biggest total you can make? What is the smallest total possible?*

Students should try to reason that the highest total will be when they roll a 6 for the first move and then another 6 for the second move to make 12 (ending on 12). The smallest total is when they roll 1 and another 1 to make 2 (ending on 2).

Reflection time

Make a giant number line in the class or outdoors using chalk. Start at zero. Choose a student to stand on zero; they are going to walk along the line. Ask two other students to each choose a number card 1–12.

Ask the walking student to take steps on the number line to match the number on one of the cards. *Do you all agree?*

Ask the walker to continue along the number line to add the second number. *Do you all agree?* Repeat, choosing other students to be walkers or to choose a number card.

Practice Book: Students can complete Practice Book page 66. This can be done directly after the main activity, as homework, or as the focus of a separate mathematics session to help students consolidate their learning and build fluency. Tell students they may use a counter or similar to move along the number track to find each total.

Differentiated outcomes	
All students	should make towers of cubes to show number sentences.
Most students	will make towers of cubes to show number sentences and record them accurately.
Some students	may make larger towers and write related number sentences.

Answers

Student Book page 63

While students work, observe how they work out what numbers they land on. Do they count along the number track or do they recall number facts, e.g. starting on 3, and adding on 5, they know that they will land on 8.

Practice Book page 66

1 I start on 4. I count on 5. I land on 9.

2 I start on 6. I count on 5. I land on 11.

3 I start on 10. I count on 3. I land on 13.

4 I start on 10. I count on 1. I land on 11.

5 I start on 10. I count on 6. I land on 16.

6 I start on 15. I count on 4. I land on 19.

7 I start on 16. I count on 4. I land on 20.

8 I start on 18. I count on 2. I land on 20.

9 I start on 12. I count on 8. I land on 20.

10 I start on 14. I count on 6. I land on 20.

Stretch zone: Check that students' '=20' statements are correct.

Explore 1 Student Book page 64 • Practice Book page 67

Specific learning focus

- Add more than two small numbers, spotting pairs to 10, e.g. $4 + 3 + 6 = 10 + 3$.
- Add a single-digit number by counting on.

Global skills

- **Creative skills:** problem solving / exploring / investigating
- **Real-world skills:** research / presenting information / interpreting information / financial literacy

Key vocabulary

- addition, add, total, +, =, number sentence

Resources

- large 0–20 number line for class use
- mini whiteboards and markers
- counters
- bean bag or beach ball

Language support

Some students may need to practise the meaning of 'counting on'. Use visual images of animals that jump to show movement. These could be frogs or kangaroos.

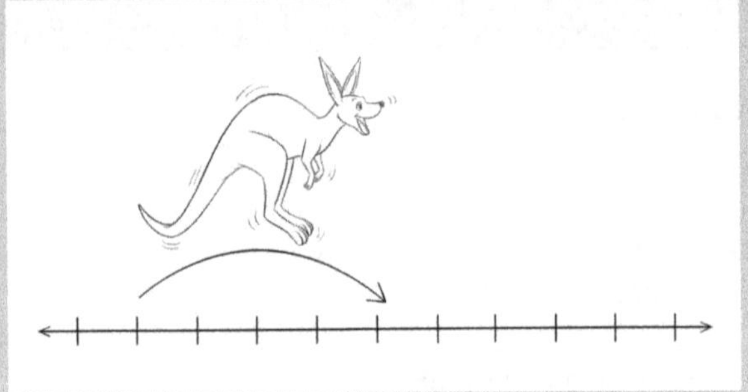

 **Introductory activity**

Arrange students in a circle. Throw a bean bag or beach ball to one of them and call out a number less than 10. The student who catches the ball has to respond with the number bond to 10. They then throw it to another student, and call out a number less than 10, and continue in the same way.

Main activity

Ask pairs to list as many number bonds to 10 as they can. Remind students that they practised this in earlier lessons.

Write the following lists of numbers on the board.

1, 8, 2

3, 4, 7

5, 2, 5

4, 6, 2

9, 1, 5

Ask a student to come up to the board and circle the two numbers in the first set which add together to make 10.

Using the class number line, start at 10 and count on the third number to find the total of all three numbers. Write the addition number sentence underneath, for example:

1, 8, 2

$8 + 2 + 1 = 11$

Ask students to complete the activities in the Student Book page 64 individually. Encourage students to circle the pairs of numbers which add together to make 10. They should use the number line to support their calculations.

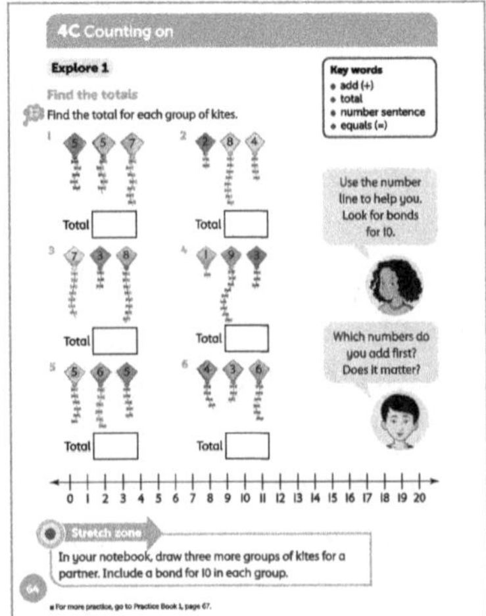

Differentiation

Supporting: Ask students to continue to use a number line to count on. Have the list of number bonds to 10 available for support.

Consolidating: Ask students to calculate in different orders to check their answers.

Extending: Encourage students to calculate mentally and then use a number line to check.

Stretch zone: *In your notebook, draw three more groups of kites for a partner. Include a bond for 10 in each group.*

Check that the numbers used include a number bond to 10 as part of the three numbers.

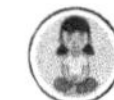 **Reflection time**

I have five. (Show a hand with five fingers.) *How many more to make ten? Count on from five.*

Ask students to respond using mini whiteboards and markers.

Choose a student to give a start and end number, e.g. *I have 7, how many more to make 13?* Repeat several times, and again ask students to respond on mini whiteboards.

Practice Book: Students can complete Practice Book page 67. This can be done directly after the main activity, as homework, or as the focus of a separate mathematics session to help students consolidate their learning and build fluency. Remind students to use number bonds to 10 to simplify the additions. Do some rapid recall of number bonds to 10 to get them started, e.g. *1 and what makes 10? 2 and what makes 10?*

Differentiated outcomes	
All students	should remember some number bonds to 10.
Most students	will find all number bonds to 10 and use this to help them calculate using a number line.
Some students	may quickly see number bonds to 10 and calculate the answers mentally.

Student Book page 64

1 17

2 14

3 18

4 13

5 16

6 13

Practice Book page 67

1 14

2 15

3 18

4 16

5 18

Stretch zone: Check that students' additions are correct.

4C Counting on

Explore 2 Student Book page 65 • Practice Book page 68

Specific learning focus

- Add more than two small numbers, spotting bonds to 10, e.g. $4 + 3 + 6 = 10 + 3$.
- Add a single-digit number by counting on.

Global skills

- **Creative skills:** exploring
- **Interpersonal skills:** communication / teamwork

Key vocabulary

- addition, add, total, $+$, $=$, number sentence, equals, jump

Resources

- number cards 0–10 (enough for one per student)
- mini whiteboards and markers
- large 0–20 number line for display, cubes

Language support

Some students may need to practise the meaning of 'counting on'. Encourage them to work with cubes to 'build' their jumps and attach them to their starting number of cubes to find the total.

Introductory activity

Allocate each student a number card from 0 to 10 (there could be three or four students with the same number). Call out a number from 0 to 10 and ask students with that number to stand up. They repeat the number together then sit down. For example, you call out 7, students who were given 7 all stand and call out '7!'

Now when you call out a number, ask students with the number bond to 10 to stand up and say the number. For example, you call out 6, and the students who were given 4 all stand and call out '4!'

Main activity

Ask students in pairs to list as many number bonds to 12 as they can on their mini whiteboard. Remind students that they have practised this for other numbers in earlier lessons. Then ask them to find number bonds for 13 and then 14.

Write the following four pairs of calculations on the board for students to look at:

$5 + 5 = 10$	$6 + 5 = 11$	$7 + 5 = 12$	$8 + 5 = 13$
$5 + 6 = 11$	$6 + 6 = 12$	$7 + 6 = 13$	$8 + 6 = 14$

Ask a student to come to the front of the class, choose one pair of sentences and describe what is the same and what is different about them. For example, they should notice that in each pair the first number is always the same, the second number increases by 1 and the answer increase by 1. Explain how making a jump that is 1 more gives an answer that is also 1 more.

Ask students to complete the activities in the Student Book page 65 individually. As they work, ask questions such as *What question did you ask yourself to find the missing number for question a?* (e.g. How many more to add to 5 to make 12?) *What question did you ask yourself to find the missing number for question d?* (e.g. How many altogether?)

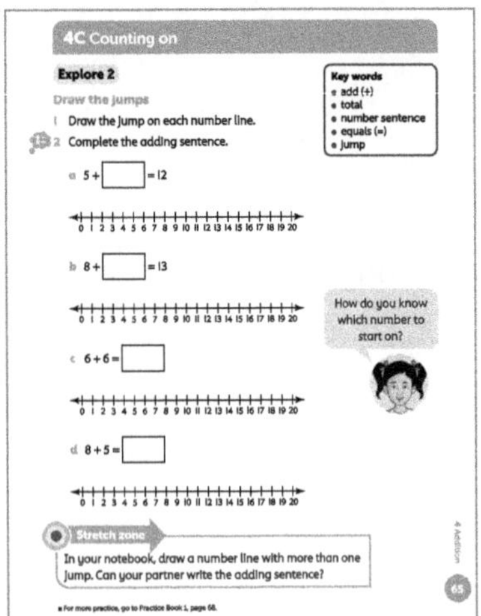

Differentiation

Supporting: Ask students to use a number line to count on. Have the list of number bonds to 10 available for support.

Consolidating: Ask students to calculate in different orders to check their answers.

Extending: Encourage students to calculate mentally and then use a number line to check.

Stretch zone: *In your notebook, draw a number line with more than one jump. Can your partner write the adding sentence?*

This activity takes students beyond adding two numbers together and makes them think about adding three numbers.

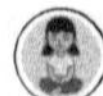 ## Reflection time

Tell students, *I have 7* (show hands with 7 fingers raised). Then ask, *How many more to make ten?* Students should be able to recall this, but if not, they can count on from 7. Ask students to respond using mini whiteboards and markers.

How many more to make 11? 12? 13? 14?

Choose a student to give a start and end number, e.g. *I have 5, how many more to make 14?* Mark this on the large number line and record the jump with an arrow. Repeat several times.

Practice Book: Students can complete Practice Book page 68. This can be done directly after the main activity, as homework, or as the focus of a separate mathematics session to help students consolidate their learning and build fluency. Ask students to think about any other number bonds to 20 that might help them to work out the total for each question, e.g. 10 and 7 make 17, so 9 and 7 will be 1 less than 17 because 9 is 1 less than 10.

Differentiated outcomes	
All students	should remember some number bonds to 10.
Most students	will find all number bonds to 10 and use these to help them calculate using a number line.
Some students	may quickly see number bonds to 10 and other totals.

Answers

Student Book page 65

2 a $5 + 7 = 12$ **b** $8 + 5 = 13$

c $6 + 6 = 12$ **d** $8 + 5 = 13$

Practice Book page 68

1 $9 + 7 = 16$

2 $9 + 5 = 14$

3 $9 + 3 = 12$

4 $17 + 2 = 19$

5 $2 + 17 = 19$

Stretch zone: Check that students have written a correct addition of three numbers to make 16, e.g. $4 + 10 + 2 = 16$.

Discover Student Book page 66 • Practice Book page 69

Specific learning focus

- Use number bonds to 10 to help with additions beyond 10.

Global skills

- **Creative skills:** problem solving / exploring / investigating
- **Real-world skills:** research / presenting information / interpreting information / financial literacy
- **Interpersonal skills:** communication / teamwork

Key vocabulary

- add (+), total, bridging 10, equals (=)

Resources

- 0–20 number lines, base-ten equipment, coloured pencils

Language support

Support students with the language of bridging through 10, counting on and totals as they complete additions using rods and cubes.

 **Introductory activity**

Take a pile of ones-cubes in one hand and another pile in the other hand, for example 8 and 6. Say, *I am going to use my number bonds to 10 to help me here. What do I need to add to 8 to make 10?* (2)

Take 2 cubes from the pile of 6, and put them with the 8 to make a pile of 10. Exchange the pile of 10 cubes for a ten-rod. You have 4 cubes left in the second pile.

Now ask how much you have altogether. Write on the board 10 + 4 = 14 and 8 + 6 = 14 so that students can see how they are related. *I add to make 10 first, then add the 4 to 10 to make 14. We can call this **bridging 10**.*

 **Main activity**

Show students the following additions on the board:

9 + 4   8 + 5   8 + 7   4 + 8

Ask a student to come forward and choose a calculation, for example 8 + 5. Ask them to make a pile of cubes for each number in the calculation. Then tell them to move some cubes from the smaller pile to the larger pile to make a pile of 10. Now exchange the pile of 10 cubes for a ten-rod.

How much do you have altogether in rods and cubes? (1 rod and 3 cubes) *What does this tell you about the total?* (It is 13)

Repeat with the other calculations.

Ask students to continue to use rods and cubes to support their calculations as they complete the activities in the Student Book page 66, working in pairs. Encourage them to explain their thinking as they find the totals.

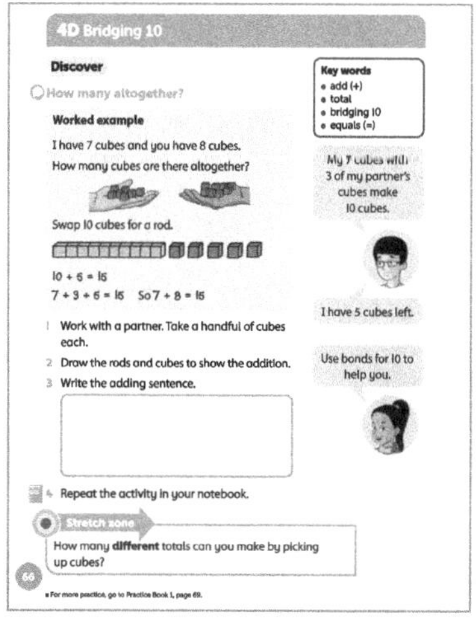

Differentiation

Supporting: Ask students to use a number line to count on. Have the list of number bonds to 10 available for support.

Consolidating: Ask students to calculate in different orders to check their answers.

Extending: Encourage students to calculate mentally and then use a number line to check.

Stretch zone: *How many different totals can you make by picking up cubes?*

Check that students are making correct totals and see what strategy they are using for finding the total.

 **Reflection time**

Ask a student to explain their strategy for adding, e.g. 3 + 9. Would they use 7 from the 9 and add it to the 3 to make 10 or would they use 1 from the 3 and add it to the 9 to make 10? Model both of these with cubes and ask, *Which is easier?*

Practice Book: Students can complete Practice Book page 69. This can be done directly after the main activity, as homework, or as the focus of a separate mathematics session to help students consolidate their learning and build fluency. Tell students to think about what filling in the first ten-frame shows them, then what filling in the second ten-frame shows them. Some students may notice that, for example, 8 add 4 is equal to 10 add 2.

Differentiated outcomes	
All students	should make a number bond to 10.
Most students	will use part of a number to make up a 10 and count how many left.
Some students	may complete additions mentally using a bridging 10 strategy.

Student Book page 66

Check that pairs are working correctly in taking cubes and totalling them using the rods and cubes.

Practice Book page 69

1 8 + 4 = 12

2 9 + 7 = 16

3 9 + 8 = 17

4 9 + 9 = 18

5 8 + 6 = 14

6 6 + 6 = 12

Check that students are completing the counters in the grids correctly

Stretch zone: Check students' number bonds to 16: 0 + 16, 1 + 15, 2 + 14, 3 + 13, 4 + 12, 5 + 11, 6 + 10, 7 + 9, 8 + 8, 9 + 7, 10 + 6, 11 + 5, 12 + 4, 13 + 3, 14 + 2, 15 + 1, 16 + 0.

4D Bridging 10

Explore — Student Book page 67–68 • Practice Book page 70

Specific learning focus

- Use number bonds to 10 to help with additions beyond 10.

Global skills

- **Creative skills:** problem solving
- **Interpersonal skills:** communication

Key vocabulary

- add (+), total, bridging 10, equals (=)

Resources

- large 0–20 number line for class use
- cubes, coloured pencils
- base-ten equipment, ten-frames, counters

Language support

Support students with the language of bridging through 10, counting on and totals as they complete additions using rods and cubes.

 **Introductory activity**

Show students the large 0–20 number line and write the calculation 8 + 5 on the board.

Mark 8 on the number line and then show a jump of 2 to 10. *How much did I jump?* (2) *If we need to add 5 to 8, how much more is there left to jump?* (3)

Jump 3 more from 10 to get to 13 and write 8 + 5 = 13.

Main activity

Create or draw a large 0–20 number line on the floor. Ask a student to come to the front of the class and stand on 9 on the number line. Write on the board 9 + 5. Ask the class, *How much of a jump will get the student to 10?* (1)

If we need to add 5, how much more of a jump is left to do? (4) Ask the student to jump on 4 to get to 14. Write 9 + 5 = 9 + 1 + 4 = 14. Highlight the 9 + 1 in the addition.

Repeat this with two more students for the calculations 8 + 7 and 9 + 7.

Ask students to complete the activities in the Student Book pages 67–68 individually. Explain that for each question you would like them to make two jumps, one to 10 and then another jump for the remaining part of the second number. Provide students with rods and cubes or ten-frames and counters to help them bridge to 10.

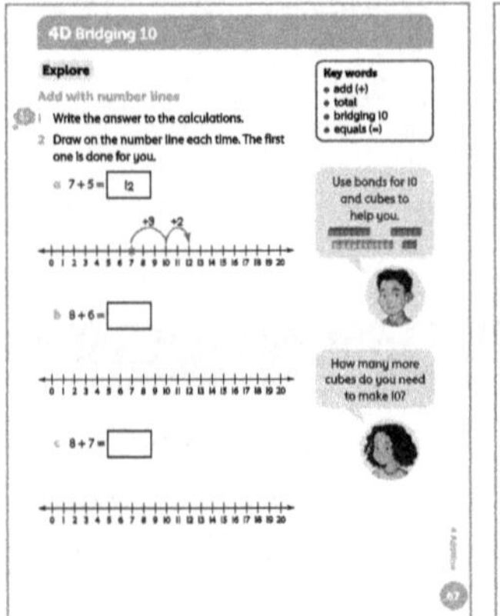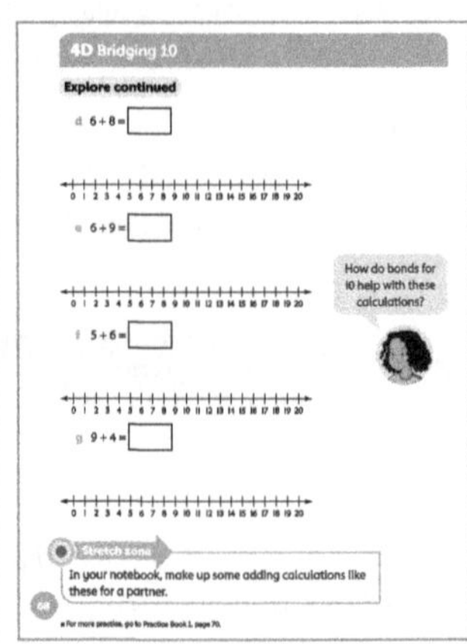

Differentiation

Supporting: Ask students to continue to use rods and cubes to help to count on. Have the list of number bonds to 10 available for support.

Consolidating: Ask students to calculate in different orders to check their answers.

Extending: Encourage students to calculate mentally and then use a number line to check.

Stretch zone: *In your notebook, make up some adding calculations like these for a partner.*

Check that students have made suitable addition calculations for a partner.

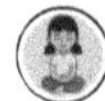 **Reflection time**

Ask students what helps them know how much to jump to 10 and how they know how much more is left to jump. Write 3 + 9 and ask how they would do this. Remind students that 3 + 9 is the same as 9 + 3, so they can jump to 10 from 9, which is easier.

Practice Book: Students can complete Practice Book page 70, using the number line and rods and cubes if needed. This can be done directly after the main activity, as homework, or as the focus of a separate mathematics session to help students consolidate their learning and build fluency. Tell students to continue to use bonds to 10 to make the additions simpler to calculate, using ten-frames and counters or rods and cubes for support.

Differentiated outcomes	
All students	should make a number bond that adds to 10.
Most students	will use part of a number to make up a 10 and count how many left.
Some students	may complete additions mentally using a bridging 10 strategy.

Answers

Student Book pages 67–68

1 and **2** Students should write the correct answer in the box and add the correct jumps on the number lines.

b $8 + 6 = 14$ **d** $6 + 8 = 14$ **f** $5 + 6 = 11$

c $8 + 7 = 15$ **e** $6 + 9 = 15$ **g** $9 + 4 = 13$

Practice Book page 70

1	15	**5**	14	**9**	13	**13**	11
2	16	**6**	15	**10**	14	**14**	12
3	12	**7**	11	**11**	10	**15**	12
4	18	**8**	17	**12**	16	**16**	13

Stretch zone: Look for answers such as: 'I jump 1 from 9 to 10 and then add what is left' or 'I add 10 instead of 9 and then take away 1 from the total.'

4E Addition word problems

Discover Student Book page 69 • Practice Book page 71

Specific learning focus

- Write addition stories from pictures and calculate them.

Global skills

- **Creative skills:** problem solving / exploring / investigating
- **Real-world skills:** research / presenting information / interpreting information / financial literacy
- **Interpersonal skills:** communication / teamwork / leadership
- **Self-development skills:** reflecting on learning

Key vocabulary

- addition, total, altogether, part, whole, addition sentence

Resources

- picture of a scene with some objects to count and add
- base-ten equipment, 0–20 number lines

Language support

Students may need support with describing the 'story' in their pictures and then interpreting how that translates to an addition sentence. Emphasise words like add, total, together.

 **Introductory activity**

Show students a picture of a scene with some objects to count and add: for example, an image of children playing in a playground.

Tell a matching short story such as, *At playtime/ breaktime, there were some children in the playground. There were 8 girls playing football and 7 boys playing on the climbing frame. How many children were in the playground altogether?*

Ask students to tell you what the addition sentence is $(8 + 7)$ and then use a number line or cubes and rods to find the total.

 **Main activity**

Explain to students that they are going to make their own pictures and addition stories. Ask them to think about some objects to draw in two groups to be added. For example, they could choose to draw two sets of toys, two groups of animals, some adults and children.

Ask them to record the addition as a number sentence and to use number lines or cubes and rods to help them calculate the total.

Students can then work individually on the Student Book page 69 and complete the addition stories. Start by discussing what they notice about the first problem. Invite a confident reader to read the problem aloud. What do students notice about the diagram? (That one part is how many cars altogether and the other two connecting circles are how many cars each person has.) Say that how many altogether is the **whole** and the other two are the **parts** and we call this a part-whole diagram. Ask them to draw a diagram and an addition sentence for the three other pictures.

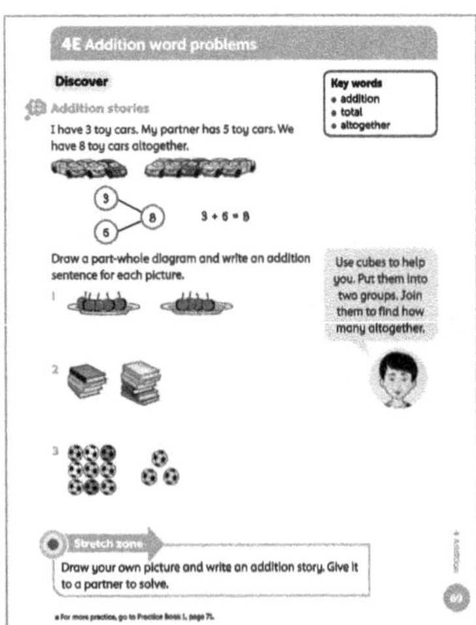

Differentiation

Supporting: Ask students to continue to use a number line and rods and cubes to help them to count. Draw a corresponding part-whole model to help students see the connection between representations.

Consolidating: Ask students to calculate using the number line.

Extending: Encourage students to calculate mentally and then use a number line to check. Students may be able to add three numbers together, for example, 6 + 4 + 3, where two of the numbers make 10.

Stretch zone: *Draw your own picture and write an addition story. Give it to a partner to solve.*

Check that the stories and addition problems are accurate and appropriate.

Reflection time

Ask a few students to share their pictures and corresponding addition stories. Ask the other students if they can see the numbers in the pictures. Ask different students to do the calculation for each story. Ask students if they can tell a different story using the picture. For example, *What story could you tell if there was one fewer houses in the picture?*

Practice Book: Students can complete Practice Book page 71. This can be done directly after the main activity, as homework, or as the focus of a separate mathematics session to help students consolidate their learning and build fluency. Read each problem aloud as a class and leave time for students to solve each addition story as you go along. Alternatively, students should ask an adult to read each problem to them.

Differentiated outcomes	
All students	should draw a picture which includes objects to be counted and added.
Most students	will describe the picture and form an addition sentence from it.
Some students	may form more complex additions that need three numbers to be added.

Answers

Student Book page 69

1 $5 + 4 = 9$ **2** $4 + 6 = 10$ **3** $9 + 3 = 12$

Practice Book page 71

1 $8 + 5 = 13$ **3** $8 + 6 = 14$ **5** $6 + 7 = 13$

2 $6 + 6 = 12$ **4** $5 + 7 = 12$

Stretch zone: Students' addition stories should use $7 + 4 = 11$.

4E Addition word problems

Explore Student Book page 70 • Practice Book page 72

Specific learning focus

- Write addition stories from pictures and calculate them using part-whole diagrams.

Global skills

- **Creative skills:** problem solving
- **Real-world skills:** presenting information
- **Interpersonal skills:** communication

Key vocabulary

- addition, total, altogether, part, whole

Resources

- objects such as small toys, marbles
- base-ten equipment, 0–20 number lines (large class one and small ones for student use)

Language support

Students may need support with describing the 'story' in their pictures and then interpreting how that translates to a part-whole model. Emphasise words like add, total, together, part and whole.

Draw an empty part-whole diagram on the board.

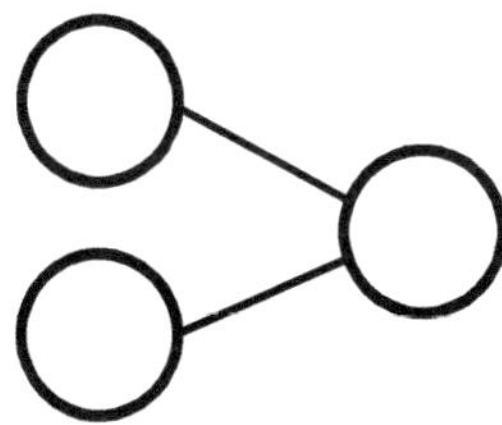

Tell students a story, for example, *I have 3 jumpers and 4 pairs of trousers. How many items of clothing is that altogether?*

Explain that the 3 jumpers and the 4 pairs of trousers are the separate parts, and the whole is when you add them together to make 7 items of clothing. Choose students to write the parts and the whole in the diagram and also write this as an addition sentence $3 + 4 = 7$.

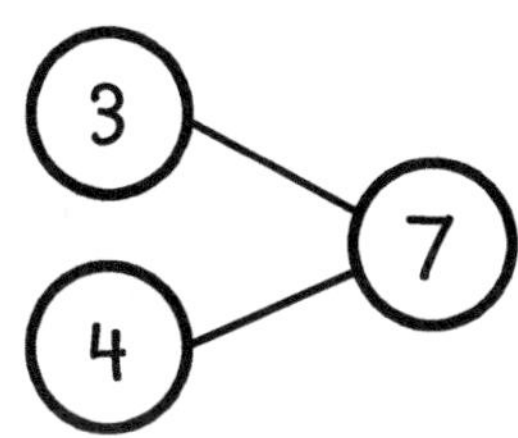

 Main activity

Show students two groups of items, for example, 8 toy animals and 5 marbles. Ask them, *How many animals are there?* (8), and say that this is one part. Now ask, *How many marbles are in the other part?* (5).

Draw the empty part-whole diagram and put the numbers in the correct places.

Now ask students what addition number sentence they need to write to calculate the total ($8 + 5 = $). Use cubes or a number line to get the answer 13 and write the total as the 'whole' in the diagram. Complete the number sentence as $8 + 5 = 13$.

Ask students to complete the problems on page 70 of the Student Book individually. Support students to read each problem as necessary.

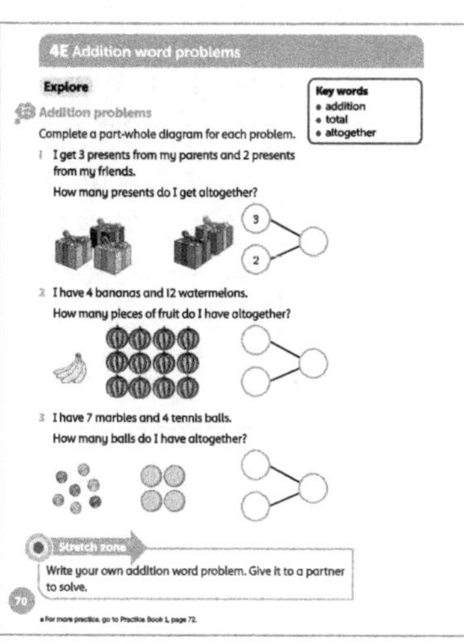

Differentiation

Supporting: Ask students to continue to use rods and cubes to help them to count on. Students can arrange the objects in a part-whole model then swap them for numerals to make the number sentence.

Consolidating: Ask students to calculate using part-whole representations.

Extending: Encourage students to calculate mentally and then use a number line to check.

Stretch zone: *Write your own addition word problem. Give it to a partner to solve.*

Check that the word problem is sensible and that the partner can solve it using a part-whole diagram.

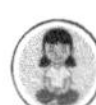 Reflection time

Ask a few students to share their part-whole diagrams and explain why they put the numbers where they did. Make sure they understand the difference between the parts and the whole and how this helps them make the number sentence.

Practice Book: Students can complete Practice Book page 72. This can be done directly after the main activity, as homework, or as the focus of a separate mathematics session to help students consolidate their learning and build fluency. For further practice representing problems using part-whole diagrams, ask students to draw a diagram for each of the problems.

Differentiated outcomes	
All students	should count how many in each part.
Most students	will add the parts to make the whole.
Some students	may complete the part-whole diagram and the number sentence accurately.

Answers

Student Book page 70

1 Parts 3 and 2, whole 5

2 Parts 4 and 12, whole 16

3 Parts 7 and 4, whole 11

Practice Book page 72

1 There are 6 helicopters. There are 4 airplanes.

There are 10 aircraft altogether. $6 + 4 = 10$

2 There are 5 ducks in one pond. There are 3 ducks in the other pond.

There are 8 ducks altogether. $5 + 3 = 8$

3 There are 6 socks with stripes. There are 5 socks with spots.

There are 11 socks altogether. $6 + 5 = 11$

4 There are 8 ladybirds. There are 4 butterflies.

There are 12 insects altogether. $8 + 4 = 12$

Stretch zone: Students should write an addition story and draw a picture to show $5 + 7 = 12$

4 Addition

Connect Student Book page 71

Big Idea

- We can use objects and number lines to add small numbers. We can explain the strategy we have used.

Global skills

- **Creative skills:** exploring / investigating
- **Real-world skills:** presenting information
- **Interpersonal skills:** communication / teamwork

Key vocabulary

- addition, add, total, +, =

Resources

- cubes or counters
- base-ten equipment

Language support

Students may need support with describing how they found the different amounts. Reinforce the language of 'add' and 'total' and emphasise the part-whole aspect of each addition.

 Introductory activity

Write the numbers 1, 2, 5 and 10 on the board. Ask students which totals they can make using two of the numbers at a time, for example, 2 + 5 = 7. Give them time to find the six different possible totals.

Check that they have found 1 + 2 = 3, 1 + 5 = 6, 1 + 10 = 11, 2 + 5 = 7, 2 + 10 = 12 and 5 + 10 = 15.

 Main activity

Now tell students they are going to try to make up different postage costs using the numbers for 1¢, 2¢ and 5¢ stamps, but this time they can use stamps more than once. For example, to make 8¢, they could use 5¢ + 2¢ + 1¢, or they could use 2¢ + 2¢ + 2¢ + 2¢.

Ask them to try to make every amount from 1¢ to 10¢ but using as few stamps as possible each time. For example, making 8¢ (above) uses fewer stamps if they use 5¢ + 2¢ + 1¢.

Students can then complete page 71 in the Student Book in pairs. Students may choose to make towers of single, two, five and ten cubes to represent the different stamp values and combine them to make towers to represent the postage needed for each letter, e.g. a tower of 9 cubes.

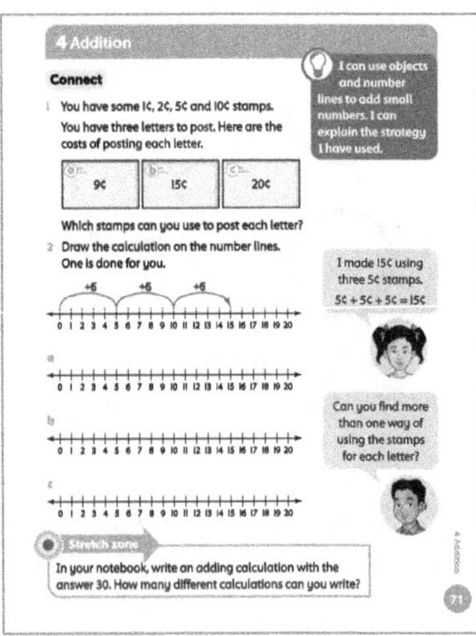

Differentiation

Supporting: Let students use cubes or counters to support their addition.

Consolidating: Ask students to check their answers aloud, explaining to you how they carried out the calculations.

Extending: Ask students to check the calculations by adding the amounts in a different order. Encourage them to calculate mentally.

Stretch zone: *In your notebook, write an adding calculation with the answer 30. How many different calculations can you write?*

There are many solutions to this question. Check that students have written calculations that total 30.

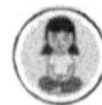 Reflection time

Ask questions such as, *What amounts were easy to make? Were they ones that only needed one or two stamps?* Get two students to share their answers to how to make 14¢ using the stamps they had. Ask the class to look at the ways they used and ask what is the same and what is different about their answers. For example, *Does one answer use more stamps than the other?*

Differentiated outcomes	
All students	should be able to calculate using cubes or counters to model the addition.
Most students	will find all the amounts using the stamps available.
Some students	may write addition sentences for three and four small numbers, for example, 2¢ + 2¢ + 2¢ = 6¢.

4 Addition

Global skills

- **Creative skills:** problem solving
- **Real-world skills:** presenting information
- **Interpersonal skills:** communication
- **Self-development skills:** reflecting on learning

Student Book

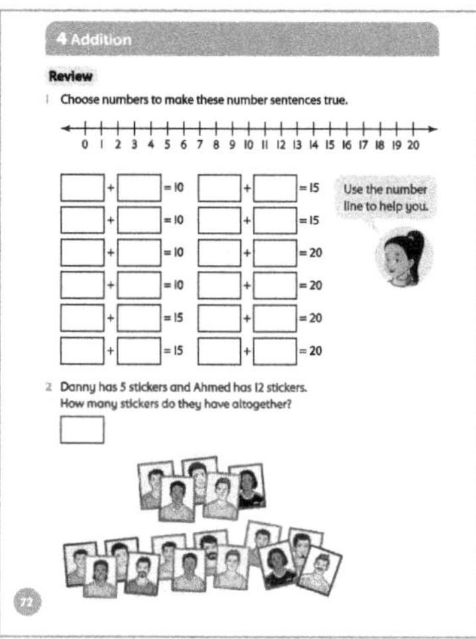

With young children, assessment activities are most effective when carried out as an everyday classroom activity. Students should have number lines with both the digits and the numbers in words available, and tens-rods and ones-cubes, so they can refer to these to support them.

Students complete page 72 of the Student Book independently. Watch as students complete addition pairs with different number bonds. Listen to them as they count to make sure that they say the number words in order and do not omit any numbers. Listen to check that they understand the last number they say is the total. If there are errors, add the numbers for the students so that they can hear the correct sequence.

It may help to create flash cards with the numbers and key vocabulary that are in the review. You can use these to support for some students.

Encourage students to use the number line for question 2. Extend by asking a range of similar questions or asking the students to make their own adding problems. For example, ask, *Can you make three different addition sentences that have a total of 14? Can you find all the number bonds that make 17?*

Answers

Student Book page 72

1 There are many possible answers to these sentences, so check that students have competed them correctly.

2 17 stickers.

Practice Book

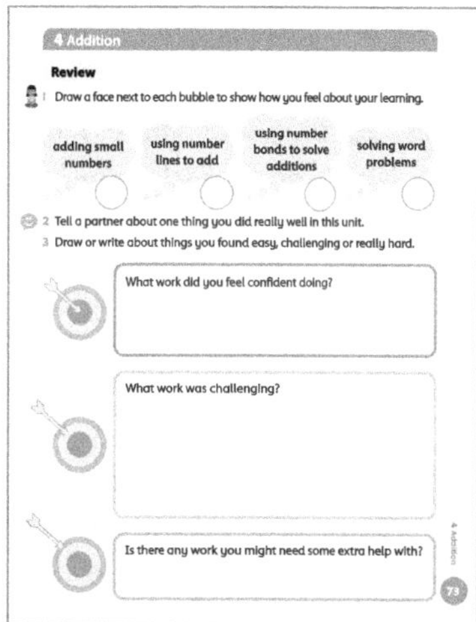

With young children it is appropriate to complete this as a whole-class discussion. You may choose to keep a record of the class discussion or a copy of the review page for your own records. Use the Student Book to briefly remind the students of the areas of mathematics that they have worked on in this unit.

Give out the Student Book to support pairs of students as they discuss and answer the questions in the Practice Book. Allow students plenty of time for discussion before asking them to share their responses with the rest of the class. If students complete this assessment at home, encourage them to discuss this with adults. Make a note of areas that they still feel unsure about.

As counting is at the heart of many of the other units you can revisit these areas regularly. You can also build counting into everyday practice. For example, counting how many students are in the class each day; counting objects as you give them out, and so on.

Additional material

There are additional end-of-unit assessments available on the *Oxford Owl* website.

Overview

Big Idea

It is important that students understand the connection between addition and subtraction.

- Addition and subtraction of whole numbers are based on sequential counting with whole numbers.
- Subtraction has an inverse relationship with addition. Students can use their understanding of addition to develop recall of subtraction and related subtraction facts.

Subtraction is used to represent and solve many different kinds of problems.

- Students need to understand the difference between 'partitioning', 'reduction' and 'comparative difference'.
- Reduction is reducing the value of one quantity. This is often called 'take away'.
- Comparative difference involves the comparison of two quantities and finding the numerical difference between them.

Students learn how to solve number problems by applying their understanding of models of addition and subtraction (such as combining or separating sets or using number lines), relationships and properties of number (such as place value), and properties of addition (commutativity and associativity). Students need to develop, discuss and use effective methods for subtraction. Subtraction is not commutative or associative for whole numbers.

Basic number facts are best developed through a process of noticing patterns and relationships between addition and subtraction. The development of subtraction facts comes from the relationship of subtraction and addition. For example, to work out $15 - 7 = ?$, students might think

$7 + ? = 15$. Students need activities so that they can see and apply these patterns.

Look out for

- **Students who are unsure of what is required because of the different ways in which subtraction tasks can be phrased**, e.g.

3	minus	2	equals	1
3	subtract	2	is	1
3	take away	2	leaves	1
2	less than	3	is	1

- **Students who are not using language precisely.** It is important to use correct mathematical language when talking to students.

Possible misconceptions

- **Students may confuse the symbols for addition and subtraction.**
- **Students may think that subtraction is commutative because addition is commutative**: they are 'opposites'.
- **Students may think that subtraction always means 'take away'** and so do not know how to solve a problem involving finding the difference.

Key vocabulary

- take (away), how many are left/left over?
- 1 less, 2 less, ..., 10 less, count on, count back, jump on, jump back, subtract
- how many fewer/less, how many more, difference, what is the difference?

Coverage in lessons

Learning focus	Learning outcomes (the ENC objectives)
Counting back	Read, write and interpret mathematical statements involving subtraction (–) and equals (=) signs.
Take away	Subtract single-digit and 2-digit numbers to 20, including 0.
Finding the difference	Subtract single-digit and 2-digit numbers to 20, including 0.
Subtraction word problems	Solve one-step problems that involve subtraction, using concrete objects and pictorial representations.

5 Subtraction and difference

Engage Student Book page 73

Big question

- What is subtraction?

Global skills

- **Creative skills:** problem solving / exploring
- **Interpersonal skills:** communication / teamwork

Key vocabulary

- take away, subtract, how many less, how many are left, 1 less than, 2 less than

Resources

- number rhymes and songs from Resource Sheet 3 (available on the Oxford Owl website)
- 0–20 number lines: a large one for the front of the classroom and small ones for each table
- counters or cubes

Language support

Make the rhymes as active as possible. Encourage all students to join in counting and finding how many are left. Some students may need to count the 'frogs' or 'monkeys' using one-to-one correspondence – one person, one number.

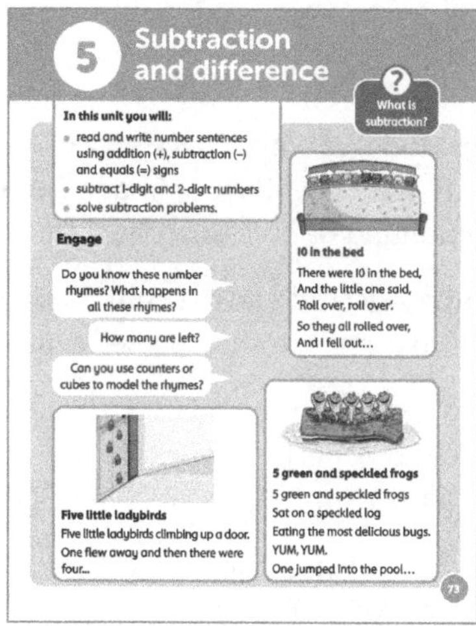

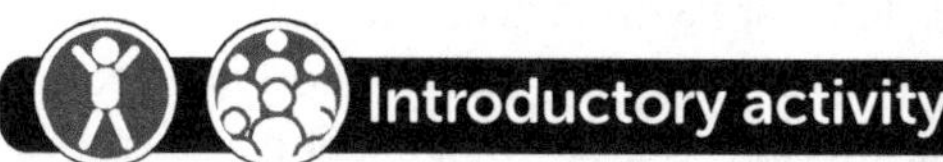

Introductory activity

Tell students they are going to learn a new rhyme about counting backwards. Choose five students to come to the front of the class to play the role of monkeys. Sing the rhyme 'Monkeys on the bed' with the five students acting it out:

5 little monkeys jumping on the bed,
1 fell off and bumped his head.
Mummy called the doctor and the doctor said,
'No more monkeys jumping on the bed'.

Repeat the rhyme and ask 'one monkey' to leave at the end. *We had five monkeys, how many have gone? One has gone. There is one less.* **How many are left?** Say the rhyme again with four monkeys jumping. At the end of the rhyme each time ask, *There is 1 less than before, how many did we start with? (5) How many have gone? (1, 2, 3, ...). How many are left?*

Main activity

Introduce to students the rhyme 'Ten in the bed'. Ask ten students to come to the front of the class and line up, facing the rest of the class. Say the first verse:

There were 10 in the bed and the little one said, 'Roll over, roll over'. So they all rolled over, and one fell out.

At this point demonstrate the action of all the students moving along the line one place and one of them 'falling out of bed' by removing one student from the end of the line. Ask that student to return to their seat. *How many are left? How can we find out?* Count the remaining students in the line and write on the board: 10 – 1 = 9. *If we take one away from 10 we have 9 left.* Repeat for 9 – 1 = 8.

Next, ask two students to 'fall out of bed' at the same time. *So they all rolled over and two fell out.* Vary the vocabulary that you use to ask how many are left, e.g. using take away, **how many less?**, how many fell out? how many are left?

Repeat, each time removing one, two or three students from the line, until there are no students left.

Differentiation

Supporting: Sit with students to encourage them to join in the rhymes.

Consolidating: Students can make pictures or posters of the rhymes.

Extending: Students can take the lead in sharing the rhymes.

Reflection time

Continue asking questions of 1 less than and 2 less than. *What is 1 less than five? What is 2 less than twelve?* Students can use a number line as well as counters or cubes to help them.

5A Counting back

Specific learning focus

- Add/subtract a single-digit number by counting back/on.

Global skills

- **Creative skills:** problem solving
- **Interpersonal skills:** communication / teamwork

Key vocabulary

- count on, count back, take away, jump on, jump back

Resources

- 0–20 number lines: large one for the classroom floor, large one for classroom display and small ones for tables
- number cards 1–6

Language support

You can help write the answers for students, saying the subtraction aloud as you record the operation. Then record the operation as students say it aloud. Finally, ask them to say aloud a subtraction sentence that you have written.

Introductory activity

Place or draw a large 0–20 number line on the classroom floor. Ask students to sit in a circle around the number line. Choose a student to be 'frog' and ask them to stand on the number 9.

Jump back two spaces. Is the 'frog' on 7?

Repeat several times, asking students to choose different start numbers, initially up to 10 and then up to 20. Choose a number card 1–6 from a shuffled set to give the number to jump back. Ask students to predict the number they will land on.

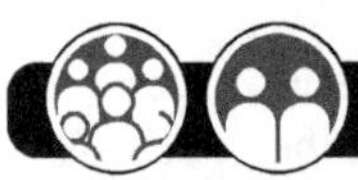

Main activity

Give each pair a table-top 0–20 number line. *We are going to **count back** along the number line to zero. Count with me.* Start at 15 and count back to zero. Point to the numbers on a large displayed number line as you say them.

Start at 6 and count back to 2. Repeat using different start and finish numbers. Show the jumps along the line and mark the start and finish numbers.

Start at 12 (circle that number). Count back to 6 (point to each number as you count back). Circle 6. *We started at 12 and counted back to 6. How many numbers did we*

jump back? (6) *We are on 6. If we **count on** 6 more we get to 12. Will this happen with other numbers?*

Repeat to show how to check the subtraction with an addition by counting on to the start number.

Choose another start number, e.g. 10. *Count back 5. 10 count back 5 is 5. We can say this a different way. We can say 10 **take away** 5 is 5.* Write on the board: 10 – 5 = 5.

Repeat with another start number, recording the subtraction on the board.

Ask students to complete page 74 in the Student Book in pairs so they can check each other's answers. Look at the number track at the top of the page together. *Think about our number line. What is the same? What is different?* (They both go up from 1 to 20 in a row but this track does not start with 0.) Together, do some quick counting on and back along the number track, encouraging students to touch each number as they count. Once they have completed the questions, ask pairs to investigate the second speech bubble together.

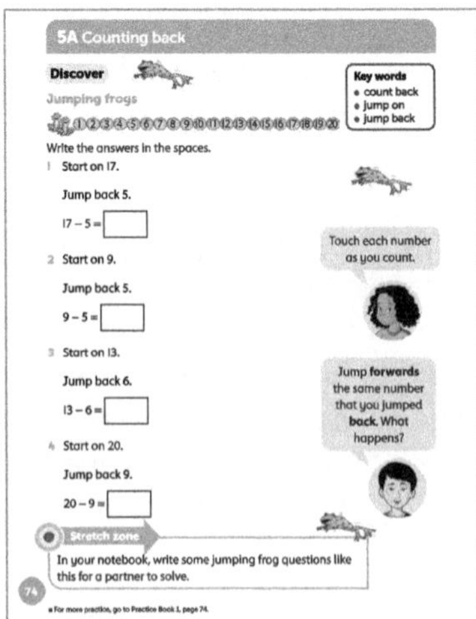

Differentiation

Supporting: Model counting back on a number line by touching each number with your finger. This supports students in counting back accurately.

Consolidating: Ask students to count forwards as well as backwards to demonstrate the inverse nature of the operations.

Extending: Ask students to write fact families based on one fact, for example 3 + 4 = 7, 4 + 3 = 7, 7 – 4 = 3, 7 – 3 = 4.

Stretch zone: *In your notebook, write some jumping frog questions like this for your partner to solve.*

Check that students are writing accurate questions that involve jumping back along the number line, and look to see if the partner can answer the question and record their answer as a number sentence.

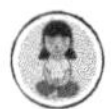 **Reflection time**

Draw a 0–20 number line on the board and select three number cards from a set of cards 1–6. Write the three numbers on the board. Ask pairs to work out the total. Mark this on the number line. Ask a student to come to the front. They select a number card and then move this number back on the number line. Ask them to read out the subtraction sentence as they write it on the board.

Practice Book: Students can complete Practice Book page 74. This can be done directly after the main activity, as homework, or as the focus of a separate mathematics session to help students consolidate their learning and build fluency. Explain that they may choose any number they like to 'jump back' but you would like them to try to use a different number each time.

Differentiated outcomes	
All students	should subtract accurately by counting back on a number line.
Most students	will predict the answer when jumping back 1 or 2.
Some students	may be able to predict the answer for larger numbers.

5A Counting back

Explore   Student Book page 75 · Practice Book page 75

Specific learning focus

- Understand subtraction as counting back and 'take away'; record related subtraction sentences.

Global skills

- **Creative skills:** problem solving
- **Interpersonal skills:** communication / teamwork

Key vocabulary

- count on, count back, take away, subtract

Resources

- sandcastle game board from Resource Sheet 4 (available on the Oxford Owl website), copied onto card, one for each group
- a set of number cards 1–6 for each group, counters in different colours (each student in a group needs a different colour)
- 0–20 number lines: a large one for the front of the classroom and small ones for each table, cubes

Language support

Instead of drawing the jumps on a number line, use practical equipment that can be moved along a line, so that the actions match the verbal instructions.

 **Introductory activity**

Use the class number line. *I start on 12 and count back 3. What number do I land on?* (9)

Student Book page 74

1 12   **2** 4   **3** 7   **4** 11

Practice Book page 74

Answers will vary as students can choose their own numbers to subtract from 10, 15 and 20. Check that students have written correct number sentences.

Stretch zone: Students should check their answers using addition.

I want to check that I have counted correctly. I can add the 3 back on. Show that, after counting back 3, you can count on 3 and end up on the same number.

Choose other numbers, count back and then count on to end on the same number.

Write 12 − 3 = 9, 9 + 3 = 12. Explain to students what you are writing as you write it.

 **Main activity**

Students work in mixed attainment groups of 4 to play the sandcastle game. They will need a game board (Resource Sheet 4). Each player in a group needs counters in a different colour and each group needs a set of number cards 1–6. Display the game board and the instructions for the game, and read them out:

- Put your counter on Start.
- Take turns to choose a number card.
- Move your counter that number of squares.
- If you land on a subtraction, find the answer and cover that number on the sandcastle with one of your counters.
- If you land on a starfish, cover any number on the sandcastle.
- Play until all sandcastle numbers are covered.
- The person to cover the last number on the sandcastle is the winner.

Tell students they can use their table-top number lines to help them work out the subtractions. Model how to play the game by playing a few rounds as a class.

Students should complete the activities on Student Book page 75 individually. Encourage them to think about how they used their number lines while playing the sandcastle game to help them complete the questions.

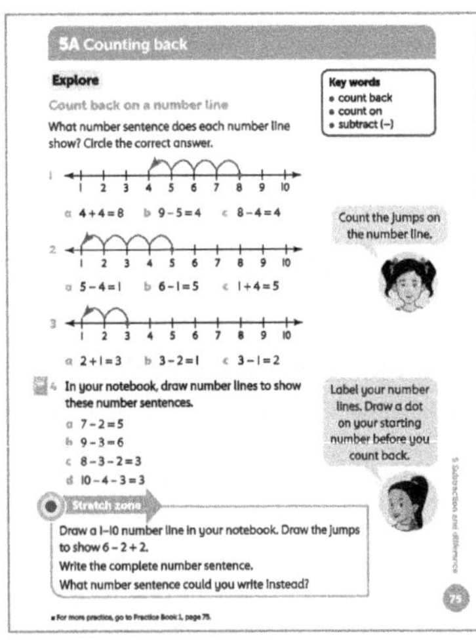

Differentiation

Supporting: Provide support to help students play the game and help them in recording subtraction calculations. Read the operation aloud as you write a subtraction sentence.

Consolidating: Ask students to explore moving back a certain number on the number line and then moving forward the same number. Ask them to explain why you get back to the same number.

Extending: Ask students to model their own addition and subtraction calculations on a number line and write down the operation.

Stretch zone: *Draw a 1–10 number line in your notebook. Draw the jumps to show 6 − 2 + 2.*

Write the complete number sentence. What number sentence could you write instead?

This activity will help students reinforce the idea that subtracting and then adding the same number leaves the original number unchanged. They should be able to write a number sentence to show this.

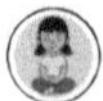 Reflection time

Model the answers to activity 4 in the Student Book on the large 0–20 number line. Pairs can check each other's books. Write one of the subtractions on the board, e.g. 7 − 2 = 5. Say, *7 **subtract** 2 equals 5*. Explain that sometimes we use the word 'subtract' instead of 'take away'.

Practice Book: Students can complete Practice Book page 75. This can be done directly after the main activity, as homework, or as the focus of a separate mathematics session to help students consolidate their learning and build fluency. This activity extends students' learning about subtraction to connect with addition. Fact families will be explored in depth in Year 2. Work through the example and first question together. If students seem confident, they can work on questions 2–6 in pairs. Otherwise, continue to work as a class to complete the questions. Students who attempt the stretch zone activity should use the number track and concrete resources such as cubes for support.

Differentiated outcomes	
All students	should play the game.
Most students	will play the game without support from an adult.
Some students	may play the game without referring to a number line.

Answers

Student Book page 75

1 **c** $8 - 4 = 4$

2 **a** $5 - 4 = 1$

3 **b** $3 - 2 = 1$

4 **a** $7 - 2 = 5$ start at 7 with 2 jumps back to 5
    b $9 - 3 = 6$ start at 9 with 3 jumps back to 6
    c $8 - 3 - 2 = 3$ start at 8 with 3 jumps back to 5 and then 2 jumps back to 3
    d $10 - 4 - 3 = 3$ start at 10 with 4 jumps back to 6 and then 3 jumps back to 3

Practice Book page 75

1 $5 + 3 = 8$ so $3 + 5 = 8$ and $8 - 3 = 5$ and $8 - 5 = 3$

2 $2 + 7 = 9$ so $7 + 2 = 9$ and $9 - 2 = 7$ and $9 - 7 = 2$

3 $1 + 5 = 6$ so $5 + 1 = 6$ and $6 - 1 = 5$ and $6 - 5 = 1$

4 $3 + 8 = 11$ so $8 + 3 = 11$ and $11 - 3 = 8$ and $11 - 8 = 3$

5 $6 + 4 = 10$ so $4 + 6 = 10$ and $10 - 4 = 6$ and $10 - 6 = 4$

6 $8 + 5 = 13$ so $5 + 8 = 13$ and $13 - 5 = 8$ and $13 - 8 = 5$

Stretch zone: Check that students have written two more fact families.

5B Taking away

Discover
Student Book page 76 • Practice Book page 76

Specific learning focus
- Understand subtraction as counting back and 'take away'.

Global skills
- **Creative skills:** problem solving
- **Interpersonal skills:** communication / teamwork

Key vocabulary
- take away, subtract, 1 less than, 2 less than, how many less? how many left?

Resources
- interlocking cubes (10 per student)
- large 0–10 number line for the front of the classroom
- number cards 1–6 for each table, counters

Language support
Drawing the calculations and modelling the vocabulary as you handle cubes will support language development. Emphasise the language of 'less than' in relation to subtraction calculations.

Introductory activity

Recite the rhyme 'Five currant buns':

Five currant buns in the baker's shop,
Big and round with sugar on the top.
A little boy/girl came with a penny one day,
Bought a currant bun and took it right away.

Choose five students to be currant buns and one student to come to the shop. Say the rhyme with the class. After the first bun has been taken away, ask: *How many are left? Count with me: 1, 2, 3, 4.* Continue with the rhyme and counting until all of the buns have been taken away. Choose different buns and a shopper and say the rhyme again.

Main activity

Say, *In my garden, I have a very tall wall.* Build a tower of 10 cubes. *Imagine this is my wall. I like to look over my wall, but it is too high. There are ten bricks in my wall. Count the cubes. I need to take away one brick from the top.* Take a cube off and put it on a table. *How many bricks are in my wall now? Count with me 1, 2, 3, ..., 9.* Show how to write this: 10 – 1 = 9. Say, **1 less than** 10 is 9.

My wall is still too high. I will take two bricks away. Take two cubes off. *I had nine bricks in my wall. I have taken two more away. How many are left? Count with me 1, 2, 3, ..., 7.* Show how to write this: 9 – 2 = 7. Say, **2 less than** 9 is 7. Model it on a large 0–10 number line.

My wall is still too high. I will take five more bricks away. Take five cubes off the wall. I had seven, I took five away. How many bricks are left on my wall? Count with me 1, 2. Show how to write this, 7 – 5 = 2, and model it on the large number line. *Now my wall is just right!*

Ask students to make a wall of 10 cubes. Then ask them to pick a number card from a shuffled pack and take that number of cubes away. They write: 10 – ? = ? Then ask them to put the cubes together again. They repeat this six more times.

Ask students to complete page 76 in the Student Book individually. Students can continue to use cubes or counters to model the subtractions.

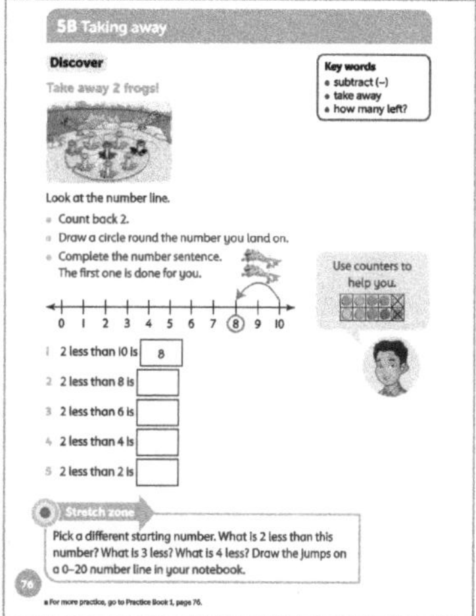

Differentiation

Supporting: Give students concrete materials such as number lines, counters and cubes to support them.

Consolidating: Ask students for the answer if they add back again the number they had taken away. This will begin to help them to make the link between subtraction and addition.

Extending: Students can write facts families from a subtraction. For example:
10 – 3 = 7 leads to: 10 – 7 = 3, 7 + 3 = 10, 3 + 7 = 10.

Stretch zone: *Pick a different starting number. What is 2 less than this number? What is 3 less? What is 4 less? Draw the jumps on a 0–20 number line in your notebook.*

This activity will show students the effect of taking away bigger numbers. As the number being subtracted gets bigger, the answer gets smaller.

Reflection time

Ask students to sit in mixed-ability pairs. Ask questions such as:

I have five pencils. I give you two, how many are left?

I have eight coins. I spend half of them. How many are left?

Ask one student to draw a picture to show each subtraction and their partner to say the answer. The more confident student should do the drawing.

Give pairs time to discuss the answer before asking them to share their answers.

Practice Book: Students can complete Practice Book page 76. This can be done directly after the main activity, as homework, or as the focus of a separate mathematics session to help students consolidate their learning and build fluency. For additional practice, ask students to record each as a subtraction sentence, e.g. 3 less than 7 is 4, $7 - 3 = 4$.

Differentiated outcomes	
All students	should understand that taking away a brick/cube is the same as counting back on a number line.
Most students	will be able to record the operation as a subtraction sentence.
Some students	may begin to understand that subtraction and addition undo one another (subtraction and addition are inverse operations).

Answers

Student Book page 76

1 8 **2** 6 **3** 4 **4** 2 **5** 0

Practice Book page 76

1 5 **2** 5 **3** 4 **4** 7 **5** 2

For each answer, check that the student has drawn the correct jumps on the number line.

Stretch zone: Check that students have drawn a number line with an arrow showing a jump from 13 to 8.

5B Taking away

Explore 1   Student Book page 77 · Practice Book page 77

Specific learning focus

- Understand subtraction as counting back and 'take away'.

Global skills

- **Creative skills:** problem solving / exploring / investigating
- **Real-world skills:** research / presenting information / interpreting information / financial literacy

Key vocabulary

- take away, subtract, less than, how many less? how many left?

Resources

- counting stick (1 metre long and marked in 10 equal divisions)
- number cards 0–20, sticky tack
- 0–20 number lines, counters
- interlocking cubes

Language support

You can help write the answers for students, saying the subtraction aloud as you record the operation. Then record the operation again as they say it aloud. Finally, ask them to say aloud a subtraction sentence that you have written. Emphasise the language of 'less than' in relation to the subtraction calculation.

 Introductory activity

Show the class the counting stick. Practise counting on and then back in ones. *I have eight sweets, but I eat three. How many do I have left?* Point to 8 on the stick, count back 1, 2, 3. *What number is this? How many sweets do I have left?* Write on the board: $8 - 3 = 5$. Say, *8 subtract 3 equals 5*.

Choose different students and different start numbers to practise using the counting stick to count back.

 Main activity

Students work in pairs with a 0–20 number line for each pair. They start with a counter on 10. Ask them to jump back 3. *What number did you land on?* (7)

If any students did not land on 7, show the jumps on the board. Ask students to record the operation, $10 - 3 = 7$, saying, *Ten subtract three equals seven* as they write it.

Ask students to secretly choose a different start number, jump back 3 and tell their partner the number they landed on. Their partner has to say what their start number was. *How did you work it out?* (Jump forward 3.)

Ask students to repeat this activity, taking it in turns to choose a start number and a number to jump back. Each time the other student works out the start number by jumping forward.

Students complete the activities on page 77 of the Student Book in pairs so that they can check each other's answers. As students work, ask questions such as *How many ducks altogether in the first group? What is the whole? How big is the part that you are taking away? How big is the part that is left?*

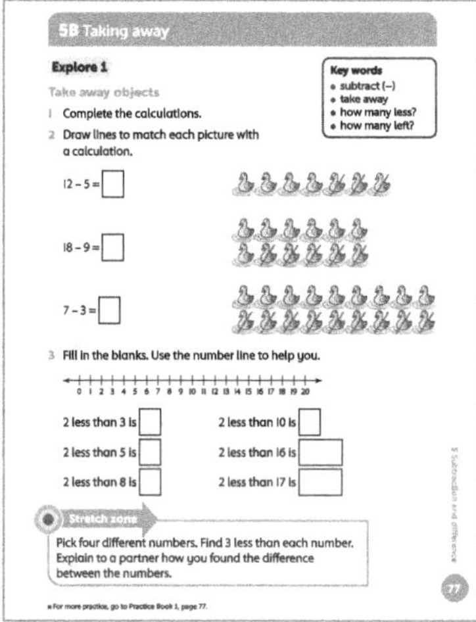

Differentiation

Supporting: Model subtractions by pointing to the numbers on a number line and linking up the idea of 'less than' with the size of the jump back. This supports students in counting back accurately.

Consolidating: Ask students to count forwards as well as backwards to demonstrate the inverse operations.

Extending: Ask students to find 2 less than 10 − 2; 2 less than 10 − 3; and so on. What pattern do they notice?

Stretch zone: *Pick four different numbers. Find 3 less than each number. Explain to a partner how you found the difference between the numbers.*

Check that students use the correct language for finding '3 less' and explain how they carried out each subtraction.

 **Reflection time**

Draw a number line on the board and mark 0 and 10 on either end, but nothing in between. Ask individual students to come to the board, write a start number (in its approximate position) on the number line and draw a curved arrow to show a jump back. Ask them to write the size of the jump above the arrow, then ask the other students to say the number they will land on.

Repeat several times.

Practice Book: Students can complete Practice Book page 77. This can be done directly after the main activity, as homework, or as the focus of a separate mathematics session to help students consolidate their learning and build fluency. Students may choose to continue to use concrete resources such as counters or cubes to model the subtractions. For additional practice, ask students to record each question as a subtraction sentence, e.g. 2 less than 10 is 8, 10 − 2 = 8.

Differentiated outcomes	
All students	should subtract by counting back on a number line.
Most students	will record the calculation accurately.
Some students	may understand the inverse relationship between subtraction and subtraction.

Answers

Student Book page 77

1 & 2 12 − 5 = 7   2nd group of ducks
        18 − 9 = 9   3rd group of ducks
        7 − 3 = 4   1st group of ducks

3 2 less than 3 is 1   2 less than 10 is 8
    2 less than 5 is 3   2 less than 16 is 14
    2 less than 8 is 6   2 less than 17 is 15

Practice Book page 77

1 8		**4** 3		**7** 5	
2 7		**5** 2		**8** 4	
3 8		**6** 1		**9** 3	

Check that students have correctly written three more subtraction sentences using 'less than'.

Stretch zone: Check that students can describe the pattern and have written a relevant sentence.

5B Taking away

Explore 2 — Student Book page 78 · Practice Book page 78

Specific learning focus

- Understand subtraction as counting back and 'take away'.

Global skills

- **Creative skills:** problem solving / exploring
- **Real-world skills:** presenting information
- **Interpersonal skills:** communication / teamwork

Key vocabulary

- take away, subtract, how many less? How many left?

Resources

- large 0–20 number line to display at the front of the class and a large number line (or number track) for the classroom floor
- cubes

Language support

Help write the answers for students, saying the subtraction aloud as you record the operation. Then record the operation as they say it aloud. Finally, ask them to say aloud a subtraction sentence that you have written, ensuring they are using the correct language for subtraction.

 **Introductory activity**

Show students a 0–20 number line on the board. Circle the number 17 and then draw a jump back of 8 to land on 9. Say, *I am going to write this as a subtracting number sentence. What will be the first number in the sentence?* (17) *What will be the next number in the sentence?* (8). Write the complete number sentence on the board: 17 − 8 = 9.

Now ask, *What will be the next number in the sentence?* (8). Write the complete number sentence on the board: 17 − 8 = 9.

Repeat with 15 and a jump of 5, leading to the sentence 15 − 5 = 10.

 **Main activity**

Place or draw a 0–20 number line on the classroom floor. Ask a student to come to the front of the class and stand on 17 on the number line. Tell them to take a jump of 9 backwards from 17. They can do this one step at a time and count as they go. What number do they land on? (8)

Now mark this on the display number line, showing a jump from 17 to 8. Write the number sentence on the board: 17 − 9 = 8.

Explain to students that they can check the final number by jumping in the opposite direction and counting on. Ask the same student, starting at 8 on the floor number line, to jump forward 9 places to land on 17. Write the addition sentence on the board: 8 + 9 = 17.

Repeat with another student and another set of numbers, for example, start at 13 and jump back 6 to land on 7, then start on 7 and jump forward 6 to land on 13.

Write the sentences 13 − 6 = 7 and 7 + 6 = 13. Ask students what they notice about the sentences.

Students complete the activities on page 78 of the Student Book in pairs so that they can check each other's answers. Draw students' attention to the second speech bubble and ask them to check their answers by counting on. Can they describe to their partner what they do?

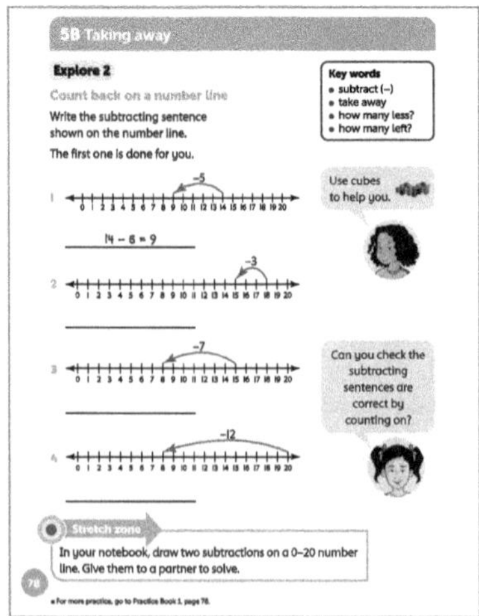

Differentiation

Supporting: Model counting back on a number line by touching each number with your finger. This supports students in counting back accurately. Students can use cubes to help them.

Consolidating: Encourage students to think about the inverse relationship by asking questions such as, *I jump back 3 and land on 7. What number did I start on?*

Extending: Ask students to write facts families based on one fact.

Stretch zone: *In your notebook, draw two subtractions on a 0–20 number line. Give them to a partner to solve.*

Check that the subtractions are appropriate and drawn correctly.

 **Reflection time**

Write the sentence 5 + 9 = 14 on the board. Tell students that this sentence has been used to check a subtraction. Ask the students if they can say which subtractions could be checked using this addition (14 − 5, 14 − 9). Mark them on a number line.

Practice Book: Students can complete Practice Book page 78. This can be done directly after the main activity, as homework, or as the focus of a separate mathematics session to help students consolidate their learning and build fluency. Take students through the steps. Explain that they should use the card they choose to find the number they must take away. Remind them which missing number box they must record it in, if necessary.

Differentiated outcomes	
All students	should subtract by counting back on a number line using cubes for support.
Most students	will record the calculation accurately.
Some students	may understand the inverse relationship between subtraction and subtraction.

5B Taking away

Explore 3 Student Book page 79 · Practice Book page 79

Specific learning focus

- Understand subtraction as counting back and 'take away'.

Global skills

- **Creative skills:** problem solving / exploring / investigating
- **Real-world skills:** research / presenting information / interpreting information / financial literacy
- **Interpersonal skills:** communication / teamwork / leadership
- **Self-development skills:** reflecting on learning

Key vocabulary

- take away, subtract, how many? how many left?

Resources

- 0–9 digit cards (set for each table), large 0–20 number line for front of class, interlocking cubes, coins or counters

Language support

Try to focus on the correct use of terms like 'take away', 'count back' and 'how many left?' so that students understand the words associated with the process they are doing.

Introductory activity

Start with a pile of 18 cubes. Ask a student to choose a number card from the 0–9 digit cards and then take

Student Book page 78

1 14 − 5 = 9

2 18 − 3 = 15

3 15 − 7 = 8

4 20 − 12 = 8

Practice Book page 78

Check that students have correctly calculated the subtraction for each number, and marked the jump correctly on the number line.

Stretch zone: Make sure students have written correct subtractions, using 2-digt and single-digit numbers.

away that number of cubes from the pile. For example, if the card is 4, then the student will take 4 cubes from the pile to leave 14.

Show this on a number line and then write it as a sentence on the board (18 − 4 = 14). Repeat with another number and student.

Main activity

Draw a set of 12 cubes. Tell students that this is the 'whole' and that they are going to take away a 'part'. Ask a student to pick a number card to decide what size the part will be, for example, 7. Say to students that 12 is the whole and 7 is a part, so what is the other part?

On the drawing of the 12 cubes, circle 7 of them and ask how many are left. (5) Write the number sentence 12 − 7 = 5. Ask students to tell you how much is the whole and how much is each part.

Repeat this with 14 cubes and 19 cubes.

Students complete the activities on page 79 of the Student Book individually. They should then come together in pairs to share some of their pictures and describe them, using the speech bubbles to help them structure their descriptions.

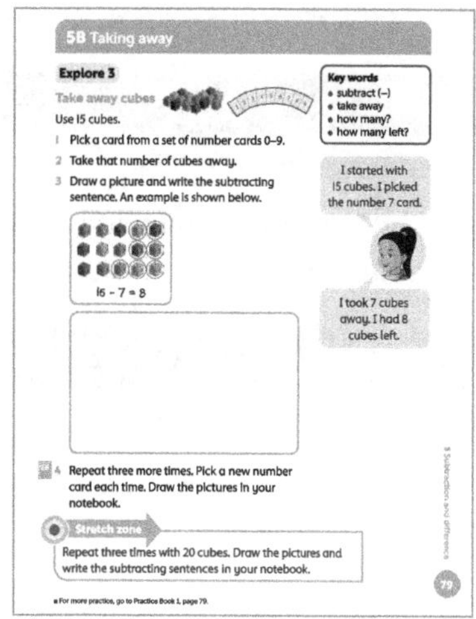

Differentiation

Supporting: Model for students how to take the whole and remove a part.

Consolidating: Help students to relate the subtraction sentence by pointing to the cubes and moving the parts or the whole as necessary.

Extending: Ask students to write fact families based on one subtraction fact, for example, from $12 - 5 = 7$, make $12 - 7 = 5$ and the corresponding additions.

Stretch zone: *Repeat three times with 20 cubes. Draw the pictures and write the subtracting sentences in your notebook.*

Check that students have drawn and written correct pictures and sentences for the numbers they used.

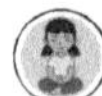 Reflection time

Draw a part-whole diagram on the board and show students a pile of cubes. Ask, *How can you know which number to put in the 'whole' on the diagram.* (Count the cubes.) Count the cubes aloud and write the number in the diagram.

Now remove some cubes and put them to one side. The students should be able to say that you have two parts. Ask a student to count the number of cubes in each part, then write the numbers in the diagram.

Reinforce the language and then relate it to the number sentence.

Practice Book: Students can complete Practice Book page 79. This can be done directly after the main activity, as homework, or as the focus of a separate mathematics session to help students consolidate their learning and build fluency. Point out to students that this is the same activity as they did in the introduction and they will need to use 18 coins or counters and digit cards to complete it.

Differentiated outcomes	
All students	should identify how many cubes are left after subtracting by counting back.
Most students	will record the calculation accurately.
Some students	may understand the inverse relationship between subtraction and subtraction.

Answers

Student Book page 79

Students will have different subtraction sentences as they are generating numbers by randomly choosing a number card. Check that each sentence is a correct version of $15 - ? = ?$

Practice Book page 79

Students will have different subtraction sentences as they are generating numbers by randomly choosing a number card. Check that each sentence is a correct version of $18 - ? = ?$

Stretch zone: Check that students have written two subtractions with the answer 12.

5C Finding the difference

Discover   Student Book page 80 • Practice Book page 80

Specific learning focus

- Understand difference as 'how many more to make'.

Global skills

- **Creative skills:** problem solving
- **Real-world skills:** presenting information
- **Interpersonal skills:** communication

Key vocabulary

- difference, how many more, how many less, what is the difference?

Resources

- interlocking cubes, counters
- large number cards 0–20, standard digit cards 1–9 (enough for one set per pair)
- 0–20 number lines: a large one for the front of the classroom and small ones for tables
- number grid per pair (as shown below)

8	8	6	4
4	9	9	7
9	5	5	4
5	7	6	8

Language support

Use games and practical resources with very little recording for some students. Understanding needs to be developed before formal recording.

Make some cube towers.

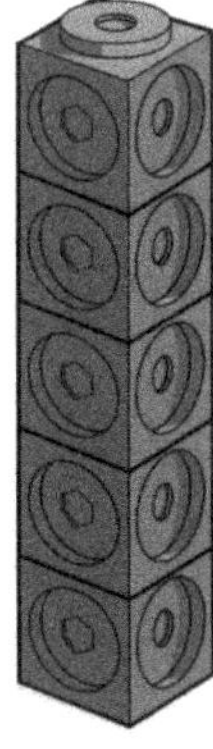 

Point at each tower, and ask, *How many are in this one?* (5) *How many are in this one?* (3) *How many more cubes to make 5?* (2)

Explain that we say 'the **difference** between 5 and 3 is 2'.

We can find the difference between two numbers by counting on from one to the other.

Ask students to work in pairs to make as many towers as they can with a difference of 2 in one minute. Encourage students to describe their towers in a similar way: *There are X cubes in this tower and Y cubes in this tower. We need 2 more cubes to make X. The difference between X and Y is 2.*

Give each student a large number card between 0 and 20. Some will have the same as others. Choose two students to come to the front of the class and ask them to show their number cards to the other students.

What is the difference between these two numbers? If you have the number that is the difference, come and stand with these two students. How can we check to see if that is right?

Some students may suggest using a number line. Demonstrate the subtraction on the class number line. Start on the lowest number and count to the highest in jumps of 1. Ask the three students to sit down and repeat the activity three more times with different sets of students.

Ask students to work in pairs. Give each pair a 4 × 4 number grid, some interlocking cubes, a set of digit cards 1–9 and counters in two different colours. Each pair builds a tower of ten cubes. Then they take turns to pick a card and build another tower with the number of cubes matching the number on the card. Ask them to find the difference between the number of cubes in the two towers. Then they cover the answer on their grid, using a counter. Play continues until all the numbers are covered. The player with the most counters on the gird wins.

Ask students to work in pairs to complete the activities in Student Book page 80. Look at the first question together. *How many ladybirds in the top row?* (5) *How many in the bottom row?* (3) *What are we trying to find?* (the difference between the number of ladybirds in the two rows) Give each pair a number line to help them, and cubes for those who need them.

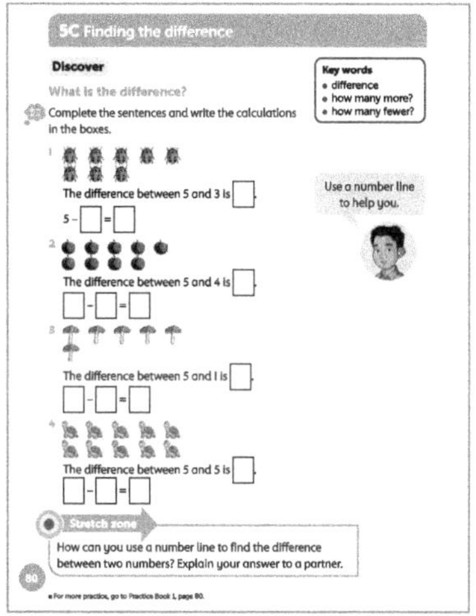

Differentiation

Supporting: Ask students to continue to build towers to find differences.

Consolidating: Ask students to model differences on a number line and to record them as subtraction sentences.

Extending: Encourage students to calculate mentally and ask them to share their strategy.

Stretch zone: *How can you use a number line to find the difference between two numbers? Explain your answer to a partner.*

Students should be able to explain correctly how they use the number line to find the difference between two numbers.

 Reflection time

Students work in pairs. On your count of three, every student holds up a number of fingers between zero and ten. Ask pairs to calculate the difference between the numbers of fingers they are each holding up.

Use this time to see and hear how students are working out the differences. Some may need more support. *How did you find the difference between the two numbers?* Ask different students to share their strategies.

Practice Book: Students can complete Practice Book page 80. This can be done directly after the main activity, as homework, or as the focus of a separate mathematics session to help students consolidate their learning and build fluency. Students can continue to make towers of cubes or rows of counters to help them find the difference.

<table>
<tr><td colspan="2">Differentiated outcomes</td></tr>
<tr><td>All students</td><td>should find the difference between cube towers by counting.</td></tr>
<tr><td>Most students</td><td>will be able to record differences on a number line.</td></tr>
<tr><td>Some students</td><td>may predict the difference between two numbers without using a number line.</td></tr>
</table>

Answers

Student Book page 80

1 2; $5 - 3 = 2$

2 1; $5 - 4 = 1$

3 4; $5 - 1 = 4$

4 0; $5 - 5 = 0$

Practice Book page 80

Check that students have completed a subtraction sentence correctly and shown the jump on the number line using the cards they have chosen. For example, if the cards are 2 and 7:

I picked 2 and 7. Show a jump starting on 7 and jump back 2 to land on 5.
The difference between 7 and 2 is 5.

Stretch zone: Check that students have written three pairs of numbers, each with a difference of 3.

5C Finding the difference

Explore   Student Book page 81 · Practice Book page 81

Specific learning focus

- Understand difference as 'how many more to make'.

Global skills

- **Creative skills:** problem solving
- **Real-world skills:** presenting information
- **Interpersonal skills:** communication / teamwork

Key vocabulary

- difference, how many more, how many less, what is the difference?

Resources

- a double-9 set of dominoes, digit cards 1–9 (several sets)
- two 20-bead strings, dice, cubes or counters
- large 0–20 number line, number cards 0–20

Language support

Use visual and practical images such as moving beads on a bead string. Encourage students to move the beads as they say the number sentence. Model this for them. As you write subtraction sentences, say them aloud.

 ## Introductory activity

Arrange students in groups of three or four and give each group two digit cards between 1 and 9 to place face up on their table. Place the dominoes face down at the front of the class. Select a domino (e.g. one showing 5 and 7) to show the class and walk around the room so that all students can see it. *One side shows 5; the other side shows 7. If I start with 5, how many more to make 7?* (2) *The difference between 5 and 7 is 2.*

I am going to show you a new domino I have chosen. Find the difference between the number of spots at each end. If the difference matches one of your cards, your group wins a point. We will play until one group has five points.

Play the game several times, changing the digit cards for each table.

 ## Main activity

Show students two 20-bead strings. Count 8 beads from one side to the other on the first string and then 14 on the second bead string. Count the difference between them by moving beads on the first string to show how many more are needed to make the same length of string as the 14 beads. *We say the difference between 8 and 14 is 6.* Display a large 0–20 number line. *Count the number of jumps on the number line with me from 8 to 14.* (1 ... 2 ... 3 ... 4 ... 5 ... 6)

Write on the board: $14 - 8 = 6$. We say, *14 subtract 8 equals 6. The difference between 8 and 14 is 6.*

Ask students to complete the activities on page 81 of the Student Book in pairs, describing the difference between the two numbers they roll and then recording their rolls in writing in their Student Books. They can continue to use cubes or counters to model each of the numbers to find the difference.

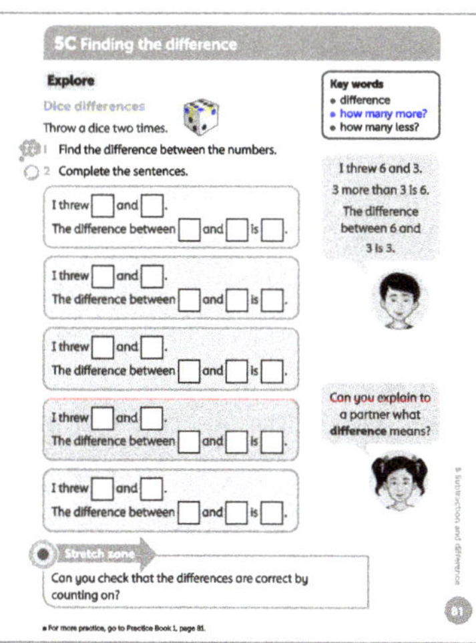

Differentiation

Supporting: Help write the answers for students as they complete the activities in the Student Book.

Consolidating: Ask students to explain how finding the difference relates to moving on a number line.

Extending: Ask students to continue to write fact families, and to use numbers that bridge 10.

Stretch zone: *Can you check that the differences are correct by counting on?*

This activity helps students understand that finding the difference between two numbers is the same as counting on from the smaller number to the larger one.

 **Reflection time**

Ask students to sit in pairs and list all the dice throws which have a difference of two, for example, a roll of 5 and a roll of 3. Ask for feedback using the phrases 'is 2 more than' and 'is 2 less than' and model the answers on a number line. Repeat looking for numbers with a difference of 3.

Practice Book: Students can complete Practice Book page 81. This can be done directly after the main activity, as homework, or as the focus of a separate mathematics session to help students consolidate their learning and build fluency. Students may benefit from using cubes or counters alongside the number track to find the difference between the pairs of numbers.

Differentiated outcomes	
All students	should be able to make towers with a difference of 2.
Most students	will record these as subtraction sentences without support.
Some students	may find differences bridging over 10 and make the link to number bonds to 10.

Answers

Student Book page 81

Students choose their own numbers and so answers will vary.

Observe students as they play the game and note who is beginning to recall the differences between their scores.

Practice Book page 81

Missing numbers in each puzzle are:

1 6		**5** 7	
2 5		**6** 5	
3 2		**7** 14	
4 9		**8** 5	

Stretch zone:

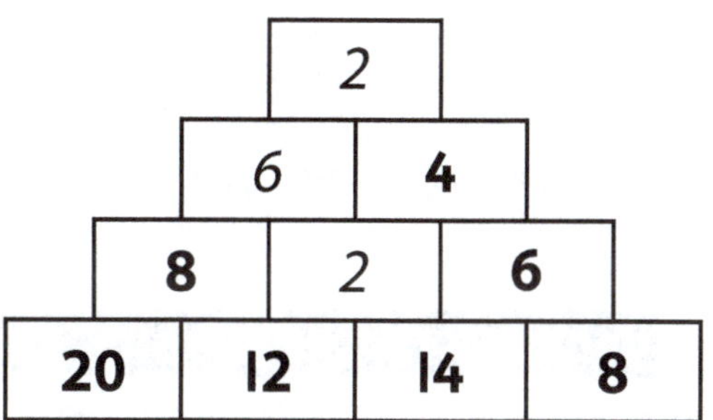

5D Subtraction word problems

Discover Student Book page 82 • Practice Book page 82

Specific learning focus

- Write subtraction stories from pictures and calculate them.

Global skills

- **Creative skills:** problem solving
- **Real-world skills:** presenting information / interpreting information
- **Interpersonal skills:** communication / teamwork

Key vocabulary

- subtraction, take away, how many are left?

Resources

- picture of a scene with objects/people to count (e.g. children playing) for display at the front of the classroom
- interlocking cubes or base-ten equipment, 0–20 number lines

Language support

Students may need support with describing the 'story' in their pictures and then interpreting how that translates to a subtraction sentence. Emphasise phrases like take away, how many left?

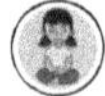 Introductory activity

Show students a picture of a scene with objects to count and subtract: for example, an image of children playing in a playground.

Tell a short story such as, *At lunchtime, there were some children in the playground. There were 17 children playing and 8 of them went inside for their lunch. How many children were left in the playground?*

Draw a part-whole diagram to represent the problem, emphasising which is the whole (17) and which are the parts, and identifying that one part is missing (the number of children left in the playground).

Ask students to tell you what the **subtraction** calculation should be (17 – 8) and then use a number line or cubes and rods to find how many are left. (9) Write the subtraction sentence 17 – 8 = 9 on the board and complete the part-whole diagram by adding 9.

Repeat with another story.

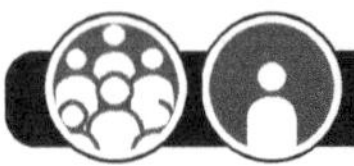 Main activity

Explain to students that they will be making their own subtraction stories using pictures on page 82 of the Student Book. Tell them to think firstly about what kind of story they could tell with the images, for example Milan gave his horse a bag of apples to eat. They can then work the numbers into it, e.g. Milan gave his horse a bag of 9 apples to eat. The horse ate 4. Ask students, *What question can you now ask?* (How many apples are left?)

Remind students to record the subtraction as a number sentence and to use number lines or cubes and rods to help them to calculate how many are left when some are removed. Support students to represent their problems as part-whole diagrams as well as write their problems out in clear sentences, on a separate piece of paper if necessary.

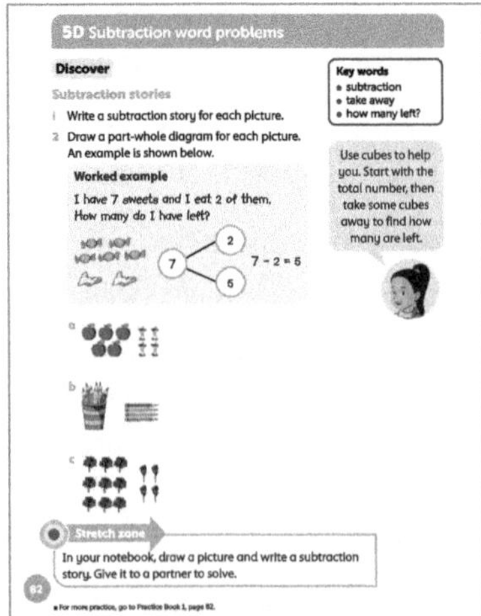

Differentiation

Supporting: Ask students to use interlocking cubes or cubes and rods to help them count. Help students to write the word probolems.

Consolidating: Encourage students to represent the problems as part-whole diagrams and ask them to explain the link to subtraction.

Extending: Encourage students to calculate mentally and then check using a suitable addition.

Stretch zone: *In your notebook, draw a picture and write a subtraction story. Give it to a partner to solve.*

Check that the stories and subtraction problems are accurate and appropriate.

Reflection time

Ask a few students to share their pictures and their subtraction stories. Ask the other students if they can see the numbers in the pictures. Ask different students to write the calculation for each story.

Practice Book: Students can complete Practice Book page 82. This can be done directly after the main activity, as homework, or as the focus of a separate mathematics session to help students consolidate their learning and build fluency. Students can continue to use counters and cubes to model the problems. Students could also represent each problem in a part-whole diagram on a separate sheet of paper.

Differentiated outcomes	
All students	should draw a picture which includes objects to be counted and some removed.
Most students	will describe the picture and form a subtraction sentence from it.
Some students	may subtract mentally to find how many are left.

Answers

Answers

Student Book page 82

1 **a** I have 5 apples. I eat 4 of them. How many are left?
$5 - 4 = 1$

   b I have 6 pencils and I give away 4 of them. How many are left?
$6 - 4 = 2$

   c I have 9 roses. 4 of them die. How many are left?
$9 - 4 = 5$

2 Check that the part-whole diagram for each picture is correct.

Practice Book page 82

1 There are 12 eggs. 5 eggs are broken. 7 unbroken eggs are left. $12 - 5 = 7$

2 Somchai has 18 bananas. He eats 4 bananas. He has 14 bananas left. $18 - 4 = 14$

3 We have 17 apples. We eat 3 apples. We have 14 apples left. $17 - 3 = 14$

4 9 frogs are on a lily pad. 3 frogs jump off. 6 frogs are left on the lily pad. $9 - 3 = 6$

Stretch zone: Check that students have drawn a picture showing 6 objects being removed from 15 objects to leave 9.

5D Subtraction word problems

Explore   Student Book page 83 • Practice Book page 83

Specific learning focus

- Write subtraction stories from pictures and calculate them using part-whole diagrams.

Global skills

- **Creative skills:** problem solving / exploring
- **Real-world skills:** presenting information
- **Interpersonal skills:** communication

Key vocabulary

- subtract, take away, how many are left?

Resources

- items to count, e.g. biscuits
- cubes or counters, 0–20 number lines

Language support

Students may need support with describing the 'story' in their pictures and then interpreting how that translates to a subtraction sentence. Emphasise words and phrases like subtract, take away, how many are left?

 **Introductory activity**

Draw an empty part-whole diagram on the board. Tell students a story, for example, *I have 12 cupcakes in a tin. I give 4 cupcakes to my friends. How many cupcakes are left?*

If the 12 cupcakes are the whole, what does the 4 you give away represent? What could you do to find the other part? Show students how to write the parts and the whole in the diagram and also write as a subtraction sentence $12 - 4 = 8$.

 **Main activity**

Show students a group of items, for example, 17 biscuits. Say that you will eat 4 biscuits and want to know how many will be left. *How many biscuits in the whole?* (17). *How many are in the part that I have eaten?* (4) *The other part will be the biscuits that are left.*

Draw the empty part-whole diagram and write the numbers 17 and 4 in the correct circles.

Now ask students what subtraction number sentence they need to write to calculate the total ($17 - 4 =$). Use cubes or a number line to get the answer 13 and write it as the other 'part' in the diagram. Complete the number sentence as $17 - 4 = 13$.

Ask students to complete the similar problems on page 83 of the Student Book individually. Continue to provide counters or cubes for students who may need the additional support.

Differentiation

Supporting: Help students by modelling the process of take a part away from a whole, naming each part and counting the objects.

Consolidating: Help students relate the subtraction sentence to the objects and interpret the whole and part.

Extending: Encourage students to calculate mentally and use subtraction fact families to help, for example, knowing $18 - 4 = 14$ to help with $18 - 14 = 4$.

Stretch zone: *In your notebook, write your own subtraction word problem. Give it to a partner to solve.*

Check that the word problem is sensible and that the partner can represent it using a part-whole diagram.

 **Reflection time**

Ask a few students to share their part-whole diagrams and explain why they put the numbers where they did. Make sure they understand the difference between the parts and the whole and how this helps them make the number sentence.

Practice Book: Students can complete Practice Book page 83. This can done be directly after the main activity, as homework, or as the focus of a separate mathematics session to help students consolidate their learning and build fluency. Students can continue to use counters and cubes to model the problems. Students could also represent each problem in a part-whole diagram on a separate sheet of paper.

Differentiated outcomes	
All students	should count how many in the whole.
Most students	will subtract one part from the whole.
Some students	may complete the part-whole diagram and the number sentence accurately.

Answers

Student Book page 83

1	Whole 5, parts 2 and 3	$5 - 2 = 3$
2	Whole 9, parts 4 and 5	$9 - 4 = 5$
3	Whole 12, parts 4 and 8	$12 - 4 = 8$
4	Whole 15, parts 6 and 9	$15 - 6 = 9$

Practice Book page 83

1	$15 - 4 = 11$	**4**	$12 - 2 = 10$
2	$9 - 4 = 5$	**5**	$8 - 7 = 1$
3	$14 - 4 = 10$		

Stretch zone: Check that students have written their own subtraction story with a total of 16 balloons.

---

5 Subtraction and difference

Connect Student Book page 84

Big Idea

- Subtraction means taking away or finding the difference between two numbers.

Global skills

- **Creative skills:** problem solving / exploring
- **Real-world skills:** presenting information
- **Interpersonal skills:** communication / teamwork

Key vocabulary

- take away, subtract, difference, how many more? how many less?

Resources

- digit cards 0–9 (enough for one per student), cubes, counters

Language support

Use practical materials or drawings for those students who still need support.

 **Introductory activity**

Give each student a digit card between 0 and 9.

When I say 'start', try to find other students with numbers so that you can make a complete number sentence. For example, cards 8, 5 and 3 will go together because 8 take away 3 equals 5, or 8 take away 5 leaves 3. When you have found the other two students in your group, sit together and work out the subtraction that your cards will give. There will be two different subtractions for each set of cards.

Play several times.

Tell students there is a song in English called 'Five Little Ducks'. Teach them the words and combine with actions to support understanding of the language:

FIVE little ducks went swimming one day, (Hold up five fingers and action swimming)
Over the hill and far away. (Action rolling hills and looking off into the distance)
Mamma duck said: 'Quack, quack, quack, quack!'
And only FOUR little ducks came back. (Hold up four fingers and make a sad face)

Repeat until there are no ducks left and then sing the final verse:

No little ducks went swimming one day, (Hold up empty hands and action swimming)
Over the hill and far away. (Action rolling hills and looking off into the distance)
Mamma duck said: 'Quack, quack, quack, quack!'
And all 5 little ducks came back! (Hold up five fingers and make a happy face)

Repeat using a different starting number, e.g. 12 and sing a couple of verses. Stop and ask questions such as, *How many ducks to start?* (12) *How many have swum away so far?* (3) *How many left now?* (9) Model with cubes or on a number line as necessary.
How could we write that as a subtraction sentence? (12 – 3 = 9) Stop several times at different points in the song and repeat the questioning. Also ask questions such as, *There are 8 ducks now and 4 have left. What's the difference?* (4)

Pairs should draw or write their own subtraction story about the ducks.

Students should work in pairs to complete page 84 in the Student Book. Explain that they are now going to write subtraction stories about fruit like they did for the ducks.

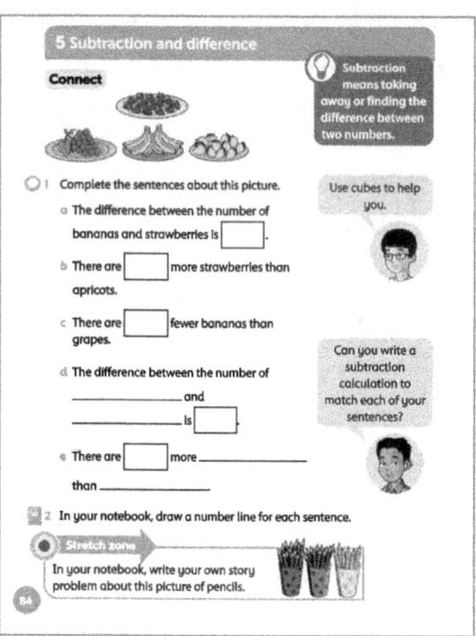

Differentiation

Supporting: Ask students to draw their story on paper, folded into four sections, so that there is an expectation of what you want (only four parts to the story). Students can use cubes or counters to help with the calculations.

Consolidating: Encourage students to bridge tens with their subtraction stories.

Extending: Encourage students to write subtraction sentences to match their story problems.

Stretch zone: *In your notebook, write your own story problem about this picture of pencils.*

This activity requires students to construct an example of a story where subtraction calculations are part of the story. Check that the story makes sense and that the subtraction sentences are accurate.

 **Reflection time**

Ask students to share their duck stories with the rest of the class. They could challenge one another to say how many are left or what the difference between two numbers is.

Differentiated outcomes	
All students	should write simple subtraction story problems.
Most students	will write subtraction stories using numbers up to 20.
Some students	may write subtraction sentences to match their stories.

Answers

1a 7 **b** 4 **c** 9

5 Subtraction and difference

Review   Student Book page 85 • Practice Book page 84

Global skills

- **Creative skills:** problem solving
- **Real-world skills:** presenting information
- **Interpersonal skills:** communication
- **Self-development skills:** reflecting on learning

Student Book

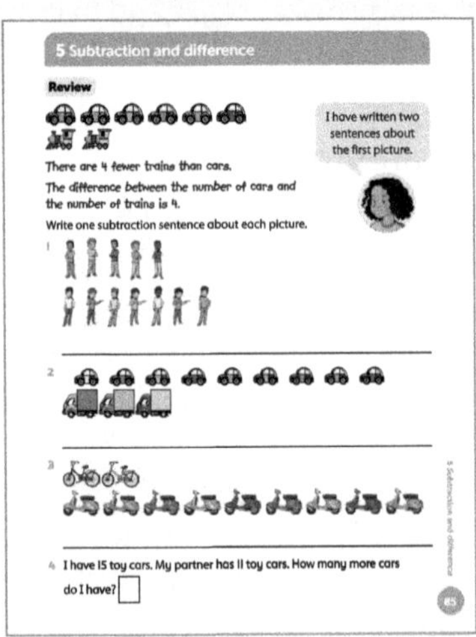

With young children, assessment activities are most effective when carried out as an everyday classroom activity. Students should have number lines available with both the digits and the numbers in words, so they can refer to this to support them.

Watch as students complete the sentences that represent subtraction and difference situations. If there are errors count on or back to help students find the difference. Check that students understand the language of difference, take away, subtract, more than and fewer than.

It may help to create flash cards with the numerals and words that are in the review. You can use these to support some students.

Encourage students to use the number line for the questions. Extend by asking a range of similar questions or asking students to write their own problems.

Answers

Student Book page 85

1 There are 2 more girls than boys (or 2 fewer boys than girls).

2 There are 6 fewer lorries than cars (or 6 more cars then lorries).

3 There are 7 more scooters than bicycles (or 7 fewer bicycles than scooters).

4 4

Practice Book

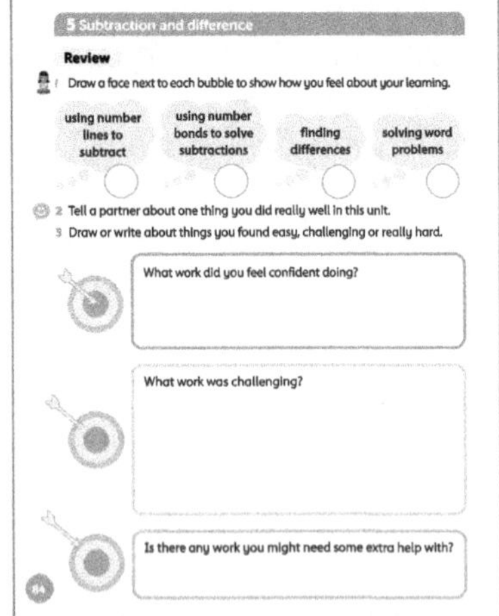

With young children it is appropriate to complete this as a whole-class discussion. You may choose to keep a record of the class discussion or a copy of the review page for your own records. Use the Student Book to briefly remind the students of the areas of mathematics that they have worked on in this unit.

Give out the Student Book to support pairs of students as they discuss and answer the questions in the Practice Book. Allow students plenty of time for discussion before asking them to share their responses with the rest of the class. If students complete this assessment at home, encourage them to discuss this with adults. Make a note of areas that students still feel unsure about.

Additional material

There are additional end-of-unit assessments available on the *Oxford Owl* website.

6 Multiplication and division

Overview

Big Idea

- Multiplication and division facts are best learned through noticing patterns and relationships between them: for example, knowing that:
  - division and multiplication are opposite operations, e.g. $4 \times 5 = 20$ leads to $20 \div 4 = 5$; $5 \times 4 = 20$ leads to $20 \div 5 = 4$
- Multiplication and division both involve equal groups.
- The development of skills in multiplication starts with patterns of repeated addition.
- The development of skills in division starts with patterns of multiplication and informal practical activities, where students are asked to carry out equal sharing tasks, distributing items among a group. Later tasks will be related to numbers, for example, *If you share the 12 sweets equally among 4 people, how many will they each get?*
- Using an array of objects allows students to represent their thinking with concrete materials, and helps them form mental pictures to support memory and reasoning.
- Students need to explore the concept with concrete materials before they notice relationships between division and the other operations.
- The inverse relationship between division and multiplication can be shown using arrays. For example, with $3 \times 5 = 15$ (3 rows of 5 make 15), looking at the array differently shows the inverse, that is, $15 \div 3 = 5$ (15 put into 3 rows makes 5 columns).
- Language plays an important role in being able to express mathematical relationships, and the physical array supports this by giving the students something concrete to talk about.

Look out for

- **Students who still count in ones to find how many there are in a collection of equal groups, and do not understand vocabulary such as 'groups of', 'multiplied by'.** Encourage counting and recognising small groups or patterns on dice.
- **Students who do not link counting up in equal steps to the operation of multiplication and do not use the vocabulary associated with multiplication.** Use a number line or 100-square to see and say patterns of counting.
- **Students who have difficulty relating multiplying or dividing by 2 to known facts about doubles and halves and record double 4 as 4 + 4.** Encourage the use of previous knowledge to find half or double a number.

Possible misconceptions

- **Students may see an array as a collection of ones only** and not focus on 'rows of' or 'columns of'.

Key vocabulary

- lots of, group, groups of, times, multiply, halve, repeated addition, array, column, row
- share, share equally, equal groups, equal parts, left over, odd, even
- multiple of 2, multiple of 5, multiplication

Coverage in lessons

Learning focus	Learning outcomes (the ENC objectives)
Equal sharing	Solve one-step problems involving multiplication and division, by calculating the answer using concrete objects, pictorial representations and arrays with the support of the teacher.
Grouping	Solve one-step problems involving multiplication and division, by calculating the answer using concrete objects, pictorial representations and arrays with the support of the teacher.
Repeated addition	Solve one-step problems involving multiplication and division, by calculating the answer using concrete objects, pictorial representations and arrays with the support of the teacher.
Multiplication word problems	Solve one-step problems involving multiplication and division, by calculating the answer using concrete objects, pictorial representations and arrays with the support of the teacher.

6 Multiplication and division

Engage Student Book page 86

Big Question

- What strategies help us multiply and divide?

Global skills

- **Creative skills:** problem solving / exploring / investigating
- **Real-world skills:** research / presenting information / interpreting information / financial literacy
- **Interpersonal skills:** communication / teamwork / leadership
- **Self-development skills:** reflecting on learning

Key vocabulary

- multiply, divide, share, lots of, array

Resources

- interlocking blocks (1 × 1, 1 × 2, 2 × 2, 3 × 2, 3 × 4, 4 × 2), large blocks for class use

Language support

Placing the mathematics into a real-life context through word problems can help with both understanding the relationships and their expression through words. Saying and writing the words clearly while doing the board work reinforces the vocabulary and understanding. Acting out problems, when possible, alongside writing or saying them, is also helpful.

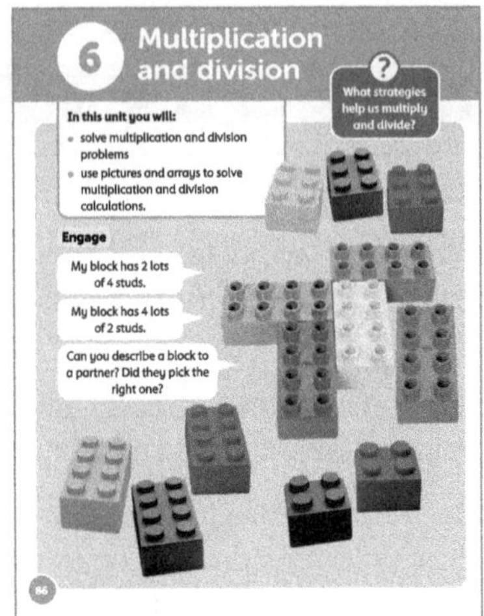

Introductory activity

Ask students to look at the pictures of blocks. Use the questions in the speech bubbles to discuss what they can see.

Put a pile of interlocking blocks on each table. Explain that the knobs on top are called 'studs'. Ask students, *Find blocks with 1, 2, 4, 6, 8 studs.* Ask them to hold up the 2 by 2 block (4 studs) and to tell you how many studs there are (4). Give the block a quarter turn. Ask students again to tell you how many studs (4). *Whichever way we turn the block, there are still 4 studs.* Write on the board 2 **lots of** 2 is 4. This block has 2 lots of 2 studs. *We can show the studs as a 2 by 2 **array**.* Draw an array of 4 dots on the board and label it 'array'.

Repeat for a 1 by 2 block. Show the class the block with the 2 studs facing them. *How many studs does this have?* (2) Write on the board 1 lot of 2 is 2. Give the block a quarter turn. *How many studs are there?* (2). *Could we also say 2 lots of 1 is 2?* Agree that this also describes the block.

 Main activity

Students work in pairs. Each pair chooses a block and shows it to the class. Each pair then finds another block of the same size. Where appropriate, they try to describe it in two different ways (e.g. 3 lots of 4 or 4 lots of 3) and how many studs there are altogether. Record what they say on the board. Repeat with other sizes of blocks.

Ask pairs to explore other blocks and write what they find out. They can record it in a table, a chart or by drawing. Give time for this exploration. *How many studs are on your largest block? Draw a picture of it. Find a square block. How many studs are there? Draw it.* Use these drawings as part of a display on arrays and multiplication.

Ask students to look at the blocks. Discuss the patterns of studs on the blocks. A block with 4 rows of 2 studs can be represented as 4 lots of 2 or 2 lots of 4, or 4 rows of 2 makes 8, or 2 rows of 4 makes 8.

Differentiation

Supporting: You can write the appropriate multiplication sentences as students describe the arrays.

Consolidating: Ask students to write the multiplication ('lots of') sentences for larger arrays made by the blocks, and recognise that the array represents two multiplications.

Extending: Ask students to make a range of arrays with 3 or 4 blocks.

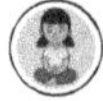 Reflection time

Put blocks side by side. *If you place a 1 by 4 block next to another 1 by 4 block, how many studs would you have?* Students can count them. Encourage them to use the correct mathematical vocabulary to describe what they are doing (1 lot of 4 add another lot of 4 is 8). Repeat using different sizes of blocks. *How are interlocking blocks similar to an array?*

6A Equal sharing

Specific learning focus

- Share numbers to 10 to find which are even and which are odd.

Global skills

- **Creative skills:** problem solving / exploring
- **Interpersonal skills:** communication / teamwork

Key vocabulary

- odd, even, share, group, equal groups, groups of, left over, share equally

Resources

- interlocking cubes or counters, ten pencils
- 100-square (one per pair), large 0–10 number line for the front of the class

Language support

Model the vocabulary of multiplication and division for a student to copy. Make posters with students to show the meanings of the words.

 ### Introductory activity

Show one pencil to the class. Choose a student to come to the front of the class. Give that student the pencil.

Choose another student and ask them to **share** the pencil equally between them without breaking it. *Is it possible?* (No) *What is the smallest number of pencils that we can share equally?* (They need one pencil each, so 2.) On the large 0–10 number line, mark 2.

Give the two students one pencil each. Continue adding one pencil at a time, saying, *two pencils – one each; three pencils – two for you, one for you; four pencils – two each; five pencils, …,* discussing whether they can be shared equally and marking the numbers that do share fairly on the number line. Discuss the pattern that is shown (**even numbers**).

 ### Main activity

Count out ten cubes. Share them into two **equal groups**. *How many cubes are in each group?* Count, as a class: 1, 2, 3, 4, 5. Write on the board, *2 **groups of** 5 = 10.* Remove some of the cubes. Share the remaining cubes between 2. Discuss what happens. *Are there any **left over**?*

Repeat several times with other numbers of cubes so that all of the numbers 1–10 are covered and record them on the board (e.g. 7 cubes, 2 groups of 3, 1 left over). Record in two lists:

1 those numbers which have no cubes left over

2 those numbers which have cubes left over (a remainder).

Label list 1 as even numbers (no cubes left over), and list 2 as **odd** numbers (1 cube left over).

Ask students to complete page 87 of the Student Book individually. Encourage them to use cubes as you did as a whole class to complete the table in their books.

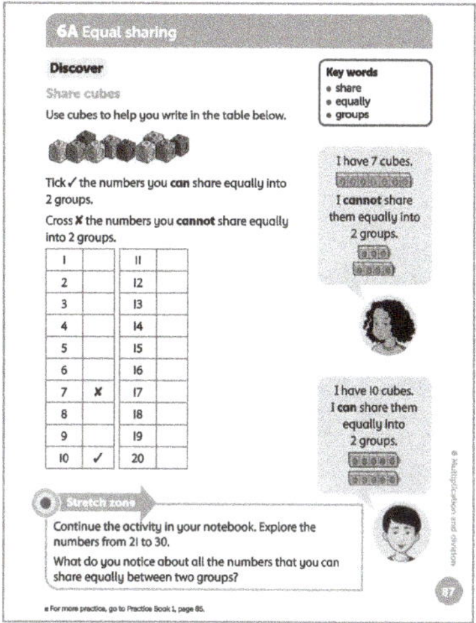

Differentiation

Supporting: Ask students to carry out the activity using concrete objects.

Consolidating: Encourage students to predict the outcomes.

Extending: Students could investigate sharing the cubes between more people (3 or 4) and recording their findings on a 100-square in a similar way.

Stretch zone: *Continue the activity in your notebook. Explore the numbers from 21 to 30. What do you notice about all the numbers that you can **share equally** between two groups?*

This activity will direct students to understand that any number that can be shared into 2 equal groups is an even number, and ends with 0, 2, 4, 6 or 8 in the ones.

 ### Reflection time

Ask students to shade in the even numbers between 1 and 20 on a 100-square. *Do you notice a pattern? What will the next even number be?* Repeat for odd numbers using a different colour. *What is the biggest even number on the 100-square? How do you know? What is the biggest odd number on the 100-square? How do you know?*

Practice Book: Students can complete Practice Book page 85. This can be done directly after the main activity, as homework, or as the focus of a separate mathematics session to help students consolidate their learning and build fluency. Before they begin the activity, ask students to tell you what equal sharing means and redefine it, if necessary. Students will be working with larger numbers; encourage them to use counters or cubes to make numbers and then try to split the cubes or counters into equal-sized groups.

Differentiated outcomes	
All students	should share cubes into equal groups.
Most students	will make the link between odd and even numbers and equal sharing.
Some students	may share larger numbers of objects.

6A Equal sharing

Explore 1 Student Book page 88 · Practice Book page 86

Specific learning focus

- Share numbers to 12 in equal groups.

Global skills

- **Creative skills:** problem solving / investigating
- **Interpersonal skills:** communication / teamwork

Key vocabulary

- odd, even, share, equal groups

Resources

- cubes, small bag, counter or cubes
- large 100-square, large 0–20 number line for the front of the class

Language support

Students might need a range of different resources to help them tackle this challenge, for example a 100-square, or some counters/cubes. Try not to pre-empt their requests by placing equipment out on tables at the start, but do make sure resources are easily accessible to students.

 **Introductory activity**

Put 8 cubes in a bag. Ask two students to come to the front of the class. Explain that you will share the cubes so that they have the same number each. Do not tell

Answers

Student Book page 87

The numbers that should be ticked are: 2, 4, 6, 8, 10, 12, 14, 16, 18, 20

The numbers that should be crossed are: 1, 3, 5, 7, 9, 11, 13, 15, 17, 19

Practice Book page 85

Blue numbers: 22, 24, 26, 28, 30, 32, 34, 36, 38, 40

Red numbers: 21, 23, 25, 27, 29, 31, 33, 35, 37, 39

Stretch zone: Odd numbers *cannot* be shared into two equal groups; even numbers *can* be shared equally.

students how many cubes are in the bag. They must solve the problem, *How many cubes could be in the bag?*

Give out the cubes one at a time. After giving one to each student, they might guess there were 2 in the bag. After giving another one each, they might revise their guess to 4. Once all the cubes are shared out, they will see they had 4 each, so there were 8 in the bag.

 **Main activity**

Ask three more students to the front of the class. Collect the cubes and put them back into the bag. Without the students seeing, add 4 more to make 12 in total in the bag. Give each student an equal number of cubes, one at a time. *How many cubes could have been in the bag?* After one each, they might think there were 3 cubes. After getting 2 each, they could say there were 6 cubes. After all the cubes are shared out, they will have 4 each, so they can see there were 12 in total at the start.

Students work in pairs to create and solve a similar problem using cubes. Allow time for feedback about what they found out and how they worked. Then ask students to move on to the activities on page 88 of the Student Book, continuing to use cubes to represent each problem.

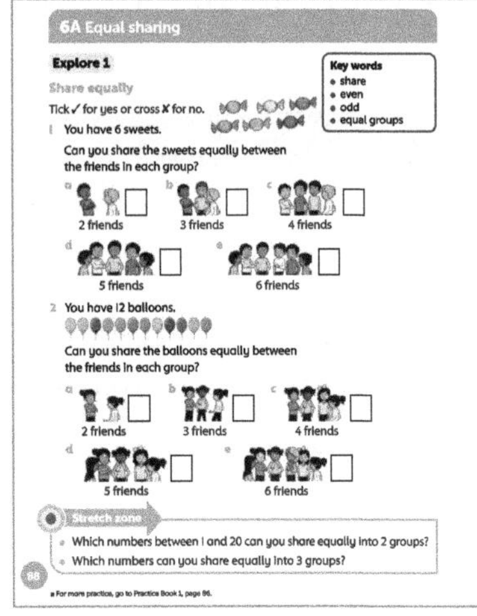

Differentiation

Supporting: Students use concrete objects for sharing amounts equally.

Consolidating: Ask students to draw pictures to represent the problems.

Extending: Ask students to draw pictures to show their own sharing problems.

Stretch zone: *Which numbers between 1 and 20 can you share equally into 2 groups? Which numbers can you share equally into 3 groups?*

This activity helps students begin to understand that not all numbers can divide equally into the same number of groups. Some numbers share equally into 2 groups (even numbers), others will share into 3 groups (3, 6, 9, 12, 15, 18), some will do both (6, 12, 18) and some do neither (1, 5, 7, 11, 13, 17, 19).

 **Reflection time**

As a class, chant all odd numbers to and from 20, and then even numbers to and from 20. Use a number line or 100-square to show the pattern.

Number students round the class, each having a different number. Play the standing/sitting game – students start counting from 1; if they are odd, they sit down, if they are even, they stand up. Use the pattern of standing/sitting to reinforce odd and even numbers.

Practice Book: Students can complete Practice Book page 86. This can be done directly after the main activity, as homework, or as the focus of a separate mathematics session to help students consolidate their learning and build fluency. Students can continue to use concrete resources to represent each item in the group and then share the items equally into different-sized groups.

Differentiated outcomes	
All students	should recognise which numbers are odd and even up to 10.
Most students	will work out the smallest and largest number problem.
Some students	may devise their own sharing problems.

Answers

Student Book page 88

1 a yes   **b** yes   **c** no   **d** no   **e** yes

2 a yes   **b** yes   **c** yes   **d** no   **e** yes

Practice Book page 86

	Objects	2 friends	3 friends	4 friends
1	12 cakes	Yes	Yes	Yes
2	9 balloons	No	Yes	No
3	5 bananas	No	No	No
4	10 eggs	Yes	No	No

Stretch zone: Check that students' chosen number is divisible by 2, 3 and 4 (e.g. 12, 24).

6A Equal sharing

Explore 2   Student Book page 89 • Practice Book page 87

Specific learning focus

- Share odd and even amounts.

Global skills

- **Creative skills:** exploring / investigating
- **Interpersonal skills:** communication

Key vocabulary

- share, equal groups, even, odd

Resources

- cubes, counters or other objects for sharing, plates

Language support

Encourage students to notice odd and even numbers by whether they share equally or not, and start to notice the digits at the end of even and odd numbers.

 **Introductory activity**

Ask 4 students to the front of the class. Tell them they need to split themselves into equal groups. Can they do it? They should be able to make 2 groups of 2.

Now ask 3 more students to come to the front of the class to make 7 students altogether. Again ask students to try to split into 2 equal groups. They will find that they cannot do it, and one student will be left over.

 **Main activity**

Show students a pile of cubes (e.g. 11) and 2 plates. Ask students whether the pile of cubes can be shared equally between the plates. Ask how they can find out. The likely response is that you can share the cubes between the plates one at a time until they have all been used.

Demonstrate this by putting one cube on each plate, then another on each plate, stopping to show the students at each stage that the plates have the same number of cubes on each. After 10 cubes have been shared, there will be one left.

Repeat this using a pile of 14 cubes. Agree that, after sharing, there will be none left over. Repeat with other sizes of piles of cubes.

What do you notice about the piles of cubes that share equally? What about the ones that do not? Allow students to discover that even numbers share equally between 2 groups and odd numbers do not.

Students can then work on the activities on Student Book page 89, using cubes or counters. Read the statements aloud before students begin, linking key vocabulary with the images.

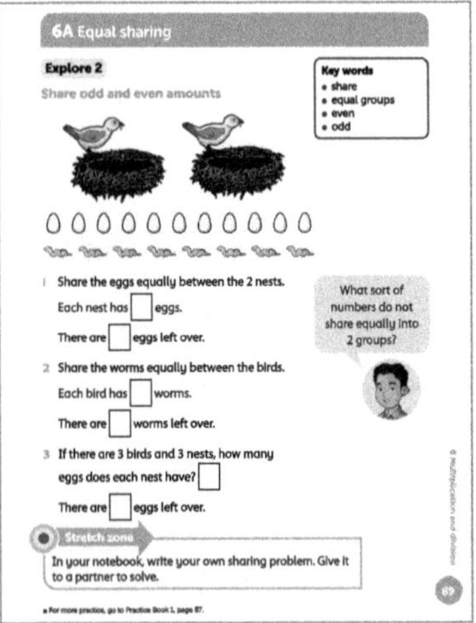

Differentiation

Supporting: Ask students to use concrete objects for sharing.

Consolidating: Ask students to draw pictures to show their calculations.

Extending: Ask students to draw pictures to show their own sharing problems.

Stretch zone: *In your notebook, write your own sharing problem. Give it to a partner to solve.*

Check that students have devised a suitable sharing problem.

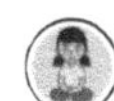 **Reflection time**

Ask students to share what they found about sharing objects equally into 2 groups. Remind them about odd numbers not sharing equally into 2 groups.

Look together at question 3 on page 89 of the Student Book. *Can odd numbers be shared into 3 equal groups? Can you use cubes to see if this is always true, always false or sometimes true?*

Practice Book: Students can then complete page 87 in the Practice Book. This can be done directly after the main activity, as homework, or as the focus of a separate mathematics session to help students consolidate their learning and build fluency.

Differentiated outcomes	
All students	should count small groups of objects and physically share them into equal groups.
Most students	will find out if an amount is odd or even by trying to share equally.
Some students	may recognise odd and even numbers up to 20.

Answers

Student Book page 89

1 Each nest has 5 eggs. There is 1 egg left over.

2 Each bird has 4 worms. There are 0 worms left over.

3 Each nest has 3 eggs. There are 2 eggs left over.

Practice Book page 87

1 4 cherries each

2 3 marbles each

3 8 sweets each

4 2 yo-yos each

5 2 apples each

Stretch zone: 24 can be shared equally into 2s, 3s, 4s, 6s, 8s, 12s.

Discover
Student Book page 90 · Practice Book page 88

Specific learning focus
- Share objects into two equal groups in a context.

Global skills
- **Creative skills:** problem solving / exploring
- **Interpersonal skills:** communication / teamwork

Key vocabulary
- share, equal groups, array, halve, half, row, column, lots of

Resources
- real-life examples of arrays (egg box, paint box, ice-cube tray), cubes, counters

Language support
As you work with students, model the phrase 'share into two equal groups'. Ask them to repeat the phrase back to you. Focus particularly on the key words array, *row* and *column*. Ask questions such as, *How many in that column? How many in that row?* Draw a poster containing a series of arrays with rows and columns labelled.

 **Introductory activity**

Give each pair of students 20 cubes. Ask one of the pair to pick an even number of cubes. They give these cubes to their partner, who decides how to share them into two equal groups. *How did you share the cubes?* Some may count out the cubes, moving one into each new group until all the cubes are used up. Some may count out in 2s or larger steps. Some may even count the total number of cubes and carry out a calculation to **halve** the number of cubes. Repeat this activity, with partners exchanging roles, until all are confident with the process.

 **Main activity**

Model two different arrangements of 8 cookies using cubes, as below, and make sure that all students understand there are 8 cookies in each arrangement but they are arranged differently in each array.

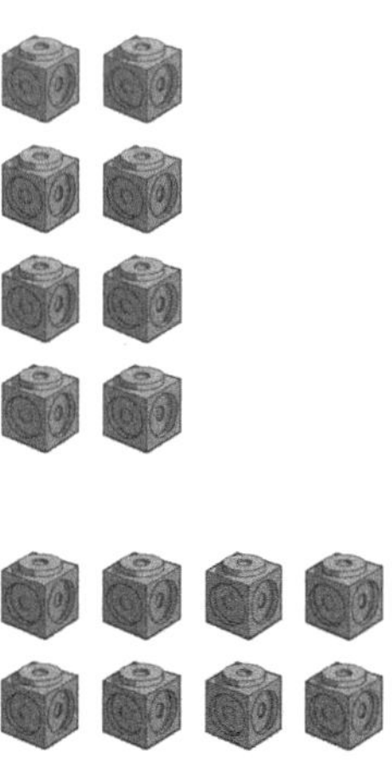

Ask a student to come to the front of the class and draw one way of sharing the 8 cookies equally into two groups. Make sure that you consistently use the phrase 'share equally into two groups'. Then ask pairs to discuss how they could arrange the 8 cookies equally into two groups in different ways. (One answer is to make two 1 × 4 arrays, the other is to make two 2 × 2 arrays.)

Ask students to complete the activities on page 90 of the Student Book in pairs.

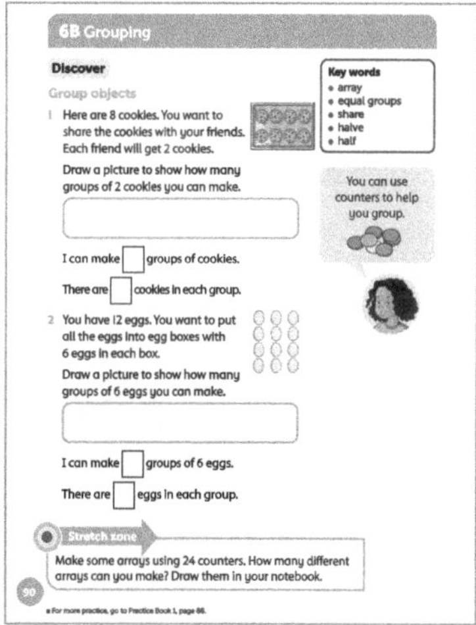

Differentiation
Supporting: Ask students to use cubes and share them into equal groups.

Consolidating: As students share into equal groups, ask them if they can find a different way to make equal groups (e.g. with 12 eggs).

Extending: Ask students to explore larger arrays.

Stretch zone: *Make some arrays using 24 counters. How many different arrays can you make? Draw them in your notebook.*

Students are being given the opportunity to make arrays for a number with several factors. Check that they have found that 24 can be represented as 1 × 24, 2 × 12, 3 × 8, 4 × 6, 6 × 4, 8 × 3, 12 × 2 and 24 × 1.

 **Reflection time**

Show the egg box (or similar) and ask students what they can see. Encourage them to use the terms **rows**, **columns**, *groups*, *lots of*. Point to each row/column as they, or you, say it. Ask questions such as, *How many in*

each row? How many rows of 2 can you see? (3) Write the number sentence that shows this (2 + 2 + 2) and draw the array. Turn the egg box so that it shows 2 rows of 3. Repeat the questioning.

Ask students what they notice about these two number sentences and arrays. (They have the same answer.)

Practice Book: Students can complete Practice Book page 88. This can be done directly after the main activity, as homework, or as the focus of a separate mathematics session to help students consolidate their learning and build fluency. Before they begin, explain that the same number of pieces of fruit need to go in each bag. Remind them to keep their drawing simple, demonstrating if necessary.

Differentiated outcomes	
All students	should make arrays and divide equally.
Most students	will begin to understand that sometimes alternative groupings are possible.
Some students	may be able to find all the possible ways to share a given number of objects equally.

6B Grouping

Explore   Student Book page 91 • Practice Book page 89

Specific learning focus

- Share objects into two equal groups in a context.

Global skills

- **Creative skills:** problem solving / exploring / investigating
- **Real-world skills:** research / presenting information / interpreting information / financial literacy
- **Interpersonal skills:** communication / teamwork / leadership
- **Self-development skills:** reflecting on learning

Key vocabulary

- share, equal groups, array, repeated addition, groups of, ×

Answers

Student Book page 90

1   The drawing should show clearly that there are 4 groups of 2 cookies.

   I can make 4 groups of cookies. There are 2 cookies in each group.

2   The drawing should show clearly that there are 2 groups of 6 eggs.

   I can make 2 groups of 6 eggs. There are 6 eggs in each group.

Practice Book page 88

	Fruit	No. in a bag	Bag 1	Bag 2	Bag 3	Bag 4	
1	Oranges	4	4	4	4	4	I can fill 4 bags of oranges.
2	Cherries	6	6	6	6	–	I can fill 3 bags of cherries
3	Bananas	3	3	3	3	3	I can fill 4 bags of bananas

Stretch zone: Students could add another row to the table for their own grouping problem.

Resources

- real-life examples of arrays (egg box, paint box, ice-cube tray, etc.), cubes, number rods (or coloured strips of squared paper stuck onto card), squared paper

Language support

Use number rods if possible to help students understand and visualise the idea of 'groups of'. If they are not available, cut coloured card to represent them, using square paper. The colours also help to reinforce the numbers in the problems. To show 3 × 4, use 3 purple (4) rods and say *3 groups of 4*. To show 4 × 3, use 4 light green (3) rods and say *4 groups of 3*.

 ### Introductory activity

Revise students' understanding of an array. Ask students how they can use an array to put any number of objects into groups of equal quantities.

Draw on the board examples of arrays such as 2 × 4, 3 × 5, 6 × 2 and so on. Ask students to come to the board and write a statement for each one, e.g. 2 groups of 4 = 8. Explain that students are going to make their own array, using an egg box (or similar) and some counters or beads.

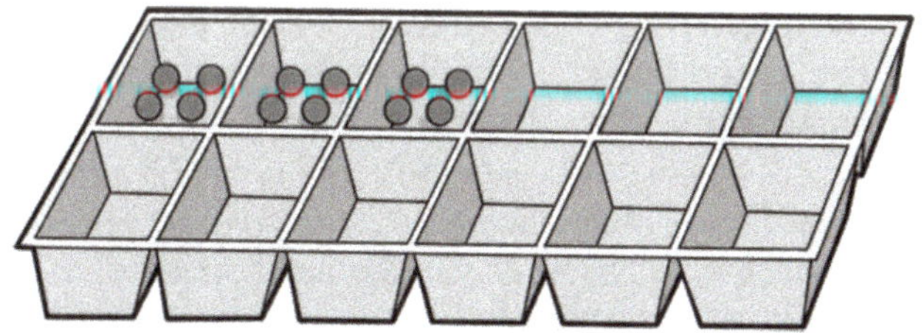

Write 3 × 4 on the board. Show the class how this can be represented by using different parts of their container. Say that the × sign can mean 'groups of', so 3 × 4 means 3 groups of 4, which is the same as 4 + 4 + 4.

Ask pairs to make their own array for each of the statements on the board.

 **Main activity**

Show two different arrays (2 × 10 and 4 × 5) and ask students to tell you what they show. For example, 2 groups of 10, 2 rows of 10, 2 × 10 = 20… (they all show 20).

Ask students to group 20 into twos. *How many groups of 2 can you make?* (10) Then ask them to group 20 into fives. *How many groups this time?* (4)

Talk students through the activities on page 91 in the Student Book. Encourage students to see that there is a link between grouping, adding groups of equal size, and the idea of multiplying.

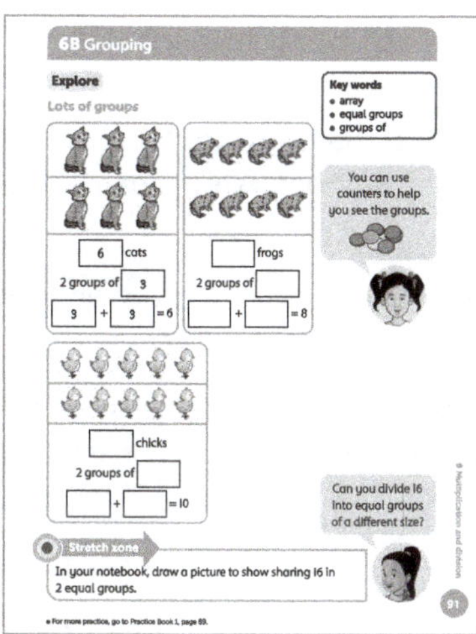

Differentiation

Supporting: You can help write grouping sentences as students describe the arrays.

Consolidating: Ask students to use concrete materials to model arrays for a range of numbers, for example, 15, 18, 24 and 30.

Extending: Ask students to find more than one way to make an array for 16, then write it as a multiplication sentence and a repeated addition.

Stretch zone: *In your notebook, draw a picture to show sharing 16 in 2 equal groups.*

Check that students have drawn 2 groups of 8.

 **Reflection time**

Turn multiplication problems into rectangles. Use squared paper and colour in squares, for example, with 3 × 4 to show 3 groups of 4. Students colour 3 rows with 4 squares in each row (4 + 4 + 4).

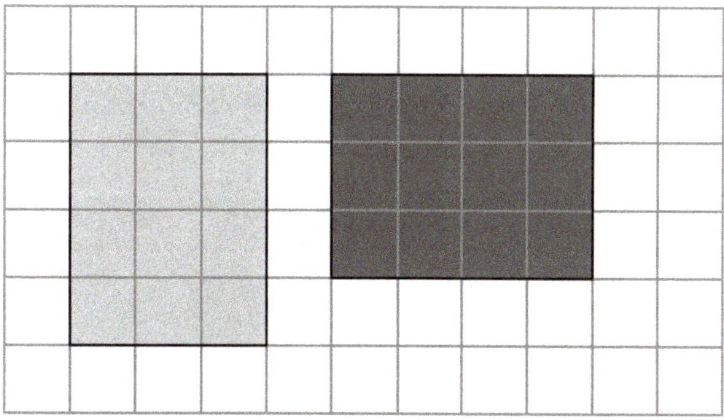

To show 4 × 3 (4 groups of 3), colour 4 rows with 3 squares in each row. Students can compare these two rectangles and see that, in both cases, 12 squares are coloured but one looks like the other turned through 90 degrees.

Practice Book: Students can complete Practice Book page 89. This can be done directly after the main activity, as homework, or as the focus of a separate mathematics session to help students consolidate their learning and build fluency.

Differentiated outcomes	
All students	should carry out the repeated addition.
Most students	will link repeated addition to multiplication.
Some students	may start to explore alternative arrays for a given number and write the related multiplication sentences.

Answers

Student Book page 91

| 8 frogs | 2 groups of 4 | 4 + 4 = 8 |
| 10 chicks | 2 groups of 5 | 5 + 5 = 10 |

Practice Book page 89

1 5 boys   5 groups of 2 equals 10

2 6 ostriches   6 groups of 2 equals 12

3 3 birds   3 groups of 2 equals 6

4 7 pairs   7 groups of 2 equals 14

Stretch zone: Check that students have created their own question, using the activity as guidance.

6C Repeated addition

Specific learning focus

- Recognise and record equal groups.

Global skills

- **Creative skills:** problem solving
- **Interpersonal skills:** communication

Key vocabulary

- equal groups, multiples of 2, multiples of 5

Resources

- cubes, counters or other matching objects, plates or paper cups

Language support

Be very clear with students about the difference between counting objects and counting equal groups of objects. The language associated with counting groups will support their multiplicative understanding.

 Introductory activity

Show students a set of identical objects, such as 8 cubes. *Count how many cubes there are.* Now ask a student to come forward and put the cubes into groups of 2. To help them with this, tell them to put 2 cubes into a paper cup, the next 2 into another cup, and so on. *How many paper cups did you use? (4) How many groups of 2 did you make? (4)* Write on the board, 4 groups of $2 = 4 \times 2 = 2 + 2 + 2 + 2$. Explain that the $\times$ sign is read as 'times', and that these are all ways of representing 8 in equal groups of 2.

Repeat using other numbers, for example, 14 cubes can be put into 7 groups of 2. Tell students that numbers that can be shared equally into groups of 2 are called **multiples of 2**.

 Main activity

Extend the introductory activity by asking a student to put a group of 10 cubes into paper cups, putting a group of 5 in each cup. *How many cups do you need? (2)*

Repeat using a pile of 15 cubes to give 3 groups of 5, then 20 cubes to make 4 groups of 5. Emphasise all the time that the group size must be equal when sharing. Write down the numbers of cubes that have been shared: 10, 15, 20, … . Tell students that because these can all be shared equally into groups of 5, these numbers are called **multiples of 5**.

Students can then work on the activities on page 92 of the Student Book. Look at the worked example as a class. Ask students to explain how the addition sentence describes the image. *How many groups of cats? How many in a group? How many altogether? How could you describe this using the multiplication sign?* Complete question 2 as a class using the same line of questioning. Then let students work on the remaining questions.

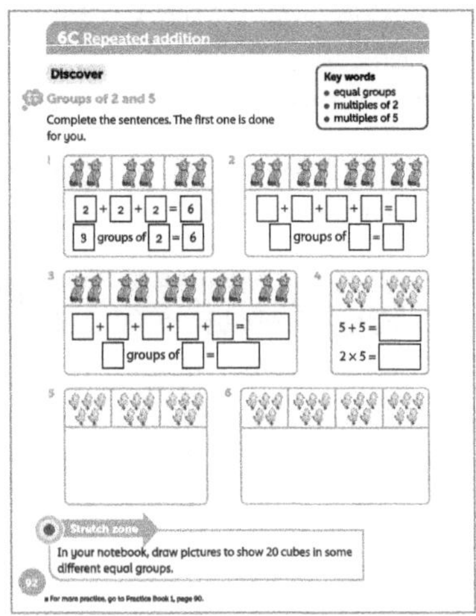

Differentiation

Supporting: You can help write linked repeated addition and multiplication sentences as students describe how many equal groups they have made using the cubes and cups.

Consolidating: Ask students to use concrete materials to model sharing larger groups into twos or fives.

Extending: Ask students to look for further multiples of 2 and 5 and then numbers which are multiples of both 2 and 5.

Stretch zone: *In your notebook, draw pictures to show 20 cubes in some different equal groups.*

Check that students' pictures represent correct groupings of 20 cubes, which can be any of the following:

1 group of 20	5 groups of 4
2 groups of 10	10 groups of 2
4 groups of 5	20 groups of 1

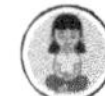 Reflection time

Show students an image of 20 identical items, for example, cats. *Can these cats be shared equally into groups of 2?* Some students should be able to recognise that 20 can be shared equally into 2s. Now ask, *Can you share 19 into equal groups of 2?* Ask students to explain their response.

Practice Book: Students can then complete Practice Book page 90. This can be done directly after the main activity, as homework, or as the focus of a separate mathematics session to help students consolidate their learning and build fluency. Check that students can describe the worked example using language such as 'groups of' and 'equal'.

Differentiated outcomes	
All students	should share objects equally into groups of 2.
Most students	will share objects into groups of 2 or 5.
Some students	may write down the number of cubes as both an addition and a 'groups of' statement.

Answers

Student Book page 92

1 $2 + 2 + 2 = 6$, 3 groups of $2 = 6$

2 $2 + 2 + 2 + 2 = 8$, 4 groups of $2 = 8$

3 $2 + 2 + 2 + 2 + 2 = 10$, 5 groups of $2 = 10$

4 $5 + 5 = 10$, $2 \times 5 = 10$

5 $5 + 5 + 5 = 15$, $3 \times 5 = 15$

6 $5 + 5 + 5 + 5 = 20$, $4 \times 5 = 20$

Practice Book page 90

1 $6 + 6 = 12$ $2 \times 6 = 12$

2 $7 + 7 = 14$ $2 \times 7 = 14$

3 $8 + 8 = 16$ $2 \times 8 = 16$

4 $9 + 9 = 18$ $2 \times 9 = 18$

Stretch zone: There are five possible ways of making 16 using equal groups: 1 group of 16; 2 groups of 8; 4 groups of 4; 8 groups of 2; 16 groups of 1. To find all the ways, students need to work systematically, starting with the smallest (or largest) number that works.

6C Repeated addition

Explore   Student Book page 93 • Practice Book page 91

Specific learning focus

- Identify a number of groups from a set of items.

Global skills

- **Creative skills:** problem solving / investigating
- **Real-world skills:** presenting information
- **Interpersonal skills:** communication

Key vocabulary

- multiples of 2, multiples of 4, repeated addition, groups, multiplcation

Resources

- cubes, counters or other matching objects, plates or paper cups, 0–20 number lines

Language support

Encourage students to begin to think about counting in groups and saying how many groups, instead of adding one group after another. Begin to use the language of 'groups of' when describing repeated groups.

 Introductory activity

Ask two students to the front of the class. Ask the first student to hold up one hand, fingers spread, palm facing forwards. Ask how many fingers there are (5). Say this is one group of 5 and write '1 group of 5 = 5'. Now ask them to hold up their other hand as well. *There are now 2 groups of 5, or 10 fingers.* Repeat with the other student, adding 2 more groups of 5.

In total there are 4 groups of 5 fingers, making 20 fingers altogether.

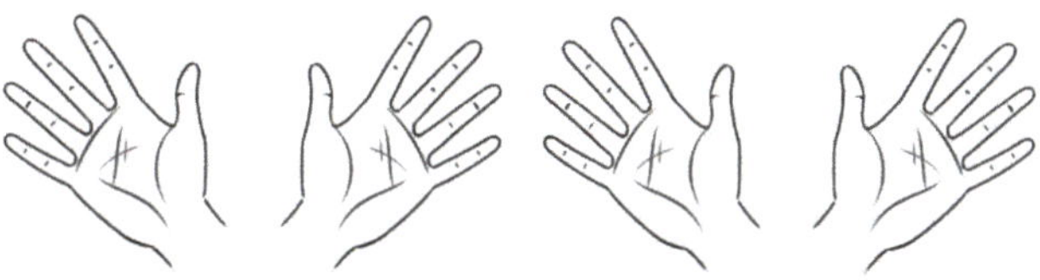

Main activity

Show students 3 paper cups that each have 2 cubes in them. Explain that each paper cup is 1 group.

How many groups are there? (3) *How many cubes in each cup/group?* (2)

Write on the board: $2 + 2 + 2 = 6$, 3 groups of $2 = 6$.

Explain to students that we use a special symbol to mean 'groups of', written ×, and that adding 2 three times (**repeated addition**) is the same as 3 groups of 2. Write on the board: $3 \times 2 = 6$, and say that we call this a **multiplication**.

Go back to the 'hand' examples from the Introductory activity and ask students to try to write an adding sentence and a multiplication sentence to describe it, e.g. $5 + 5 + 5 + 5 = 4 \times 5$. Work through another example, e.g. *Hold up both your hands with 4 fingers up on each.* Ask students to write the multiplication sentence to show this. ($2 \times 4 = 8$)

Students can now complete page 93 in the Student Book individually. As students are working, walk around the classroom and ask individuals questions such as:

How did you count the legs? Is there a faster way than counting one at a time?

Does a cat have an odd or even number of legs? How do you know?

Three chickens. How do you record how many legs as a repeated addition?

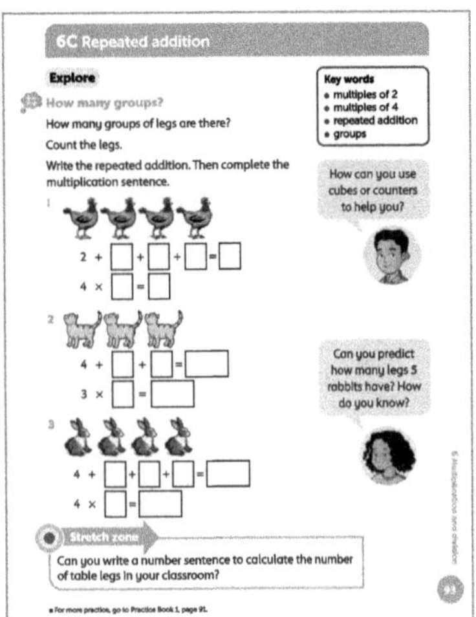

Differentiation

Supporting: Help students to write linked repeated addition and multiplication sentences as they record sentences for the pictures shown.

Consolidating: Ask students to use a number line to make equal jumps for the numbers of legs in each question.

Extending: Ask students to explore arrays linked to each question, for example, 4×2 can be 4 groups of 2, or 2 groups of 4.

Stretch zone: *Can you write a number sentence to calculate the number of table legs in your classroom?*

Before students begin, check to see how big the total will be based on the class size and then either tell students to work it out together or to build the number up by drawing an array or using cubes to represent the groups of table legs and then counting them.

Check that students have written a correct number sentence. Some students may be able to calculate the total, but it could be too large for other students, perhaps 15×4 or more.

 **Reflection time**

Write on the board $5 \times 3 = 15$. Ask students to describe what each number in the sentence means. They should be able to say that there are 5 equal groups, each group has 3 items and that makes 15 items altogether.

Repeat with $6 \times 2 = 12$ and $4 \times 5 = 20$.

Practice Book: Students can then complete page 91 in the Practice Book. This can be done directly after the main activity, as homework, or as the focus of a separate mathematics session to help students consolidate their learning and build fluency. This activity will be quite challenging for students who are not confident writers. Ask students to say their problems aloud for an adult to record, if possible.

Differentiated outcomes	
All students	should count how many in a group.
Most students	will count how many equal groups and write a sentence for this.
Some students	may write a multiplication sentence to describe a number of equal groups.

Answers

Student Book page 93

1 $2 + 2 + 2 + 2 = 8; 4 \times 2 = 8$

2 $4 + 4 + 4 = 12; 3 \times 4 = 12$

3 $4 + 4 + 4 + 4 = 16; 4 \times 4 = 16$

Practice Book page 91

Possible number sentences could include:

$2 \times 1 = 2$ (rabbits) $3 \times 1 = 3$ (ducks)

$2 \times 2 = 4$ (dogs' ears) $2 + 2 + 2 + 2 = 8$ (chicken legs)

Stretch zone: Check that students have written a word problem and also the answer.

6D Multiplication word problems

Specific learning focus

- Write multiplication stories from pictures and calculate them.

Global skills

- **Creative skills:** problem solving / exploring
- **Real-world skills:** presenting information
- **Interpersonal skills:** communication

Key vocabulary

- multiplication, groups, repeated addition

Resources

- picture of a scene with countable objects to display
- cubes or counters, number lines, number rods

Language support

Students may need support when describing the 'story' in their pictures and then in interpreting how that translates to a multiplication sentence. Emphasise phrases like 'equal groups', 'how many groups'?

 ### Introductory activity

Show the students a picture of a scene with some objects in groups. For example, an image of some cows in a field, each cow having 4 legs.

Tell a short story such as, *In the farmer's field, there are 5 cows. Each cow has 4 legs. How many legs are there altogether?*

Ask students to tell you what the multiplication calculation should be (5×4) and then use a number line or cubes to find how many there are.

 ### Main activity

Explain to students that they will be making their own pictures and multiplication stories. Tell them to think about what the objects will be in their picture, for example, some animals with the same number of legs, or vehicles with the same number of wheels.

Remind them to record the multiplication as a number sentence and to use number lines, cubes or number rods to help them calculate how many there are in total.

Students can then work on the Student Book page 94 individually. Look at the worked example and then question 1 together. Ask what size the groups could be for question 1. Agree that students could think of a row as a group or a column as a group. Point to the missing number boxes for the repeated addition and ask *How many?* (6) *How many columns of counters?* (6) Say that this suggests that students should think of each group as a column of counters.

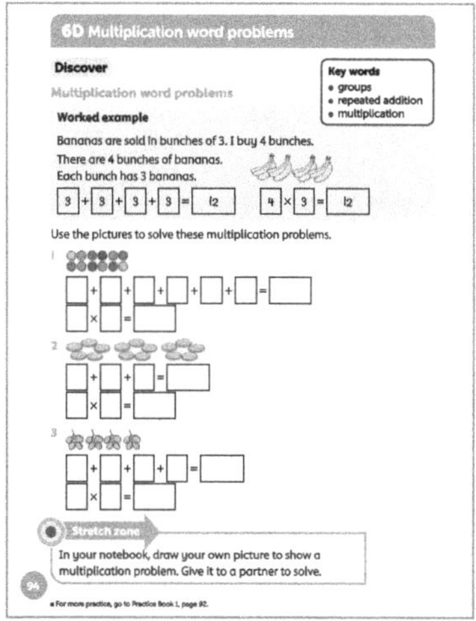

Differentiation

Supporting: Help students to identify the key information in the word problem, or read the problem where this aids access.

Consolidating: Ask students to calculate using counters or number rods.

Extending: Encourage students to calculate mentally and then use counters or an array to check.

Stretch zone: *In your notebook, draw a picture and write a multiplication story. Give it to a partner to solve.*

Check that the stories and multiplication problems are accurate and appropriate.

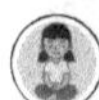 ### Reflection time

Ask a few students to share their pictures and their multiplication stories. Ask the other students if they can see the numbers in the pictures. Ask different students to do the calculation for each story.

Practice Book: Students can complete Practice Book page 92. This can be done directly after the main activity, as homework, or as the focus of a separate mathematics session to help students consolidate their learning and build fluency. Read through each question together and explain vocabulary, linking it to the drawings.

Differentiated outcomes	
All students	should draw a picture of objects in equal groups.
Most students	will describe the picture and form a multiplication sentence from it.
Some students	may multiply mentally to find how many in total.

6D Multiplication word problems

Explore 2   Student Book page 95 · Practice page 93

Specific learning focus

- Write multiplication stories from pictures and calculate them.

Global skills

- **Creative skills:** problem solving
- **Interpersonal skills:** communication

Key vocabulary

- multiplication, groups, repeated addition

Resources

- cubes or counters, number lines

Language support

Students may need support with describing the 'story' in their pictures and then interpreting how that translates to a multiplication sentence. Emphasise words like multiply, groups, repeated addition.

 Introductory activity

Tell students a story, for example, *There are 4 cupcakes in a packet. I buy 3 packets. How many cupcakes have I bought?*

Explain that the 4 cupcakes in a packet is one group, and the 3 packets you buy is how many groups. Multiplying 3×4 gives you the total number of cupcakes (12). Remind students how to write the multiplication sentence $3 \times 4 = 12$.

Main activity

Show students a group of objects, for example 5 horses. Say that you want to know how many legs there are altogether.

How many horses? (5) How many legs does each horse have? (4)

Now ask students what multiplication number sentence they need to write to calculate the total ($5 \times 4 =$). Use cubes or a number line to get the answer 20. Complete the number sentence as $5 \times 4 = 20$.

Ask students to complete the problems on page 95 of the Student Book individually.

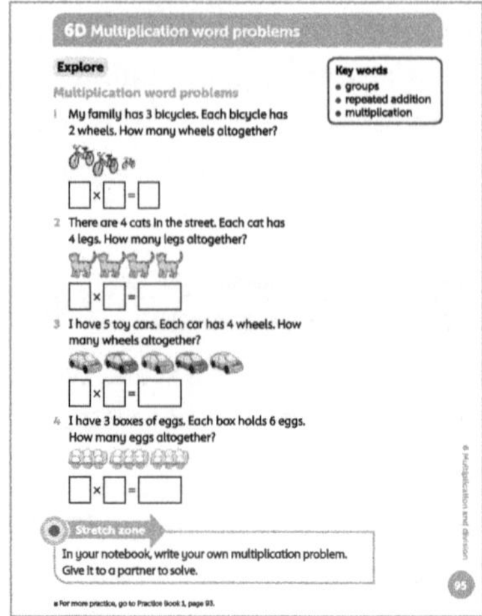

Differentiation

Supporting: Help students to identify the key information in the word problem, or read the problem where this aids access.

Consolidating: Ask students to calculate using cubes.

Extending: Encourage students to calculate mentally and then use counters or an array to check.

Stretch zone: *In your notebook, write your own multiplication word problem. Give it to a partner to solve.*

Check that the word problem is sensible and that a partner (and you!) can solve it.

Ask a few students to share their multiplication sentences and explain which numbers they put in the empty boxes, and why. Make sure they understand the difference between the number in a group and the number of groups and how this helps them make the number sentence.

Practice Book: Students can complete Practice Book page 93. This can be done directly after the main activity, as homework, or as the focus of a separate mathematics session to help students consolidate their learning and build fluency. Have pictures ready to support students to understand the vocabulary in the word problems. Students can continue to use concrete materials to solve calculations.

Differentiated outcomes	
All students	should count how many in a group.
Most students	will see how many groups there are.
Some students	may complete the number sentence accurately and calculate the total.

6 Multiplication and division

Connect Student Book page 96

Big Idea

- We can use equal grouping, sharing and arrays to help us multiply and divide.

Global skills

- **Creative skills:** problem solving
- **Interpersonal skills:** communication
- **Self-development skills:** reflecting on learning

Key vocabulary

- share, whole, equal parts

Resources

- two chocolate bars (with 24 small squares), 0–20 number lines, cubes or counters

Language support

Reinforce throughout the discussion that sharing means dividing into parts that are the same as each other (equal). Emphasise this in ways that students can understand, e.g. it wouldn't be 'fair' if one 'half' was bigger than the other 'half', or if one student in a group of five had a larger share than everyone else.

Answers

Student Book page 95

1 $3 \times 2 = 6$

2 $4 \times 4 = 16$

3 $5 \times 4 = 20$

4 $3 \times 6 = 18$

Practice Book page 93

1 $2 \times 3 = 6$

2 $5 \times 2 = 10$

3 $3 \times 8 = 24$

4 $5 \times 4 = 20$

Stretch zone: There are several answers, for example, 2 goats (8), 8 children (16); 5 goats (20), 2 children (4).

 **Introductory activity**

Show students two wrapped chocolate bars, each with 24 squares in a 4×6 array. Give one student a chocolate bar and say, *If I have one student and one chocolate bar, they can have one chocolate bar all to themselves. What if there are four students but I only had one chocolate bar? How could I share it fairly between them?*

Show the unwrapped chocolate bar and break it into unequal parts.

Do you think this is sharing fairly? Why not? Use the feedback from the class, using the vocabulary of sharing equally. Agree the pieces must be the same size as each other to make it fair.

Listen to suggestions and then agree that when you share something between four people, you need to share it into four **equal parts**.

 **Main activity**

Draw a 4×6 array of squares on the board and ask students what they notice about the array and the chocolate bar. They should notice that the array has 4 rows of 6 squares and so it matches the squares on the chocolate bar.

How can we share the chocolate bar equally between 4 students? Encourage them to look at the number of rows. They might notice that there are 4 rows of 6 squares and so suggest that each student gets 1 row of 6 squares.

Now say that you want to share the chocolate between 6 students. Ask if they can see that the bar also has 6 rows of 4 squares.

How does this help us share the chocolate between 6 students? They should be able to see that each student would get a row of 4 squares.

Can you write multiplication sentences about this array? Write students' suggestions on the board: $4 \times 6 = 24$ and $6 \times 4 = 24$.

Explain that students are going to work on another chocolate bar problem and this bar will be even bigger! Students complete Student Book page 96 individually. Check that all students have calculated the number of squares of chocolate correctly before they start working on the questions. Encourage them to continue to use concrete resources such as cubes or counters to solve each problem.

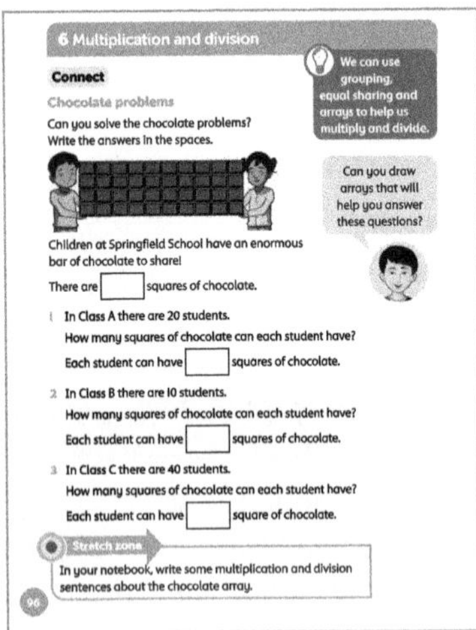

Differentiation

Supporting: You can help students by writing down what they tell you as an answer.

Consolidating: Ask students to jump in multiples of 4 on a number line to get to 24, and ask how many jumps they made.

Extending: Ask students to think about how they would share other rectangular bars, for example, *How would you share a bar that has 5 rows of 8 squares between 5 students? 8 students? 4 students?*

Stretch zone: *In your notebook, write some multiplication and division sentences about the chocolate array.*

Check that students have completed sentences accurately.

 Reflection time

Ask students to discuss what they have learned about arrays and sharing. Can they describe an array in terms of its rows in two directions? They should be able to notice that a 3×5 array has the same number of objects as a 5×3 array.

Ask students to tell you how much 5 lots of 3 are (15). Then ask, *How much is 3 lots of 5?* (15) Discuss what they notice and help them to see that 3×5 is the same as 5×3.

Differentiated outcomes	
All students	should understand that sharing involves equal parts.
Most students	will use the array to help them share equally.
Some students	may extend to use arrays for sharing other numbers of objects.

Answers

40 squares

1 2

2 4

3 1

6 Multiplication and division

Review — Student Book page 97 • Practice Book page 94

Global skills

- **Creative skills:** problem solving / investigating
- **Interpersonal skills:** communication
- **Self-development skills:** reflecting on learning

Student Book

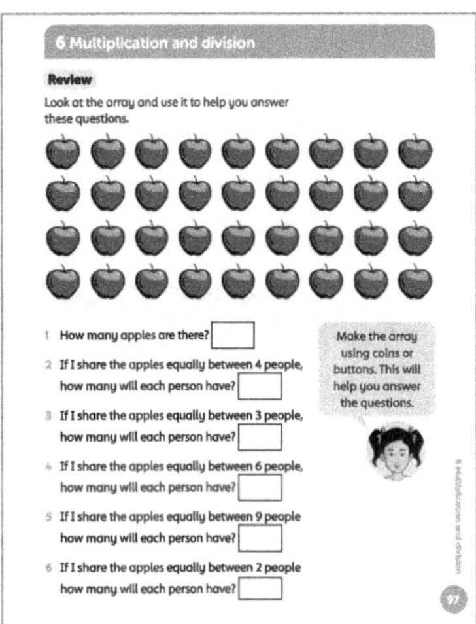

With young children, assessment activities are most effective when carried out as an everyday classroom activity. Students should have interlocking cubes or number rods available to support them.

Watch as students complete the sentences that represent repeated addition and multiplication situations. If there are errors, help students find the number of groups and the number in each group. Check that students understand the language of repeated addition, multiplication, equal groups.

Encourage students to use cubes for answering questions. Extend by asking a range of similar questions or asking students to make up their own problems.

Answers

Student Book page 97

1 36 apples

2 9 apples each

3 12 apples each

4 6 apples each

5 4 apples each

6 18 apples each

Practice Book

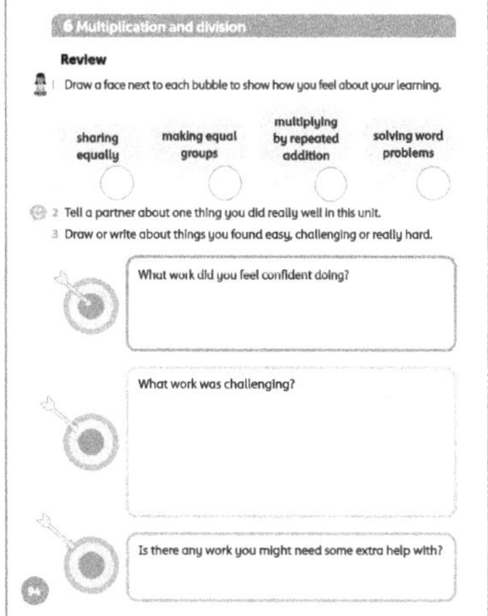

With young children it is appropriate to complete this as a whole-class discussion. You may choose to keep a record of the class discussion or a copy of the review page for your own records. Use the Student Book to briefly remind students of the areas of mathematics that they have worked on in this unit.

Give out the Student Book to support pairs of students as they discuss and answer the questions in the Practice Book. Allow students plenty of time for discussion before asking them to share their responses with the rest of the class. If students complete this assessment at home, encourage them to discuss this with adults. Make a note of areas that students still feel unsure about.

Additional material

There are additional end-of-unit assessments available on the *Oxford Owl* website.

7 Fractions

Overview

Big Idea

Fractions are numbers which students need to be able to understand, recognise and calculate with. As parts of whole numbers, they are useful in contexts where whole numbers are not accurate enough, for example when measuring out $2\frac{1}{2}$ cups of flour. Fractions are often the result when one whole is divided by another, and we need a way of recording the result, for example, 12 divided by 5 is $2\frac{2}{5}$.

Whereas whole numbers represent an amount and are written in one way (for example, 4 apples is written as 4 and no other whole number representation will do), with fractions there are many different representations that mean the same fraction, for example $\frac{1}{2}$ and $\frac{2}{4}$ mean the same number.

Finding fractional amounts of numbers or measures is an important everyday skill, and an awareness from an early age of what a fraction represents is crucial.

Look out for

- **Students who do not realise that, when dividing something into quarters, the quarters must all be the same size as each other.** Emphasise 'equal parts' when speaking and provide non-examples alongside examples of quarters.

- **Students who are unsure how to 'read' fractions and say, e.g. $\frac{1}{2}$ = *one-two or one-halves*.** Provide lots of opportunities to read and say fractions, modelling clear and accurate pronunciation for students.

Possible misconceptions

- **Students may think larger digits mean larger fractions.** One of the key misconceptions for students is that because the digits used in a fraction are large, that makes the fraction large. It is not uncommon for students to think $\frac{4}{8}$ is a bigger fraction than $\frac{1}{2}$.
- **Students may think all fractions are numbers less than 1.** Students often struggle to understand that, for example, $\frac{7}{5}$ is a fraction that represents a number between 1 and 2. Experience of fractions greater than 1 will overcome this misconception.

Key vocabulary

- double, half, halve, $\frac{1}{2}$
- equal parts, equal shares, share equally, matching
- whole, fraction
- quarter, $\frac{1}{4}$, one out of 4

Coverage in lessons

Learning focus	Learning outcomes (the ENC objectives)
Doubles and halves	Recognise, find and name a half as one of two equal parts of an object, shape or quantity.
Halves	Recognise, find and name a half as one of two equal parts of an object, shape or quantity.
Quarters	Recognise, find and name a quarter as one of four equal parts of an object, shape or quantity.

7 Fractions

Engage Student Book page 98

Big question

- How can I describe parts of a whole?

Global skills

- **Creative skills:** problem solving / exploring / investigating
- **Interpersonal skills:** communication / teamwork

Key vocabulary

- share equally, equal parts, equal shares

Resources

- food to share, e.g. apples, melon, pizza, cake

Language support

From the start, focus on ensuring students understand the vocabulary of equal parts or equal shares. Reinforce that two halves of something are always the same size as each other; avoid the misconception of describing a bigger or smaller half.

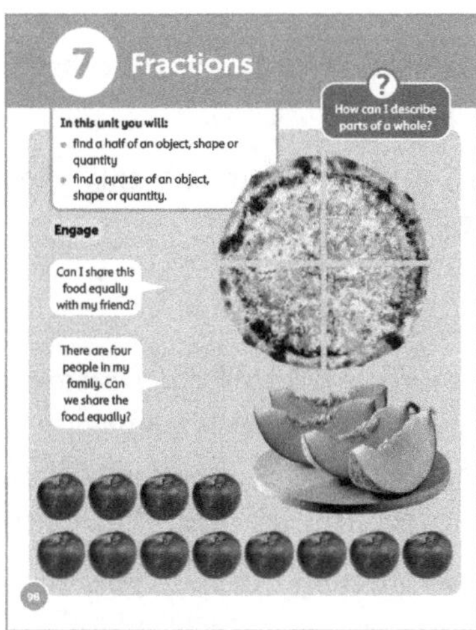

Introductory activity

If you have access to an IWB, you could display the Student Book Engage page and discuss the Big idea of the unit with students.

Prepare some food samples for the class that can be cut up to **share equally** – this could be a pizza, melon or cake.

Ask students what would be the fairest way to share a cake between 2 people. Their answers will probably include something about cutting it in half. Ask them to describe what they mean by a half. *Where do they find halves of things in their everyday life?*

 Main activity

Ask students how they might share something (e.g. a pizza) fairly between 4 people (Students can be prompted to look at the picture of a pizza on page 98 of the Student Book, or on the IWB). Ask for strategies for sharing the pizza and encourage language of **equal shares**, same size pieces and so on.

Extend this to sharing between 8 people using the same arguments. Ensure that students understand the importance of equal size pieces, by showing them some shapes where some are split into **equal parts** and others that are not, and get them to identify the ones that have equal parts.

Differentiation

Supporting: You can help students use the correct language for sharing by using sentence frames, for example, 'I shared the _____ equally between _____ people. There are _____ equal parts.'

Consolidating: Ask students to draw a picture to show how they would share something fairly.

Extending: Ask students to share a given item between different numbers of people, e.g. a cake to share between 3 and then 6 people equally.

Reflection time

Show the students half of an apple and also half of a melon. Ask them which half is larger. Discuss the idea that the two halves of the apple are the same size as each other, but half an apple is smaller than half a melon because an apple is smaller than a melon.

7A Doubles and halves

Discover
Student Book page 99 • Practice Book page 95

Specific learning focus

- Know doubles to at least double 5.
- Double any single-digit number.
- Find halves of even numbers to 10 to find which are even and which are odd.

Global skills

- **Creative skills:** problem solving / exploring / investigating
- **Interpersonal skills:** communication

Key vocabulary

- double, half, halves, matching, equal parts

Resources

- interlocking cubes, counters or countable objects

Language support

Be sure to reinforce throughout the discussion that parts like halves are the same as each other (equal). Explain the link between doubles and halves, for example that double 3 is 6, so half of 6 is 3.

 Introductory activity

Show students a group of 5 objects, cubes or counters. Say they are going to find **double** this number.

How many objects are here? (5) *To make double we have to get another 5.* Put 5 more objects with the first group and ask, *How many are there now?* (10) Then say, *Double 5 is 10.*

 Main activity

Give each student a set of interlocking cubes and ask them to make a tower of 3 cubes. Then ask them to make a **matching** tower of 3 cubes.

How many cubes do you have altogether in the two towers? (6) So then say, *Double 3 is 6.*

Now ask students to build a tower of 8 cubes. Once done, students should break their tower into 2 equal parts. Get them to see that the parts are the same by putting them side by side to show they match.

How many in one of your parts? (4) *Two matching parts are called **halves**, so 4 is **half** of 8.*

Students complete Student Book page 99 individually. Draw student's attention to the text in the speech bubble and read it aloud. Give them a sheet of paper to draw their picture on. They may want to build the towers first to copy. After they have completed their drawing discuss what they notice. (Half of 8 is 4 and double 4 is 8.)

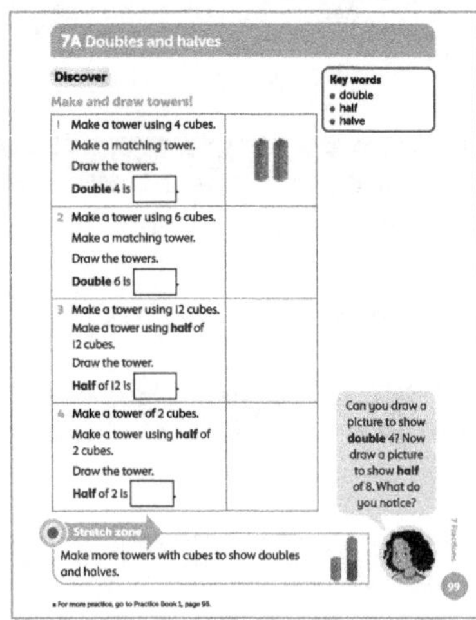

Differentiation

Supporting: Help students by modelling the building of towers and matching towers to make doubles.

Consolidating: Ask students to check their answers by using the inverse, for example if they double 4 to get 8 then they should halve 8 to find 4.

Extending: Ask students to mentally find doubles or halves up to 20.

Stretch zone: *Make more towers with cubes to show doubles and halves.*

Check that students have correctly coloured and recorded the doubles and halves.

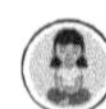 Reflection time

Ask questions such as, *I divided an apple into two equal parts. What is each part called?* (A half.) *How many halves in one orange?* (2) *How many halves in two apples?* (4) *How many apples in two halves?* (1) *How many oranges in four halves?* (2)

Practice Book: Students can complete Practice Book page 95. This can be done directly after the main activity, as homework, or as the focus of a separate mathematics session to help students consolidate their learning and build fluency. Encourage students to practise counting the shaded squares in twos to check how many each double is.

Differentiated outcomes	
All students	should make a matching tower to form a double.
Most students	will make doubles and halves up to 20 using cube towers.
Some students	may carry on doubling or halving numbers above 20.

Student Book page 99

1 8   **2** 12   **3** 6   **4** 1

Practice Book page 95

1 18   **2** 16   **3** 4   **4** 8   **5** 10

Stretch zone: Check that the students' doubles statements are correct. For example: Double 6 is 12.

7A Doubles and halves

Explore Student Book page 100 • Practice Book page 96

Specific learning focus

- Know doubles to at least double 5.
- Double any single-digit number.
- Find halves of even numbers to 10 to find which are even and which are odd.

Global skills

- **Creative skills:** problem solving / exploring
- **Interpersonal skills:** communication / teamwork

Key vocabulary

- double, half, halve

Resources

- dominoes, number cards 1–10

Language support

Listen to the vocabulary and language structures students use when discussing doubles and halves. Model, using whole sentences, using the appropriate vocabulary – half, halve, fraction, part, whole, double.

Introductory activity

Ask students to sit in two or more equal-length lines, either on the floor or on chairs. Ask the front person on every line a 'double' or 'half' question such as double 4, double 6, half 12, half 2 and so on. If they are correct, they go to the back of their line.

If they are incorrect, a member of the team can answer but the student stays at the front. Continue until all students in one line have answered and the first person is back at the front. Make sure you ask questions appropriate to students' current understanding.

Main activity

Tell the class they are going to play a doubles and halves game, using dominoes. Give each pair a set of dominoes (or 10 each depending on time available), which they place face down on the table.

Students in each pair take turns to turn over one domino and work out the total number of spots. If the total is *even*, they *halve* the total and write that as their score. If the total is *odd*, they *double* it and write that as their score. Then they place the domino face up so that it can't be chosen again. A pair who turns over a *double*, *doubles* their *total* score.

Each domino score is added to the previous total. The first pair to reach or exceed 50 wins.

Students then complete page 100 in the Student Book individually. Watch which strategies students are using for counting the objects, particularly the larger groups. Encourage them to move beyond counting in ones and model more efficient strategies, e.g. counting pairs of ladybirds or seeing two groups of 4.

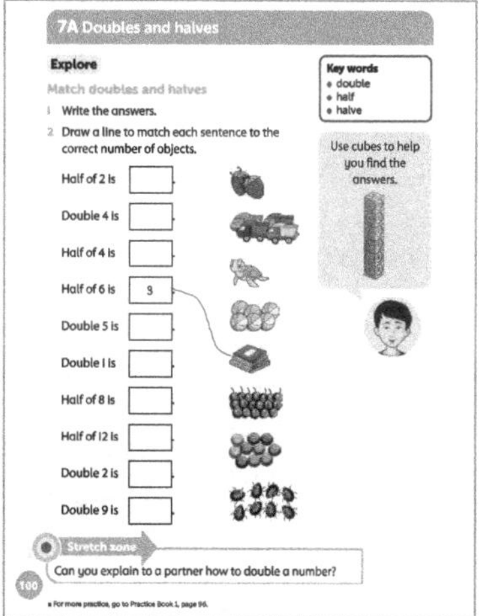

Differentiation

Supporting: Students continue to use cubes to double and halve, and refer to a list of odd and even numbers.

Consolidating: Ask students what halves and doubles they can remember without building towers.

Extending: Ask students how they can halve an odd number.

Stretch zone: *Can you explain to a partner how to double a number?*

Check that students are able to talk about making two lots of the number, or adding the number twice.

Reflection time

Draw a 5 × 5 grid on the board. Write in the numbers 2, 4, 6, 8, 10, 12, 14, 16, 18, 20, repeating them until the grid is full.

2	4	6	8	10
12	14	16	18	20
2	4	6	8	10
12	14	16	18	20
2	4	6	8	10

Group students into teams. Ask one student from each team to come to the front of the class and to choose a card from a set of number cards 1–10. Each student decides whether to double or halve the number on their chosen card, then covers that number on the grid and returns to their place.

Continue, choosing different students each time. If no number can be covered, the student returns to their seat without covering anything. The student who covers the last number on the grid wins for their team.

Practice Book: Students can complete Practice Book page 96. This can be done directly after the main activity, as homework, or as the focus of a separate mathematics session to help students consolidate their learning and build fluency. Introduce any new vocabulary as appropriate and practise it when checking any answers, *How many spider webs? How many is half the number of train engines?*

Differentiated outcomes	
All students	should recognise odd and even numbers up to 6.
Most students	will remember doubles and halves up to 10.
Some students	may know doubles and halves beyond 20.

Answers

Student Book page 100

1 & 2 Check that the matching lines have been drawn correctly.

Half of 2 is 1	Turtle
Double 4 is 8	Ladybirds
Half of 4 is 2	Strawberries
Half of 6 is 3	Books
Double 5 is 10	Marbles
Double 1 is 2	Strawberries
Half of 8 is 4	Trucks
Half of 12 is 6	Balls
Double 2 is 4	Trucks
Double 9 is 18	Blueberries

Practice Book page 96

Check that the correct number of objects has been circled each time.

1 3

2 1

3 2

4 5

5 6

7B Halves

Discover Student Book page 101 • Practice Book page 97

Specific learning focus

- Find half of simple shapes and small amounts.

Global skills

- **Creative skills:** problem solving
- **Real-world skills:** interpreting information
- **Interpersonal skills:** communication

Key vocabulary

- half, halves, equal parts, fraction, whole

Resources

- cubes, counters
- paper squares, rectangles and isosceles triangles
- coloured pencils
- shapes such as stars, wheels, hearts that can be halved

Language support

Relate the word 'half' to the shapes and amounts and help students to be precise about finding half, recognising that two halves must be equal to each other.

Introductory activity

Draw an even number of counters on the board and ask students to count how many there are. Ask students whether there is an even number or an odd number. As there is an even number, tell students they are going to share them into two equal parts.

Get students to put the same number of cubes on their tables, then share them into 2 piles, making sure there is the same number in each pile. Tell students that after sharing into 2 equal piles, one pile is half of the total number of counters or cubes.

Main activity

Draw a square on the board and ask students what shape it is (a square). Draw a straight line vertically to divide the square into halves. Say, *I drew a line to divide the square into 2 equal parts.* Point to one of the halves and say, *This is a half.* Point to the second half. *How much is this?* (a half)

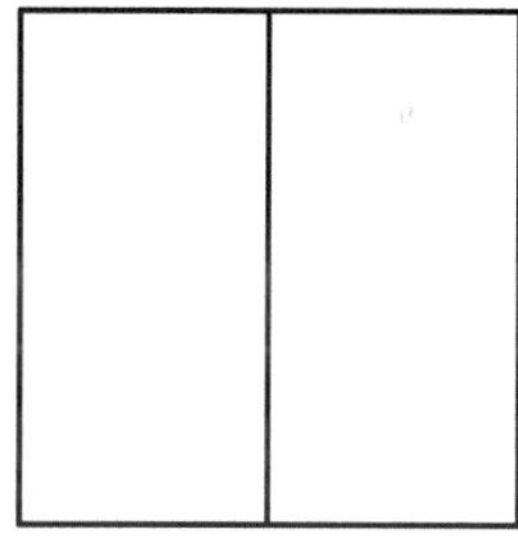

Explain that we use the word **fraction** to mean an equal part of a **whole** and that we can write the fraction 'a half' as $\frac{1}{2}$. Write this on the board.

Now draw a triangle, make it isosceles with a horizontal base at this stage, then cut it into halves vertically from the apex to the base.

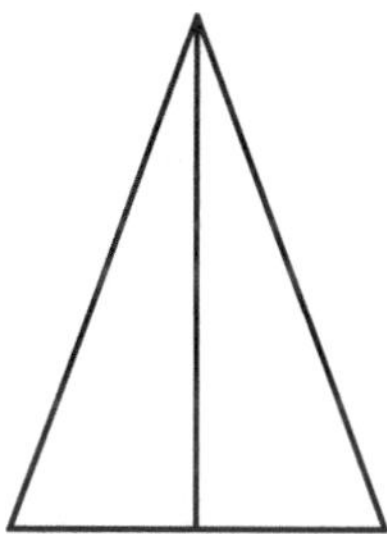

Again, ask the students how much of the whole each part is, and agree that each part is half of the triangle.

Reinforce the idea that both parts must have the same amount, either in piles of cubes, or parts of a shape. With a paper cut-out square, show students that folding it in half shows where the line needs to be to make halves, and shows them that the two parts are the same as they are able to be folded onto one another. Repeat with a paper triangle.

Now ask students to complete the activities on page 101 of the Student Book. Encourage students to use counters to find half of groups of objects.

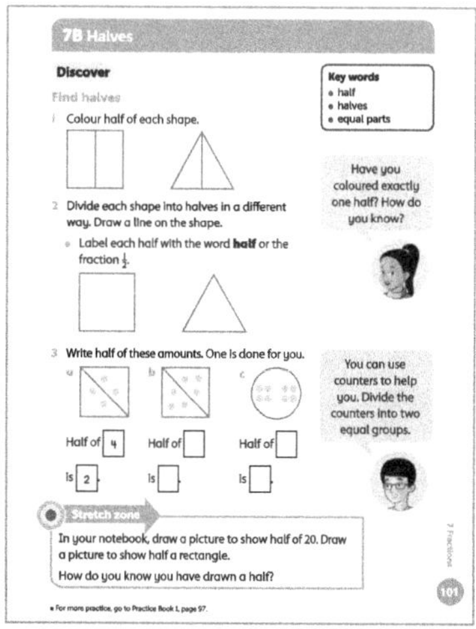

Differentiation

Supporting: Students should continue to use cubes to find half of an amount, as well as pieces of paper to fold in half.

Consolidating: Ask students what number halves they can remember.

Extending: Ask students to explain how they can halve a shape.

Stretch zone: *In your notebook, draw a picture to show half of 20. Draw a picture to show half a rectangle.*

How do you know you have drawn a half?

Students should be able to describe the two parts as being the same size, perhaps that one part will fit exactly onto the other part.

Check that students have drawn a picture to show that half of 20 is 10, and that they have coloured half the rectangle correctly.

 Reflection time

Show students some everyday shapes such as a heart, star and wheel. For each shape ask students to try to work out how they can divide the shape into halves, emphasising that the halves must be equal. Ask students how they knew they had drawn half of a shape.

Practice Book: Students can then complete page 97 of the Practice Book. This can be done directly after the main activity, as homework, or as the focus of a separate mathematics session to help students consolidate their learning and build fluency. Encourage students to use cubes or counters to help them to find half of each number by making the amounts and then splitting them into two equal-sized groups.

Differentiated outcomes	
All students	should see half of a shape and label it.
Most students	will halve a shape or an amount and label it.
Some students	may work out half of larger amounts.

7B Halves

Explore Student Book page 102 • Practice Book page 98

Specific learning focus

- Find half of a set of items by dividing them into 2 equal groups.

Global skills

- **Creative skills:** problem solving / exploring
- **Interpersonal skills:** communication / teamwork

Key vocabulary

- half, halves, $\frac{1}{2}$, equal parts

Resources

- interlocking cubes, counters, countable objects – sweets, toy animals, etc.

Language support

Students may need help with understanding that halving the number of items in a group does not have to mean cutting each item in half, as it is the amount that is being halved.

Introductory activity

Show students a square grid and draw a rectangle on the grid that is 6 squares long and 4 squares wide. Ask students, *How many squares are in the rectangle?* (24) *I want to find half of the rectangle, so how many squares will that be?* (12)

Ask students to suggest how you could cut the rectangle in half. They might suggest vertically or horizontally. Draw the cut line and then ask students to count the squares in each part to make sure they are the same number. Repeat this with a different-shaped rectangle.

Main activity

Ask 8 students to come forward and arrange themselves in 2 equal groups (2 groups of 4). Tell them that each group is half of the 8 students and record it on the board as 'half of 8 is 4' and '$\frac{1}{2}$ of 8 is 4'

Now ask 2 more students to join them.

How many students are there now? (10)

Ask which group the 2 students should join so that each group is half of 10. *How many students is half this time?* (5) Remind students that each time they find half of an amount, it is the same as finding how many in one of 2 equal groups, whether it is students, cubes, sweets or grid squares in a shape.

Students can now complete the activities on page 102 of the Student Book. Students can continue to use counters or cubes to physically split groups into two equal groups to find $\frac{1}{2}$.

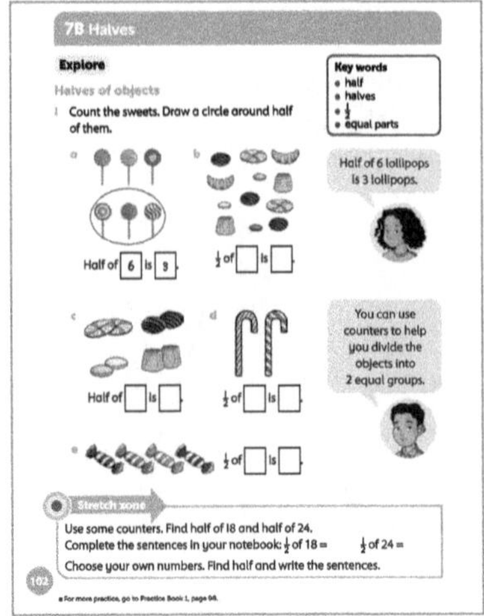

Differentiation

Supporting: Students should continue to use cubes to find half of an amount and record it as $\frac{1}{2}$ of the number. Support them to write $\frac{1}{2}$ correctly.

Consolidating: Ask students what number halves they can remember.

Extending: Ask students how they can halve a shape.

Stretch zone: *Use some counters. Find half of 18 and half of 24. Complete the sentences in your notebook:* $\frac{1}{2}$ *of 18 =* $\frac{1}{2}$ *of 24 =*

Choose your own numbers. Find half and write the sentences.

Students should be able to use counters to help them find the halves of 18 and 24. Check that they have written correct sentences for the number they chose.

 Reflection time

Show students a group of 7 apples. Ask them how they could find half of this group of apples.

What makes this a bit tricky? (7 is an odd number so 7 whole apples cannot be shared evenly.)

If we can't share whole apples evenly, what could we do to find half of 7?

Some students may say to share the apples into 2 groups of 3 and then cut the remaining apple in half, others may say to cut all the apples in halves and then share them into 2 equal groups.

Remind students that even numbers will divide into 2 groups without having to cut anything.

Practice Book: Students should now complete page 98 in the Practice Book. This can be done directly after the main activity, as homework, or as the focus of a separate mathematics session to help students consolidate their learning and build fluency. Model how to draw a rectangle on the grid and find half. Point out to students that they will need to fit three shapes on their grids, so the shapes should not be too large.

Differentiated outcomes	
All students	should see half of a shape and label it.
Most students	will halve a shape or an amount and label it.
Some students	may work out half of larger amounts.

Answers

Student Book page 102

1 a Half of 6 is 3.

  b Half of 12 is 6.

  c Half of 4 is 2.

  d Half of 8 is 4.

  e Half of 2 is 1.

Stretch zone: $\frac{1}{2}$ of 18 = 9 and $\frac{1}{2}$ of 24 is 12

Practice Book page 98

1 Check that students have drawn three different rectangles and coloured half correctly.

2 $\frac{1}{2}$ of 10 = 5

3 $\frac{1}{2}$ of 12 = 6

4 $\frac{1}{2}$ of 14 = 7

5 $\frac{1}{2}$ of 16 = 8

Stretch zone: The answers increase by 1 each time.

7C Quarters

Discover Student Book page 103 • Practice Book page 99

Specific learning focus

- Find a quarter of a shape or an amount

Global skills

- **Creative skills:** problem solving / exploring / investigating
- **Interpersonal skills:** communication

Key vocabulary

- quarter, $\frac{1}{4}$, one out of 4, equal parts

Resources

- squared paper, rectangular pieces of paper, coloured pencils, counters or cubes, rulers

Language support

Students will need to be reminded that a quarter is connected to four, and that there are 4 quarters in a whole amount. Also, that each quarter is the same size the others.

Introductory activity

Show students a square of paper. Ask what can be done to the paper to show two halves. Students may say to fold it in half. Fold the paper and then fold it again in half before opening it to reveal 4 equal parts, or **quarters**. The two folds could be in the same orientation to give 4 parallel quarters, each being a rectangle, or the folds could be at right-angles which will give 4 smaller squares as quarters.

Ask students how many of the parts should be coloured to show a quarter. (1)

Give students rectangular pieces of paper and ask then to fold them in half carefully. Then ask students to fold them again so that the piece of paper shows quarters.

Ask students which part shows a quarter and to place their finger on it. They should point to one of the parts. Then ask them to look at the parts that other students are pointing to, and they will see that different students are pointing to different quarters. This helps reinforce that each of the quarters is the same size.

Ask students to complete question 1 on page 103 of the Student Book. After they have completed it, ask them to compare with other students to notice that some have coloured different parts. As long as only one part has been coloured, it shows one quarter.

Return to the shapes they folded earlier. *Look at your shape. How many quarters can you see?* The students could point to each part with their fingers to see there are 4 quarters. *How many quarters in the whole shape?* (4). *There are a four quarters in a whole. A quarter is* **one out of 4** *equal parts.* Explain that we often write the fraction 'one quarter' as $\frac{1}{4}$.

Use different folds to make quarters of different shapes, for example, folding so that sides touch, or folding corner to corner.

Students can then complete the remaining questions on Student Book page 103.

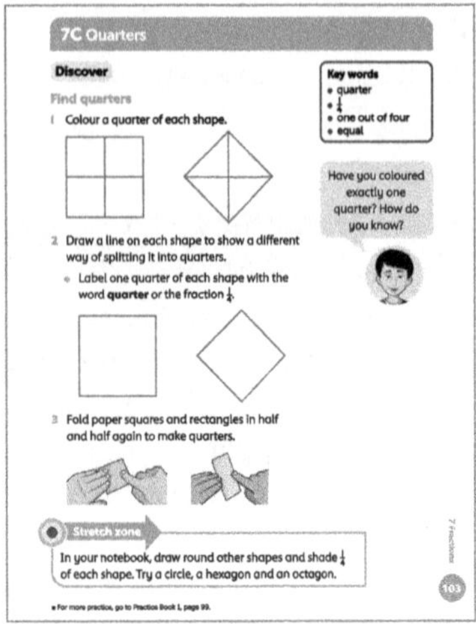

Differentiation

Supporting: Help students to find a quarter of a shape through drawing or folding.

Consolidating: Ask students to show how they make sure the 4 pieces are all equal in size to help them find quarters of the shapes.

Extending: Ask students how they can quarter shapes other than squares in more than one way, for example, rectangles and circles.

Stretch zone: *In your notebook, draw round other shapes and shade $\frac{1}{4}$ of each shape. Try a circle, a hexagon and an octagon.*

Check that students have managed to shade $\frac{1}{4}$ of each shape correctly.

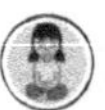 Reflection time

Ask students to show the ways they have coloured a quarter of the shapes as part of the stretch zone activity. *How do you know you have coloured a quarter? Can you think how you might find a quarter of a number by doing something similar to what you did with the paper?*

Practice Book: Students can now complete page 99 of the Practice Book. This can be done directly after the main activity, as homework, or as the focus of a separate mathematics session to help students consolidate their learning and build fluency. Encourage students to use a straight edge or ruler to so that they can split their shapes into equal parts as accurately as possible.

Differentiated outcomes	
All students	should identify 1 part out of 4 equal parts.
Most students	will divide a shape into 4 equal parts.
Some students	may find more than one way of splitting a shape into quarters.

Student Book page 103

1 Check that the students have coloured 1 of the 4 parts of the shapes.

2 Check that they have drawn equal quarters on the shapes.

3 Check that they can fold paper shapes twice to make quarters.

Practice Book page 99

Check that students have found correct ways to colour $\frac{1}{4}$ of each square.

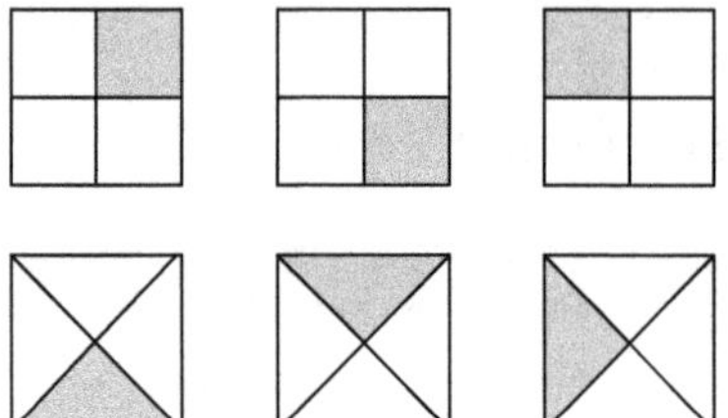

Stretch zone: For the first rectangle, students will probably divide it into quarters by using vertical lines to halve and then halve again, or by using a horizontal line to halve and then a vertical line to halve again.

For the second rectangle, students may draw diagonal lines to form four quarters.

7C Quarters

Specific learning focus

- Find a quarter of a shape or an amount.

Global skills

- **Creative skills:** problem solving / exploring / investigating
- **Interpersonal skills:** communication

Key vocabulary

- quarter, $\frac{1}{4}$, one out of 4, equal parts

Resources

- cubes or counters, paper plates
- large 0–20 number line for class use

Language support

Students will need to be reminded that quarter is connected to four, and that there are 4 quarters in a whole amount. Also, that each of the 4 quarters is the same amount.

 ### Introductory activity

Show students a pile of 12 cubes or counters. Ask a student to come to the front of the class to count them to check that there are 12. Now ask the student to share the cubes into 4 piles, sharing them out one at a time so that the piles are equal. When they are all shared out ask, *How many piles are there?* (4) *Are they the same size?* (Yes) *How many in each pile?* (3)

Explain that the 4 piles are all quarters of 12, so 3 is a quarter of 12.

 ### Main activity

Show students a set of objects (20) and have a paper plate available, marked into 4 equal parts. Ask a student to come forward and share the objects equally by placing them on the 4 parts of the plate. Check there are the same number in each. Ask, *how many are in each?* (5) Write on the board: $\frac{1}{4}$ of 20 = 5.

Repeat using 24 objects, this time leading to quarters that are 6.

Students can then complete the activities on page 104 of the Student Book. Students can continue to use counters or cubes to split amounts into equal-sized groups. Ask students how sharing to find quarters is different from sharing to find halves.

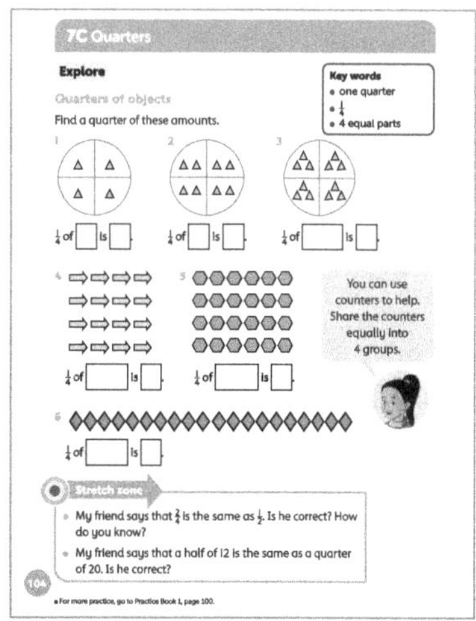

Differentiation

Supporting: Students use cubes to find a quarter of an amount.

Consolidating: Ask students to use number halves to help them find quarters.

Extending: Ask students how they can quarter a shape.

Stretch zone:

- *My friend says that $\frac{2}{4}$ is the same as $\frac{1}{2}$. Is he correct? How do you know?*
- *My friend says that a half of 12 is the same as a quarter of 20. Is he correct?*

Mention to students they may want to use counters to represent the fractions of amounts. Check that students understand that the same fraction can be represented in different ways and that halves and quarters of amounts can only be compared if the wholes are the same amount.

 ### Reflection time

Write '$\frac{1}{4}$ of 12 is 3' on the board. Show students a 0–20 number line with 12 circled on it.

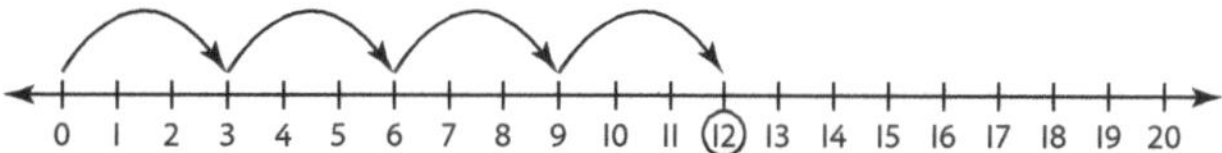

Now start from 0 and do 4 jumps, each jump being 3 places. Show that it ends on 12.

Ask students to describe what they see with the number line and how it relates to what they were doing with finding a quarter of groups of objects.

Practice Book: Now ask students to complete page 100 of the Practice Book. This can be done directly after the main activity, as homework, or as the focus of a separate mathematics session to help students consolidate their learning and build fluency. Students can continue to use counters or cubes to physically split groups into four equal groups to find $\frac{1}{4}$ of amounts.

Differentiated outcomes	
All students	should count 4 quarters making up a whole.
Most students	will find a quarter of an amount.
Some students	may know how many in the whole if they know how much a quarter is.

Answers

Student Book page 104

1 $\frac{1}{4}$ of 4 is 1

2 $\frac{1}{4}$ of 8 is 2

3 $\frac{1}{4}$ of 12 is 3

4 $\frac{1}{4}$ of 16 is 4

5 $\frac{1}{4}$ of 24 is 6

6 $\frac{1}{4}$ of 20 is 5

Practice Book page 100

1 $\frac{1}{4}$ of 8 = 2

2 $\frac{1}{4}$ of 4 = 1

3 $\frac{1}{4}$ of 24 = 6

4 $\frac{1}{4}$ of 20 = 5

5 $\frac{1}{4}$ of 16 = 4

6 $\frac{1}{4}$ of 28 = 7

Stretch zone: Students should work out $\frac{1}{4}$ of their own chosen number.

7 Fractions

Connect Student Book page 105

Big idea

- I can describe equal parts of a whole as halves and quarters.

Global skills

- **Creative skills:** problem solving / exploring / investigating
- **Real-world skills:** research / presenting information / interpreting information / financial literacy

Key vocabulary

- fraction, half, halves, quarter, equal parts

Resources

- paper shapes (e.g. squares, circles, triangles)
- counters, interlocking cubes, 2 circular pieces of paper or card cut into quarters

Language support

Encourage students to talk, within their groups, about what they are doing and why. Some students may take on different roles within the group, such as recorder, cutter, counter and so on. Make sure that students are not volunteering for a role to avoid one they find difficult.

 Introductory activity

Ask students to look at the pink cake on page 105 in the Student Book. Display on the IWB, if possible. Show students a circular piece of paper or card that has been cut into 4 pieces. Explain that this represents the pink cake.

What fraction of the cake is each piece? $\left(\frac{1}{4}\right)$ *If each person can have one piece of cake, how many people can have a piece?* (4)

Now say that there will be 8 people who need a piece of cake each.

If they each get a quarter of a cake, how many cakes are needed? (2) Model the cakes with circular pieces of paper and show how 2 cakes cut into quarters will make 8 pieces.

 Main activity

Talk through the problems in the Student Book on page 105 with students. Explain that they might like to use paper shapes, counters or cubes to help them. They could also use two different colours or shapes, one to represent sandwiches and the other to represent cakes.

Arrange students in small mixed-attainment groups to work on the problems, so that less confident language learners hear the appropriate mathematical vocabulary modelled by their peers.

As students work, ask them questions such as, *How did you work out how much an equal share was?*, *Does it matter that half a sandwich is bigger than half a cake?*

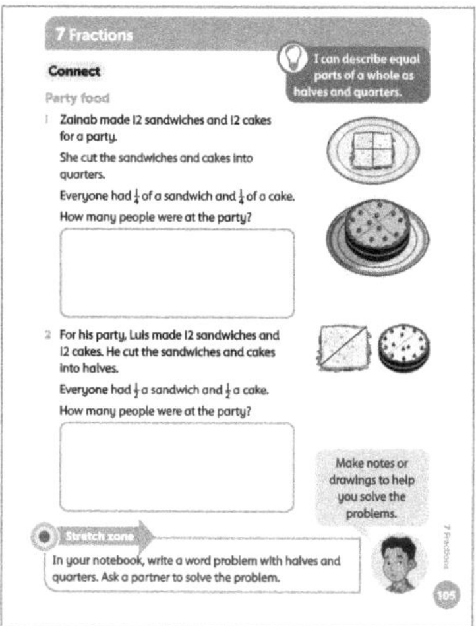

Differentiation

Supporting: Make sure students take part and use practical materials to understand the process.

Consolidating: Encourage students to share their thinking so that they say their strategies clearly.

Extending: Ask students to explain the process.

Stretch zone: *In your notebook, write a word problem with halves and quarters. Ask a partner to solve the problem.*

Check that students have written a suitable problem that can be solved.

 Reflection time

Share students' solutions with the class, including their visual representations. Did groups reach the same answer but in different ways? Ask them to share their strategies.

Differentiated outcomes	
All students	should take part in the activity suggesting solutions.
Most students	will understand the concept of three-quarters.
Some students	may confidently explain the process to other students.

7 Fractions

Global skills

- **Creative skills:** problem solving / investigating
- **Interpersonal skills:** communication
- **Self-development skills:** reflecting on learning

Student Book

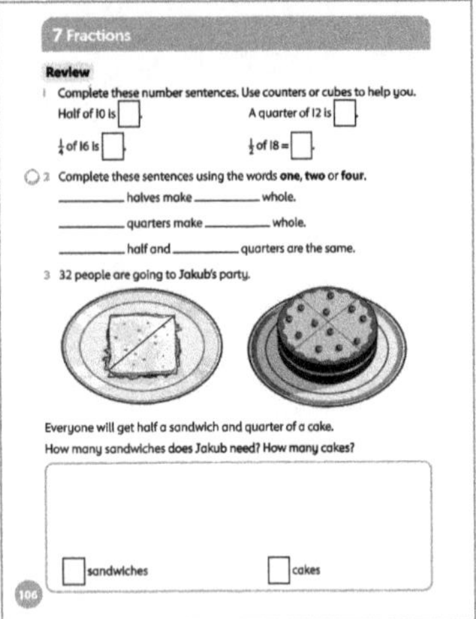

With young children, assessment activities are most effective when carried out as an everyday classroom activity. Students should have paper shapes available to support them.

Watch as students complete the sentences that represent fraction situations involving halves and quarters. Check that students understand the language of fractions with wholes, halves, quarters and equal parts, and that they can write $\frac{1}{2}$ and $\frac{1}{4}$ correctly.

Encourage students to use their paper shapes for the questions. Extend by asking a range of similar questions or asking the students to make up their own problems, for example, *How many quarters are the same as a half?*, *If you eat 3 quarters of a cake, how much is left?*

Answers

Student Book page 106

1 Half of 10 is 5.   A quarter of 12 is 3.

  $\frac{1}{4}$ of 16 is 4. $\frac{1}{2}$ of 18 = 9.

2 Two halves make one whole.
  Four quarters make one whole.
  One half and two quarters are the same.

3 16 sandwiches, 8 cakes

Practice Book

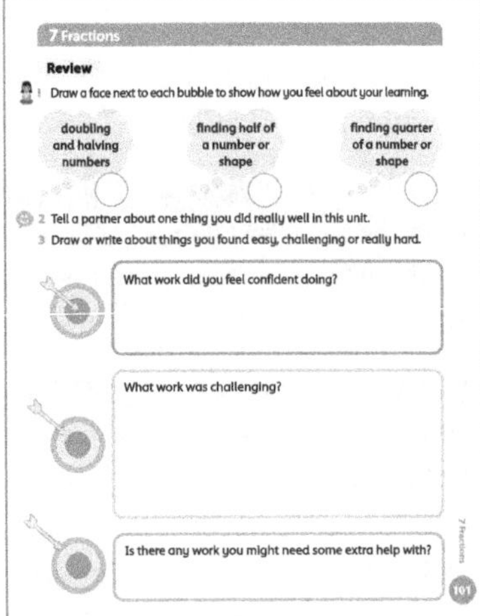

With young children it is appropriate to complete this as a whole-class discussion. You may choose to keep a record of the class discussion or a copy of the review page for your own records. Use the Student Book to briefly remind the students of the areas of mathematics that they have worked on in this unit.

Give out the Student Book to support pairs of students as they discuss and answer the questions in the Practice Book.

Allow students plenty of time for discussion before asking them to share their responses with the rest of the class. If students complete this assessment at home, encourage them to discuss this with adults.

Make a note of areas that students still feel unsure about. You can build fractions into everyday practice. For example, finding half of amounts or splitting the class into halves or quarters.

Additional material

There are additional end-of-unit assessments available on the *Oxford Owl* website.

8 Length, mass and capacity

Overview

Big Idea

Many students will be familiar with some measurements such as the distance to school or measures in food items or cooking. One of the Big Ideas in this unit is that there are standard units for measuring length, capacity and mass (weight): the centimetre and metre, litre and millilitre, kilogram and gram, respectively. These metric units of measure use our base-ten number system and can therefore be added and subtracted in the same way as with numbers.

Students also need to be able to read scales and measuring instruments. Previous experience with non-standard units is essential. Students should explore and compare non-standard and standard units before they move on to consider which unit is appropriate in a given situation. A focus on the units rather than the act of measuring will ensure that students recognise and use the appropriate unit. Estimating lengths (and weights and capacities) before measuring them will help to develop a sense of measure – 'big' does not have to mean heavy, nor does 'small' have to mean light. Students need to develop a feel for the size of standard units such as 1 kilogram and 1 litre to help them use these as a standard that they compare to.

Students also need to understand that the length of something does not change when it is placed in a different position:

- The height of a student is still the same when they lie down.
- A piece of string that looks long when it is laid out straight is the same length when in a spiral.
- Moving an object to a different position does not change its length.

Look out for

- **Students who do not remember that you need to line up the ends of two objects to compare their length.** This includes the object you are measuring and the measuring instrument you are using.
- **Students who do not read a scale accurately because they do not understand what the space between each scale mark represents.** Students need practice at reading the numbered divisions on a variety of scales, including a ruler. Using a ruler as a number line will ensure that students are already familiar with its layout when they begin to use it for measuring.
- **Students who make errors when reading scales.** With liquids, they may bring the container up to their eye rather than putting it down on a flat surface. Demonstrate correct use and ensure lots of practice.

Possible misconceptions

- **Students may think that a taller container will hold more than a shorter one**, even when the shorter container is much wider. Practically measuring, such as finding out how many cupfuls fill different containers can help to correct this misconception.

Key vocabulary

- compare, size, most, least, about the same, estimate, guess
- length, ruler, centimetre
- measure, long, longer, longest, short, shorter, shortest, wide, wider, widest, tall, taller, tallest
- weight, mass, weigh, pan balance scales, heavy, heavier, heaviest, light, lighter, lightest, gram
- capacity, litre, millilitre, full
- holds the most, holds the least, depth, width

Coverage in lessons

Learning focus	Learning outcomes (the ENC objectives)
Length	Compare, describe and solve practical problems for lengths and heights, for example, long/short, longer/shorter, tall/short, double/half.
	Measure and begin to record lengths and heights.
Mass	Compare, describe and solve practical problems for mass/weight, for example, heavy/light, heavier than, lighter than.
	Measure and begin to record mass/weight.
Estimating capacity	Compare, describe and solve practical problems for capacity and volume, for example, full/empty, more than, less than, half, half full, quarter.
	Measure and begin to record capacity and volume.

8 Length, mass and capacity

Engage Student Book page 107

Big question

- How do we measure different things?

Global skills

- **Creative skills:** problem solving / exploring / investigating
- **Real-world skills:** research / presenting information / interpreting information
- **Interpersonal skills:** communication / teamwork

Key vocabulary

- more, less, most, least, tallest, shortest, widest, full, measure, size, compare, about the same, estimate, guess

Resources

- empty containers of different heights and capacities
- water for liquid capacity or dried pules (chick peas)
- bucket or suitable container for measuring
- interlocking cubes

Language support

Listen to how students use the language and vocabulary of measures. Encourage them to use comparative language, for example, 'taller than', 'shorter than', while giving examples from the selection of containers.

Introductory activity

If you have access to an IWB, you could display the Student Book Engage page and discuss the Big idea of the unit with students. Show students a collection of containers or the picture on page 107 in the Student Book. *All of these are different **sizes**. We need to **compare** them to find which holds the **most**. Which do you think will hold the most? Which do you think will hold the **least**? Are there any that you think will hold **about the same** as each other?*

Ask students to look at each one and **estimate** (**guess**) which will hold the most. Get students to talk to their partner. Choose some students to put the containers in order from the one that holds the most to the one that holds the least, putting together any that they think will hold about the same amount.

Does everyone agree with this? Does anyone have a different idea?

Main activity

Give students a selection of containers of different sizes, and a larger bucket for measuring. *How could we find out which container holds the most and which container holds the least?* Suggest that, as the containers are all empty, students could fill them.

Show the water or the dried material you will be using to fill them. *We can pour this into each container until they are **full**.* Choose some students to fill the containers one at a time until they are full. After each one is full, pour its contents into the bucket and mark the level, then empty the bucket and fill it with cubes up to the marked level. Explain that this is a **measure** of the amount the container holds. *This container holds _____ cubes from the bottom to the top.* Keep a record of the number of cubes for each container.

Which container held the most? Which container held the least? Use the results to put the containers in order from the one that holds the least to the one that holds the most.

Differentiation

Supporting: Ask students to use the language of comparison after you have modelled the use of the vocabulary.

Consolidating: Ask students to explain the strategies they are using.

Extending: Ask students to explain why tall thin containers do not have large capacities.

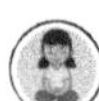
Reflection time

Ask students if they were surprised by what they found out. Ask them to share with the class what surprised them. For example, tall thin containers do not hold as much as some shorter, fatter containers.

8A Length

Specific learning focus

- Estimate, measure and compare lengths, choosing and using suitable uniform non-standard and standard units and appropriate measuring instruments.

Global skills

- **Creative skills:** exploring / investigating
- **Real-world skills:** research / presenting information / interpreting information
- **Interpersonal skills:** communication / teamwork

Key vocabulary

- long, longer, longest, short, shorter, shortest

Resources

- lengths of ribbon, string or similar

Language support

Help students with the vocabulary of comparison of lengths. Emphasise and model the use of long, longer, longest and short, shorter, shortest. Encourage students to make estimates, e.g. *I think the blue ribbon is longer than the red ribbon.*

 ## Introductory activity

Show students a piece of ribbon. Lay it out as straight as you can and ask whether they think it is **long** or **short**. Ask how they decided. Explain that, to describe something as long or short, means that it is long or short compared to other things.

Now show another piece of ribbon and ask, *Is this piece long or short compared to the first one? How can you tell?* Students may suggest that the two ribbons are laid side by side to check. Make sure they understand that one end of each ribbon must be lined up side by side to make the comparison. Encourage responses such as, *The first ribbon is **shorter** than the second ribbon. The second ribbon is **longer** than the first.*

 ## Main activity

Give each pair of students some lengths of different materials, such as ribbons, string, paper strips or lengths of pasta. Say that they are going to compare two pieces and say which is longer and which is shorter. Ask what they need to check before they compare. (That the ends are lined up.) They should record their measurement as a sentence, for example, *The pasta is shorter than the string.*

Now ask them to take three pieces and compare them all. They should be able to line them all up at one end then say which ones are longer or shorter than others.

*Which is the **longest**? Which is the **shortest**? How do you know?*

Ask students to choose new ribbons (more than three) and use them to complete the activity on page 108 in the Student Book.

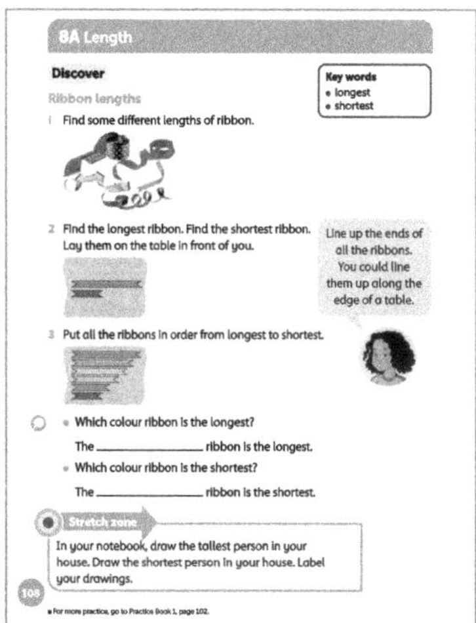

Differentiation

Supporting: Help students to line up the pieces of different lengths so they can see which is longer or shorter.

Consolidating: Ask students to estimate which is longer or shorter before they line them up to compare.

Extending: Encourage students to sort three or more items into length order from shortest to longest.

Stretch zone: *In your notebook, draw the tallest person in your house. Draw the shortest person in your house. Label your drawings.*

Check that students' drawings are correctly labelled.

 ## Reflection time

Ask students to describe some of the things they compared, for example, *My ribbon was longer than my pasta.* Some students will be able to give you two comparisons, such as, *My ribbon was longer than the pasta but shorter than the string.*

Practice Book: Students can complete Practice Book page 102. This can be done directly after the main activity, as homework, or as the focus of a separate mathematics session to help students consolidate their learning and build fluency. Look at the instructions on how to measure accurately and ask students to model accurate and inaccurate measuring for you as in the diagram. Do they see the difference?

Differentiated outcomes	
All students	should describe a length as longer or shorter than another.
Most students	will estimate and find which is longer or shorter of two things by comparing.
Some students	will sort three or more items into order by length.

Student Book page 108

Walk around the classroom while students are working and check that they are measuring accurately. They should place the ends of the objects together at one end and compare the lengths by looking at the other ends of the objects.

Ask students to explain their estimates. Observe whose estimates improve the more items they compare – they are learning to judge length.

Practice Book page 102

Check that the objects chosen are likely to be shorter than 1 m in length, and that students have completed the sentences correctly.

Stretch zone: Make sure students can measure in finger widths and record their measurements correctly.

8A Length

Explore 1 Student Book page 109 • Practice Book page 103

Specific learning focus

- Estimate, measure and compare lengths, choosing and using suitable uniform non-standard units.

Global skills

- **Creative skills:** exploring / investigating
- **Real-world skills:** research / presenting information / interpreting information
- **Interpersonal skills:** communication / teamwork

Key vocabulary

- long, longer, longest, short, shorter, shortest, tall, taller, tallest, measure, length

Resources

- a variety of uniform objects which can be used as non-standard units of measure, e.g. cubes, straws, paperclips

Language support

Draw up a grid of measuring words and talk through how the word changes from when we are talking about a single object, to comparing two objects and when comparing three or more.

Add the grid to the class display.

	Comparing 2	Comparing 3 or more
long	longer	longest
short	shorter	shortest
tall	taller	tallest
thin	thinner	thinnest
thick	thicker	thickest
narrow	narrower	narrowest
wide	wider	widest

 Introductory activity

Write on the board: long, longer, longest, short, shorter, shortest, **tall**, **taller**, **tallest**. Ask students to take it in turns to say a sentence about themselves that includes some of these words. Give a few examples such as, *I am taller than my sister. My arm is shorter than my leg.* Students should look at objects around the room and make up similar sentences using the words.

 Main activity

Ask students to measure some objects in the classroom using their hand span. Explain that when we measure something we find out how long or wide it is. This is its **length**.

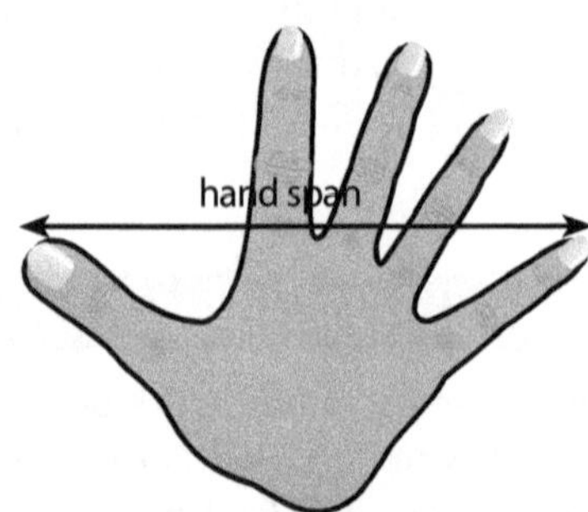

For example ask, *How long is a desk? How wide is the door?* After students have measured some objects, compare different students' measures for one item, for example, the length of a table.

Did you notice that you had different measures for the same object? Why do you think this is?

Talk about why this could be. For example, that different students have different lengths of hand spans, or that having to keep moving your hands to measure makes it hard to measure accurately.

Now ask students to choose an item to use as a unit of measure: cubes, straws or a similar non-standard uniform measure. They should measure the objects with their chosen unit. After they have finished, ask, for example, *How long is the chair?* Different students who measured the chair with cubes by laying them end to end should notice that they have the same answer because they used the same unit to measure, and all the cubes are the same size.

Students should continue the activity in the Student Book page 109, measuring more objects, working in pairs so that they can check and compare each other's measurements.

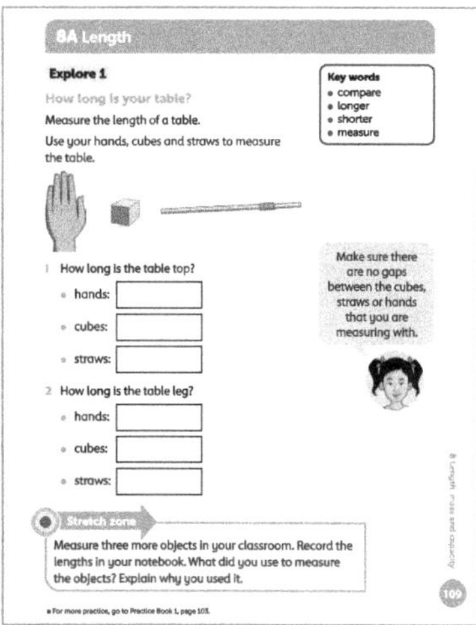

Differentiation

Supporting: Ask students to estimate to the nearest handspan, cube or other unit before they measure.

Consolidating: Ask students to check their measurement, perhaps by starting measuring from the opposite end.

Extending: Ask students to estimate the lengths of a range of objects using a chosen unit. They should then check by measuring.

Stretch zone: *Measure three more objects in your classroom. Record the lengths in your notebook. What did you use to measure the objects? Explain why you used it.*

Check that students have made sensible choices of non-standard units, for example, not choosing paper clips to measure the width of the classroom door.

 **Reflection time**

Remind students that they need to use words ending in '-er' such as shorter, taller, longer when comparing two things, and words ending in '-est' such as shortest, tallest and longest when comparing more than two things. Show students a selection of classroom items (some very close in length) and ask them to compare pairs and then groups of objects. Invite students to measure objects where necessary.

Practice Book: Students can complete Practice Book page 103. This can be done directly after the main activity, as homework, or as the focus of a separate mathematics session to help students consolidate their learning and build fluency. Talk about using a pace (stride) as a unit of measure. *What type of things would you measure with a pace? If two people measure the length of a field in paces will the number of paces be exactly the same? Why or why not?*

Differentiated outcomes	
All students	should estimate how many of the chosen unit are needed to measure the length of an item.
Most students	will estimate and measure accurately with a chosen non-standard measure.
Some students	may estimate and measure accurately using a non-standard uniform measure.

Answers

Student Book page 109

Students measure a table using non-standard units such as hands, cubes and straws. Check that they have reasonable answers for each unit they use.

Practice Book page 103

Students measure different objects using non-standard units such as hand spans and paces. Check that they have reasonable answers for each unit they use.

8A Length

Explore 2   Student Book page 110 • Practice Book page 104

Specific learning focus

- Estimate, measure and compare lengths and heights, choosing and using suitable uniform non-standard units.

Global skills

- **Creative skills:** problem solving/ exploring / investigating
- **Real-world skills:** research / presenting information / interpreting information
- **Interpersonal skills:** communication / teamwork

Key vocabulary

- long, longer, longest, short, shorter, shortest, tall, taller, tallest

Resources

- cubes, various items to compare and measure for length and height

Language support

Continue to use the grid of words associated with lengths and their comparisons. Encourage students to use the correct vocabulary as they estimate and measure.

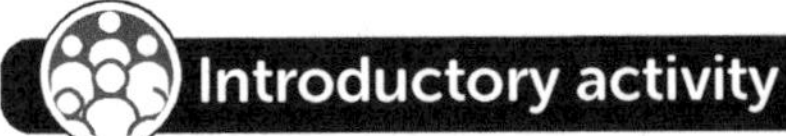 **Introductory activity**

Play a game that gets students using the vocabulary. For example, say, *A giraffe is tall*. The first student might say, *A giraffe is taller than a car*. The next student might say, *A car is taller than a cat. A cat is taller than …* and so on until the class cannot think of anything shorter. Start with other pieces of vocabulary, for example, *A river is long*.

 **Main activity**

Have a range of items on the table of different heights. Pick up two items to compare. Ask students which is the taller and which is the shorter. Ask students to say a sentence to describe this, for example, *The book is taller than the pencil*.

Ask a student to come forward and choose an item to measure using cubes. Ask, *How long is this?* The student measures with the cubes or another non-standard unit. Repeat with other students and different items. Check that they line up the ends of the items so that they have the same start point, then they can see which is longer or shorter.

Students can then complete the activities on page 110 of the Student Book individually. When students have finished, ask them to read through each of their answers in full sentences to practise using the language of comparison.

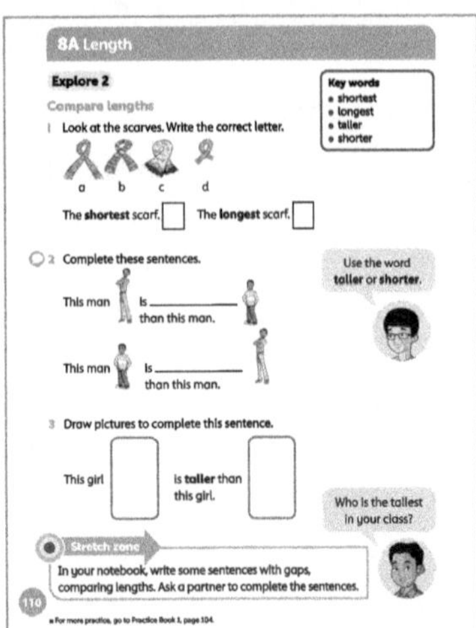

Differentiation

Supporting: Help students to estimate in cubes before they measure. For example, ask them to look at one cube, five cubes and ten cubes to get a sense of the length of each.

Consolidating: Ask students to estimate to the nearest number of cubes before they measure.

Extending: Ask students to estimate the lengths and heights of a range of objects using cubes and straws. They should explain why the answers are different.

Stretch zone: *In your notebook, write some sentences with gaps, comparing lengths. Ask a partner to complete the sentences.*

Check that students' sentences make sense and are appropriate for a length context.

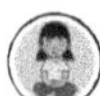 **Reflection time**

Ask students to describe how they know which item is longer/taller and which is shorter when comparing lengths.

Practice Book: Students can complete page 104 of the Practice Book. This can be done directly after the main activity, as homework, or as the focus of a separate mathematics session to help students consolidate their learning and build fluency. Remind students they must use one type of object to measure and that the objects must all be the same size.

Differentiated outcomes	
All students	should compare to find which is longer/shorter.
Most students	will measure to the nearest non-standard unit.
Some students	may find the difference in length in non-standard units.

Answers

Student Book page 110

1   shortest d; longest a

2   taller; shorter

3   Drawings of two girls, the taller girl in the first box.

Practice Book page 104

Check that students have recorded a range of objects and measured their lengths accurately using the non-standard unit of their choice.

Stretch zone: Students should select the longest and shortest from the objects they chose.

8A Length

Explore 3 Student Book page 111 • Practice Book page 105

Specific learning focus

- Estimate, measure and compare lengths using a cm-ruler.

Global skills

- **Creative skills:** exploring / investigating
- **Real-world skills:** research / presenting information / interpreting information
- **Interpersonal skills:** communication / teamwork

Key vocabulary

- centimetre, cm, ruler, long, estimate

Resources

- rulers: large one to display (if possible) and standard rulers for student use (with 0 as the first mark if possible)
- various items to compare and measure for length (e.g. pencils, books, leaves, ribbons).

Language support

Continue to use the grid of words associated with lengths and their comparisons. Encourage students to use the correct vocabulary as they estimate and measure. Explain what an abbreviation is and get them used to recording centimetres as 'cm'. Add this vocabulary to the poster/working wall and add 'ruler'.

 **Introductory activity**

Show the class a ruler marked to 30 centimetres (display on an IWB, if available). *This is a **ruler**. Do you know what we use this for? What do you notice about it?* (It has tick marks and numbers.) *What's the biggest number you can see?* (30) *What's the smallest? What is the first number?* (0) *Is it right at the edge of the ruler?* (No)

Say that the space between each pair of numbers is called 1 **centimetre**. Explain that we often shorten the word 'centimetre' to '**cm**'. Write '1 centimetre' and '1 cm' on the board. *We can find out how many centimetres long, tall or wide something is by placing the object against the ruler and counting along the ruler.*

Use the ruler to measure the length of a pencil. Place one end of the pencil against 0 cm and read off the length from the other end of the pencil against the nearest centimetre marking.

 **Main activity**

Show students a collection of objects, e.g. pencils, books, leaves, ribbons. *We are going to measure the length of the items in centimetres.* Show the 30-centimetre ruler and remind students about lining up the end of each object with the 0 cm mark.

Take one object, such as a book, and ask students to estimate how many centimetres long they think it is. Compare it with the ruler and ask whether the book will be less than 30 centimetres, or more than 10 centimetres and try to get a reasonable estimate. Model how to measure the book and record the length.

Repeat with one other object. Estimate first, then measure and record.

Give students their own rulers and a small selection of objects to measure. Ask them to work in small groups or pairs to measure items, supporting each other to estimate first, then measure accurately and then compare results.

Students can then complete the activities on page 111 of the Student Book individually. Before they begin, read the speech bubble together. *What does 'to the nearest centimetre' mean?* Make sure students understand that this means the centimetre mark on the ruler that the length of the object is closest to. Model on the IWB how to measure to the nearest centimetre, to reinforce if necessary.

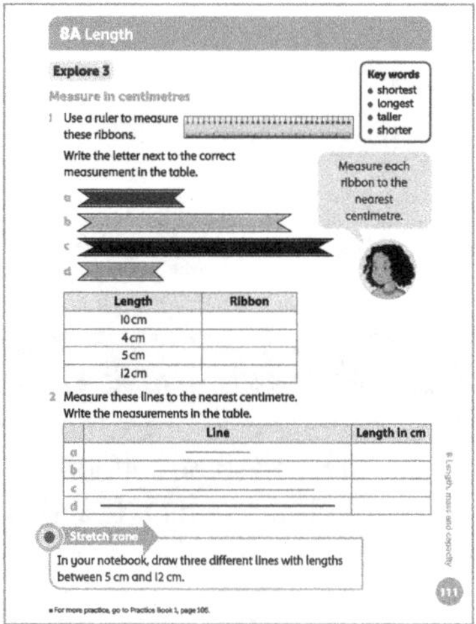

Differentiation

Supporting: Help students to line up the ruler accurately, with the '0' rather than the end of the ruler, before they measure. Help them 'read' the length on the ruler.

Consolidating: Ask students to estimate to the nearest centimetre before they measure.

Extending: Ask students to estimate and measure the lengths of a range of objects. They should then check by measuring.

Stretch zone: *In your notebook, draw three different lines with lengths between 5 cm and 12 cm.*

Check that students have drawn lines with lengths in the required range.

 ## Reflection time

Ask students how good their estimates were. *If you were measuring something that is 15 cm, would 5 cm be a good estimate? What is the longest thing you measured? Was your estimate a good one?*

How could you measure something longer than 30 cm, such as your height? Accept any reasonable answer here. Some students may suggest larger measuring tools that they are familiar with in an everyday context, e.g. metre sticks or measuring tape; others may suggest non-standard measures such as large building blocks or strips of paper.

Practice Book: Students can then complete Practice Book page 105. This can be done directly after the main activity, as homework, or as the focus of a separate mathematics session to help students consolidate their learning and build fluency. Tell students that if they do not have an actual ruler, they can use the one on the page to measure the object they find in their home. If they are unsure of the object's name, they can ask an adult or draw a picture of it.

Differentiated outcomes	
All students	should measure reasonably accurately with a ruler to the nearest centimetre.
Most students	will estimate and measure reasonably accurately with a ruler to the nearest centimetre.
Some students	may estimate and measure accurately with a ruler to the nearest centimetre.

Answers

Student Book page 111

1 10 cm = ribbon b   5 cm = ribbon a
4 cm = ribbon d   12 cm = ribbon c

2 Check that the students have measured each line accurately to the nearest centimetre.

Practice Book page 105

1 pencil: 15 cm   **4** screw: 4 cm

2 straw: 12 cm   **5** cube: 1 cm

3 pen: 10 cm

8B Mass

Discover Student Book page 112 • Practice Book page 106

Specific learning focus

- Estimate, measure and compare the mass of objects, choosing and using suitable uniform non-standard and standard units and appropriate measuring instruments.

Global skills

- **Creative skills:** exploring / investigating
- **Real-world skills:** research / presenting information / interpreting information
- **Interpersonal skills:** communication / teamwork

Key vocabulary

- heavy, heavier, heaviest, light, lighter, lightest

Resources

- collection of objects in different sizes and masses, e.g. toys, packets and tins of food.
- variety of boxes of different shapes and sizes (enough for each student to have access to at least five different boxes)

Language support

Continue to use the grid of words associated with measures and their comparisons. Add word associated with mass to the grid, such as heavy, heavier, heaviest, light, lighter and lightest.

Encourage students to use the correct vocabulary as they estimate and measure.

 ## Introductory activity

Set up what **heavy** and **light** mean with actions that are exaggerated. For example, try to lift your desk huffing and puffing and say, *The desk is heavy.* Then lift a pencil and say, *The pencil is light.* Write these phrases on the board, underlining the key vocabulary.

Show students a set of four or five packets of different size, shape and mass, for example, a cereal box, box of chocolates, tin of beans. Ask a student to come forward and choose two of the packets. They should pick each up in turn and then try to say which is **heavier** and which is **lighter**. If needed, they can pick one up in each hand to feel their masses simultaneously.

Repeat with two more students choosing different pairs of packets from the set.

Is a larger packet always heavier? Why? Why not? For example, compare a tin of beans with a large bag of popcorn.

Main activity

Give students sets of objects of different mass and ask them to work in pairs to sort the objects from **lightest** to **heaviest**. Suggest that they keep comparing different pairs of objects to get the correct order.

Students should repeat the activity, again working in pairs, using a set of five different-sized boxes, and record their findings on page 112 in the Student Book.

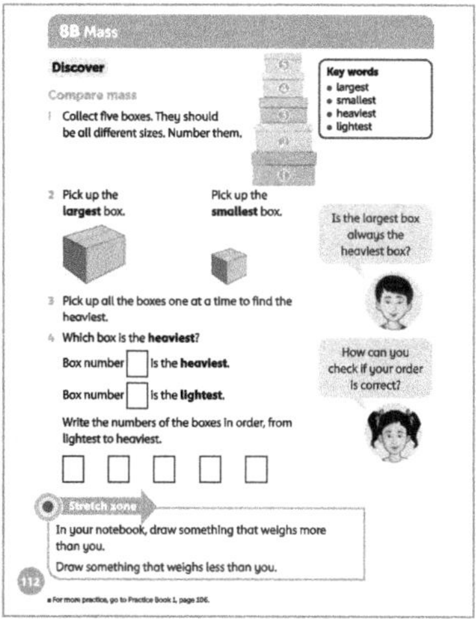

Differentiation

Supporting: Ask students to estimate and describe the objects as heavy or light. Model comparison sentences:

The _________ is heavier than the _________.
The _________ is lighter than the _________.

Consolidating: Ask students to compare by using the language of heavier and lighter.

Extending: Ask students to put the whole set of objects in order by making repeated comparisons of two objects at a time.

Stretch zone: *In your notebook, draw something that weighs more than you. Draw something that weighs less than you.*

Check that students have drawn a suitable item heavier and lighter than themselves. Ask them to explain how they know.

 Reflection time

Discuss with students the order of the boxes as they put them from lightest to heaviest. *How did they know they had the correct order at the end?*

Practice Book: Students should complete the activities on page 106 of the Practice Book. This can be done directly after the main activity, as homework, or as the focus of a separate mathematics session to help students consolidate their learning and build fluency. You can suggest that some students choose items that are quite different in mass to make the comparison easier.

Differentiated outcomes	
All students	should compare two objects to see which is heavier.
Most students	will compare and order two or more objects by mass.
Some students	may order similar masses from lightest to heaviest.

Answers

Student Book page 112

Check that students have identified the heaviest and lightest boxes correctly. Check that they are estimating the mass and not the size of the box by asking, *Which is the biggest? Which is the heaviest? How did you know?*

Practice Book page 106

Check that students have ordered their objects correctly.

Stretch zone: Students should have chosen items that are clearly heavier or lighter than they are, and also one item of about the same mass.

8B Mass

Explore 1 Student Book page 113 • Practice Book page 107

Specific learning focus

- Estimate, measure and compare the mass of objects, choosing and using suitable uniform non-standard and appropriate measuring instruments.

Global skills

- **Creative skills:** exploring / investigating
- **Real-world skills:** research / presenting information / interpreting information
- **Interpersonal skills:** communication / teamwork

Key vocabulary

- largest, smallest, heaviest, lightest, pan balance scales, weighs, mass, weight

Resources

- collection of similar objects in different sizes and masses, e.g. toys, boxes, packets of food, water bottles, books, pencils, scissors, jugs, apples
- cubes and marbles, or similar, for using as weights
- sets of pan balance scales

Language support

Continue to use the grid of words associated with measures and their comparisons. Encourage students to use the correct vocabulary as they estimate and measure. Focus on the word 'weighs' and link with a pan balance image and questions about mass, for example, *How heavy? What does it weigh? What is the mass of _______?*

 Introductory activity

Show students a set of **pan balance scales**.

Say, for example, *I want to know how heavy this book is. Another way to say this is that I want to know how much this book **weighs**. I will measure its **mass** in cubes. We need to put the book in one pan and cubes in the other.* Demonstrate how the pans move up and down, and balance. *When there are enough cubes to weigh the same as the book, the pans will be level (balance).*

Invite a student to put the book in one pan. *We put cubes in the other pan.*

How many cubes do you think will we need to make the pans balance? Make an estimate as a class and then have a student add one at a time until the pans balance.

How close were we to our estimate?

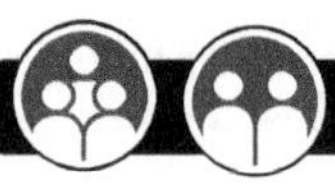 Main activity

Ask students to work in small groups. Give each group a set of objects of different masses and a set of pan balance scales. Ask students to estimate the mass of each object first in cubes and then weigh the object on the pan balance scales by placing cubes in the other pan until the pans are level.

Students should repeat the activity to find the masses of the objects listed on page 113 of the Student Book, working in groups and recording their answers individually.

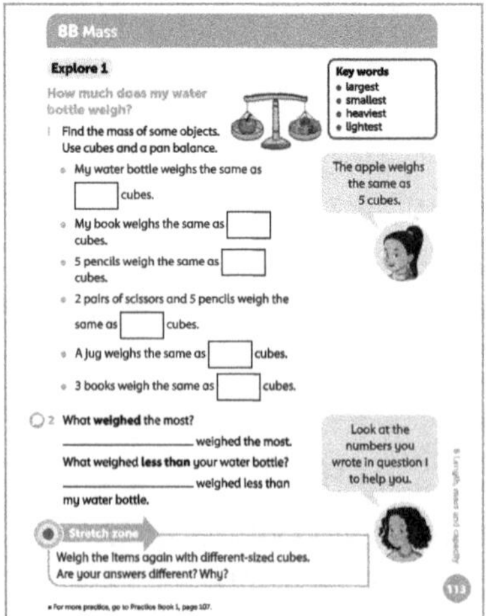

Differentiation

Supporting: Help students to estimate mass by giving them one cube, five cubes, ten cubes to hold to get a feel for the mass of each.

Consolidating: Ask students to estimate to the nearest number of cubes before they measure.

Extending: Ask students to estimate the mass of objects using cubes and marbles, and to explain why the answers are different.

Stretch zone: *Weigh the items again with different-sized cubes. Are your answers different? Why?*

Check that students understand that using a heavier unit to measure with will give a smaller mass for each object.

Which is lighter, a cube or a marble? How many of each did one of your objects weigh? Why was one measurement more than the other?

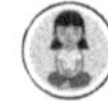 Reflection time

Ask students to describe how they estimated the mass of objects in cubes and then checked using the pan balance. Were they able to balance every object exactly with a number of cubes?

*Did you think about the **weight** of other objects to help you? Why not?*

Practice Book: Students can now complete Practice Book page 107. This can be done directly after the main activity, as homework, or as the focus of a separate mathematics session to help students consolidate their learning and build fluency. This activity requires students to measure with digital scales in grams. You should consider whether this activity is suitable for your students to do at home, and whether you would prefer them to complete it after you have finished the Explore 2 lesson.

Differentiated outcomes	
All students	should weigh objects using cubes.
Most students	will estimate and measure reasonably accurately with scales to the nearest number of cubes.
Some students	may estimate and measure accurately using uniform, non-standard units.

8B Mass

Explore 2   Student Book page 114 • Practice Book page 108

Specific learning focus

- Estimate, measure and compare the mass of objects, choosing and using suitable uniform non-standard and standard units and appropriate measuring instruments.

Global skills

- **Creative skills:** exploring / investigating
- **Real-world skills:** research / presenting information / interpreting information
- **Interpersonal skills:** communication / teamwork

Key vocabulary

- largest, smallest, heaviest, lightest

Resources

- collection of similar objects in different sizes and masses, for example, boxes, toys, packets of food and so on
- sets of gram weights: 10 g, 20 g, 50 g, 100 g, 200 g (one set per group)
- sets of pan balance scales (one per group)

Language support

Continue to use the grid of words associated with measures and their comparisons. Encourage students to use the correct vocabulary as they estimate and measure.

Answers

Student Book page 113

Check that students have recorded the weight of their items against a number of cubes with some accuracy. You should see that they have identified the heaviest object, and also something that weighs less than their water bottle.

Practice Book page 107

Check that students have written some reasonable statements about the weights of objects that weigh about 100 g.

Stretch zone: Check that the students' responses are reasonable.

 **Introductory activity**

Show students a set of gram weights – 10 g, 20 g, 50 g, 100 g, 200 g. Explain that these are used to measure mass (or weight) in **grams**. Pass these around so that students can get a feel for what each one feels like, especially the lightest and the heaviest ones. Have objects ready that match each weight and a couple that are multiples of 10 g (e.g. 40 g). Show how to weigh with one of these objects. Ask students to use this object as a benchmark to estimate which of the objects has a mass of 10 g, 20 g, 100 g etc. Then they come to the front of the class to weigh the object using the scales to see if they are correct.

 **Main activity**

Ask students to work in groups. Give each group a set of objects of different mass, each one being close to 10 g, 20 g, 50 g, 100 g or 200 g, and a set of pan balance scales. Ask students to estimate the mass of each object first, in grams, and then weigh the object on the pan balance scales by placing weights in the other pan until the pans are level.

Students can then choose a weight from the set and place it in one pan.

Can you find an object that weighs the same as that?

Students should continue to work in their groups on the activities on page 114 of the Student Book. Encourage students to think about the masses of the objects they have already weighed to help them choose five objects that would be suitable to weigh with the pan balance scales and weights.

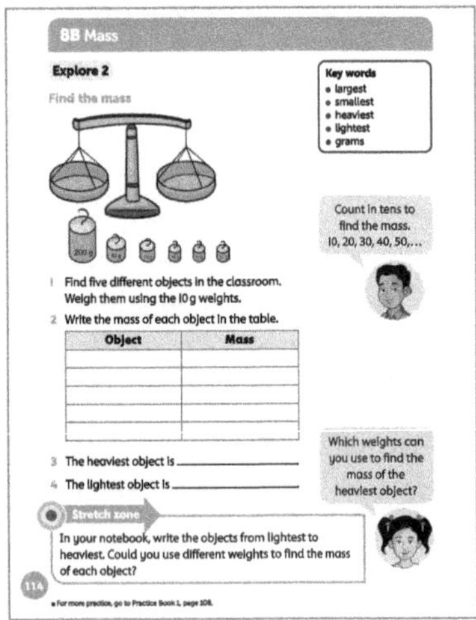

Differentiation

Supporting: Ask students to estimate which object is the lightest or heaviest.

Consolidating: Ask students to estimate to the nearest 50 g before they measure.

Extending: Ask students to estimate the mass of a range of objects in grams to the nearest 10 g. They should then check by measuring using the set of weights.

Stretch zone: *In your notebook, write the objects from lightest to heaviest. Could you use different weights to find the mass of each object?*

Check that students have ordered the objects correctly. Look to see if they have suggested weighing the objects in different gram weights.

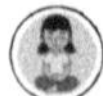 **Reflection time**

Ask students to describe how they estimated the mass of objects in grams and then checked using the pan balance. Were they able to balance every object exactly with the 10 g weights?

Practice Book: Students can now complete Practice Book page 108. This can be done directly after the main activity, as homework, or as the focus of a separate mathematics session to help students consolidate their learning and build fluency. Both this activity and the activity on page 107 can be adapted so that students continue to work with pan balance scales if you feel it is more appropriate for your setting.

Differentiated outcomes	
All students	should measure with scales to the nearest weight available.
Most students	will estimate and measure reasonably accurately with scales to the nearest 50 g.
Some students	may estimate and measure accurately with scales to the nearest 10 g.

Answers

Student Book page 114

Check that students have weighed the mass of their objects accurately and recorded the mass. They should have identified the heaviest and lightest items.

Practice Book page 108

Check that the students have weighed the mass of their objects accurately and recorded the mass. They should have identified the heaviest and lightest items and written their masses in grams.

Stretch zone: Make sure the three items all weigh less than 50 g.

8C Estimating capacity

Discover   Student Book page 115 • Practice Book page 109

Specific learning focus

- Estimate and compare capacities by direct comparison, then by using uniform non-standard units.

Global skills

- **Creative skills:** problem solving / exploring / investigating
- **Interpersonal skills:** communication / teamwork

Key vocabulary

- capacity, holds the most, holds the least, how many, less, more, width, depth, deep, wide

Resources

- interlocking cubes, containers of different shapes and sizes, including a cup as the smallest (six containers per group, labelled 1–6 in random order)

Language support

Make prompts cards which use the key vocabulary as well as corresponding illustration cards. Ask students to match the word prompts and pictures. Make posters that display the vocabulary for the lesson.

 **Introductory activity**

Organise students into pairs. Tell them, *Look at the containers on the table. I want to fill them with cubes. Which container will **hold the most** cubes?* Give a few minutes for discussion in pairs and then take feedback. *Which do you think would hold the most cubes? Why?* Some students may think the tallest one will, because it is the tallest.

Others may look at other qualities of the containers such as **width** and **depth**. You will need to introduce these words as part of a practical demonstration. Ask different students/pairs for their thoughts.

Choose a container. *Who estimates that this one will hold the most?* Then choose a different container. *Who thinks this one will hold the most? How could we find out?*

Wait for a response, e.g. 'We could fill each of them with cubes and then count the number of cubes to see which holds most.'

*The amount that a container can hold is called its **capacity**. A container that holds more has a larger capacity.*

Which container had the largest capacity?

 **Main activity**

Students work in pairs. Put six containers of different shapes and sizes on each table, along with a large number of cubes. *On your table, you have some containers and some cubes. Some containers are tall and thin, some are short and fat (or wide). Work with your partner to find which has the largest capacity (holds the most cubes) and which container has the smallest capacity (**holds the least** cubes). You will need to fill all the containers with cubes Then arrange them in a row from the highest capacity to the lowest.*

Ask students to make sure they don't put in too many cubes, or too few. Each container must be full. Demonstrate the meaning of 'too few' and 'too many' by over-filling and under-filling a container. Also demonstrate how to tip out the cubes from each container one at a time and then make the cubes into a long 'train' to find the capacity of a container. Students compare the lengths of the cube trains to order the containers.

Students should use page 115 of Student Book to record their findings. (Water can be used instead of cubes, if suitable.)

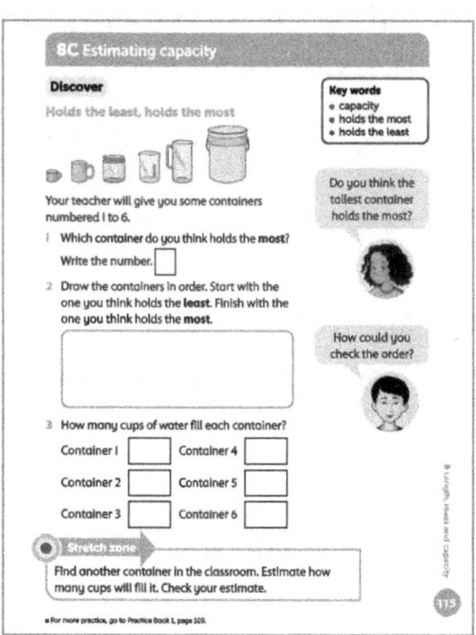

Differentiation

Supporting: Encourage students to check the measurements they are making. *Which holds the most? Which holds the least?*

Consolidating: Ask students to estimate capacities using their experience of previous measurements.

Extending: Ask students to use different-sized cubes to compare and explain their results.

Stretch zone: *Find another container in the classroom. Estimate how many cups will fill it. Check your estimate.*

Check that students have made reasonable estimates and help them see how to be more accurate if their estimates are very inaccurate. For example, show them how much space is taken up by one cupful, 5 cupfuls and 10 cupfuls to give them a sense of the different capacities.

Reflection time

Ask students to verbally complete some sentences such as:

This container holds fewer cube than this one. So it has a (greater/smaller) capacity.

If a container holds fewer cubes than another it means that the capacity is … . (smaller)

If two containers have the same number of cubes it means … ? (the capacities are the same)

Practice Book: Students can complete Practice Book page 109. This can be done directly after the main activity, as homework, or as the focus of a separate mathematics session to help students consolidate their learning and build fluency. If students are doing this activity at home, tell them to ask an adult to help them to set it up and choose appropriate containers, measuring cup and so on.

Differentiated outcomes	
All students	should make estimates about which container will hold the most.
Most students	will articulate the reasons for their choices.
Some students	may understand that two containers may have different shapes but have the same capacity.

Answers

Student Book page 115

Observe students as they work through the activities. Listen to the language they are using when they compare the containers. Note who can explain why one container must have a greater capacity than another and justify their choice for which ones hold the most and least.

Practice Book page 109

Ensure that students estimate before they count the chickpeas. Note whose estimates improve as they move on to different containers – these students are using their previous results to help them make new estimates.

Stretch zone: Make sure students use the results of the activity to check that their chosen container has a smaller capacity.

8C Estimating capacity

Explore 1 Student Book page 116 • Practice Book page 110

Specific learning focus

- Estimate and compare capacities by direct comparison, then by using uniform non-standard units.

Global skills

- **Creative skills:** problem solving / exploring / investigating
- **Real-world skills:** research / presenting information
- **Interpersonal skills:** communication / teamwork

Key vocabulary

- capacity, holds most, holds least, full, empty

Resources

- containers of different shapes and sizes, including the ones from the previous lesson and some new ones
- large bottle of water and some paper cups
- cubes

Language support

Model key vocabulary such as *full, empty, holds* and *container* in practical sessions. Use the words in other contexts such as, *Your lunch box is full* or *The vase is empty*.

 **Introductory activity**

Ask the class, *How many things can you hold in your hand?*

Show students the cubes. Pick up a small number but don't count them. *I think I am holding 12 cubes. Is that correct? Do I know how many? No, that's just what I think, it's my idea. I can count them.* Count them out loud.

Was my idea, my estimate, a good one? Estimate how many cubes you can hold in your hand. Choose some students to tell you their estimates. Ask them to come and fill their hand with cubes. Count the cubes.

Ask each student, *How close to your estimate were you? You estimated and you counted. Was your estimate close? Was it bigger or smaller than the number of cubes?*

 **Main activity**

Show students a large bottle of water and some cups.

How many cups do you think I can fill using the water from this bottle?

Take some suggestions and ask students to say why they chose that number. Pour out water from the bottle to fill one cup, so they can see how much is left in the bottle. Do they want to change their estimate? Continue to fill cups from the bottle until the bottle is empty. You can repeat this using a jug of water which has either greater or smaller capacity than the bottle.

Ask students to work through the activities on page 116 of the Student Book with a partner. They should write what they find out and then explore the capacity of other containers. Remind students to estimate their capacity before they measure.

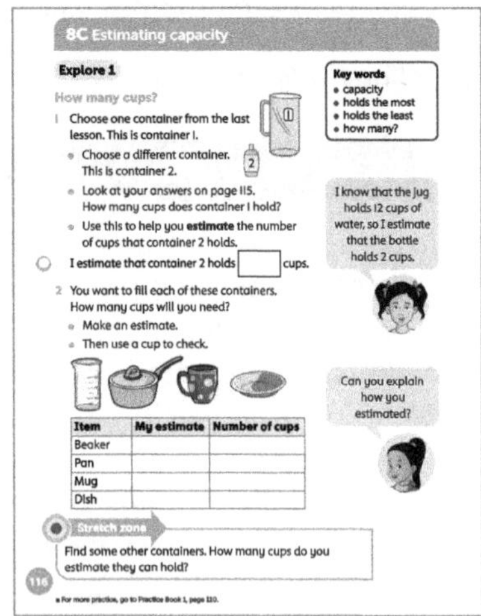

Differentiation

Supporting: Help students to describe the relative capacities of the different containers. *Which holds the most? Which holds the least? How do you know?*

Consolidating: Ask students to estimate capacities using their previous experience of estimating measurements.

Extending: Ask students to predict what the results would be if they repeated the activities with different-capacity containers, and then carry out the activities.

Stretch zone: *Find some other containers. How many cups do you estimate they can hold?*

Check that students have made reasonable estimates and help them see how to be more accurate if their estimates are very inaccurate.

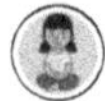 **Reflection time**

Ask pairs of students to work together to explain to the rest of the class what they found out. Choose two students for each activity in the Student Book. *Did all of you find the same thing? Did anyone have a different answer? Why do you think this was?*

Revise the meaning of 'capacity' and discuss how different-shaped containers have different capacities.

Practice Book: Students can complete Practice Book page 110. This can be done directly after the main activity, as homework, or as the focus of a separate mathematics session to help students consolidate their learning and build fluency. Tell students that if they are unsure of the names of the objects they choose they can ask an adult for help or draw a picture of the object.

Differentiated outcomes	
All students	should make estimates about which container will hold the most.
Most students	will make accurate estimates of how many of a smaller container can be filled from a larger one.
Some students	may link their estimates for different containers, for example, if four cups fill a jug and three jugs fill a bottle, how many cups fill a bottle?

8C Estimating capacity

Explore 2 Student Book page 117 • Practice Book page 111

Specific learning focus

- Estimate and compare capacities by direct comparison, then by using uniform non-standard units.

Global skills

- **Creative skills:** problem solving / exploring / investigating
- **Real-world skills:** research / presenting information
- **Interpersonal skills:** communication / teamwork

Key vocabulary

- capacity, holds most, holds least, how many, millilitres (ml), litre

Resources

- containers of different shapes and sizes, measuring jugs in millilitres (one per pair), large bucket of water for filling containers

Language support

Make prompts which contain the important words and pictures to illustrate them. Cut them into cards and ask students to match words and pictures. Make posters that display the vocabulary for the lesson. Focus on the relationship between millilitres and ml.

 **Introductory activity**

Organise students into pairs. Tell them, *Look at the containers on the table. I want to fill them with water. Which container holds the most water?* Give a few minutes for pair discussion and then take feedback. Ask students to explain their choice.

Student Book page 116

Observe students while they are estimating and ask them to explain how they estimated how many cups a new container would hold based on a previous estimate for another container. Check their estimates for accuracy.

Practice Book page 110

Ask students to predict whether they will be able to hold more or fewer large objects than small objects. (More small objects)

Some students may think the tallest one will hold most because it is the tallest. Ask different students/pairs for their thoughts. Choose a container and fill it with water. Now pour the water from the container into a measuring jug. Show students how to read the scale to see how many **millilitres** of water are in the container. Explain that we usually measure capacity of containers in millilitres and the short way to write it is **ml**. Repeat this with another container.

Main activity

Ask students to work in pairs. *On your table, you have some containers. Work with your partner to find which container has the largest capacity and which has the smallest capacity. Fill all the containers with water and then use the measuring jug to find out how many millilitres of water each holds. Each container must be full.*

Demonstrate the meaning of 'too little' and 'too much' by over-filling and under-filling a container.

Students should use page 117 in the Student Book to record their findings.

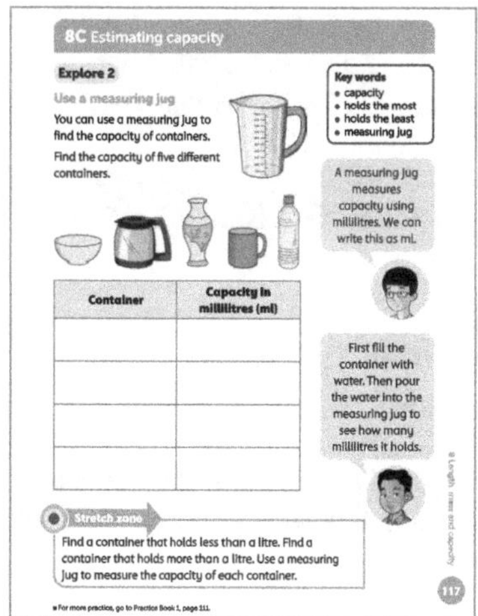

Differentiation

Supporting: Help students to measure how much water a container holds: help them to fill a container, pour the water into a measuring jug and then read how many ml.

Consolidating: Ask students to estimate capacities using their previous experience of estimating.

Extending: Ask students to sort different-sized containers according to their capacity.

Stretch zone: *Find a container that holds less than a* **litre**. *Find a container that holds more than a litre. Use a measuring jug to measure the capacity of each container.*

Check that students have measured the capacity carefully.

 **Reflection time**

Ask students to verbally complete some sentences such as, *This container holds more millilitres of water than this one. So it has a _________ capacity.* (larger)

If a container holds fewer millilitres of water than another it means that the capacity is ______. (smaller)

If two containers hold the same number of millilitres of water it means _______. (the capacities are the same)

Practice Book: Students can complete Practice Book page 111. This can be done directly after the main activity, as homework, or as the focus of a separate mathematics session to help students consolidate their learning and build fluency. Students will need access to a measuring cup with a millilitre scale to complete this activity.

Differentiated outcomes	
All students	should make estimates about how much water each container can hold.
Most students	will make accurate measures of how many ml of water a container can hold.
Some students	may sort a set of containers into order of capacity recorded in ml.

Student Book page 117

Observe students as they work through the activities. Listen to the language they are using when they compare the containers and the amount of water each holds.

Practice Book page 111

Ensure that students estimate before they fill the containers with water. Observe their estimating and note whose estimates improve as they move on to different containers — these students are using their previous results to help them make new estimates.

Stretch zone: Suggest that students use the results of the activity to try to find a container with a capacity of 500 ml.

8 Length, mass and capacity

Connect   Student Book page 118

Big idea

- To measure length, mass and capacity, we can use non-standard units, such as hands, cubes and cups. We can also use standard units such as centimetres, grams and millilitres.

Global skills

- **Creative skills:** problem solving / exploring / investigating
- **Real-world skills:** research
- **Interpersonal skills:** communication / teamwork

Key vocabulary

- taller, shorter, longer, wider

Resources

- toy animals (one per small group)
- pictures of different shelters suitable for small animals (from books or the internet)
- large sheets of paper or card, metre rules, rulers, cubes, cardboard boxes

Language support

Working in mixed-attainment groups will help students hear good models of English. Ask questions as the students are making the models, e.g. How long is the arm? How wide is the tall person? Is the little person shorter than you?

 **Introductory activity**

Tell students that an animal centre is planning to make shelters for their smallest animals (e.g. guinea pig, hamster, hedgehog, squirrel). Show students pictures of what these animal shelters might look like. *Can you give the centre an idea of how big the shelters may need to be? What type of water or food dish might the animals need? How much would these dishes hold?* Focus on using language of comparison: *Which animal is heavier/taller/wider?* Together, discuss or find out some facts relating to small animals that would help with decisions about the size of the shelter for example, *How much water does a dog drink every day compared to a mouse?*

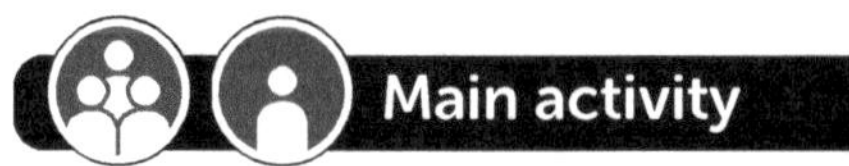

Ask students to work in small groups and decide for themselves how they are going to design and draw a shelter for one of the small animals they have found out about. Encourage them to think about the size of the shelter – height, length, width – so that the animal has enough room to move comfortably. Think about what it will have inside, such as feeding and drinking bowls, an area to sleep. Have a selection of rulers, metre rules and cubes available for students to measure with.

Students should complete the activity in the Student Book on page 118. Tell students that the toy animal is just a stand-in for a real animal.

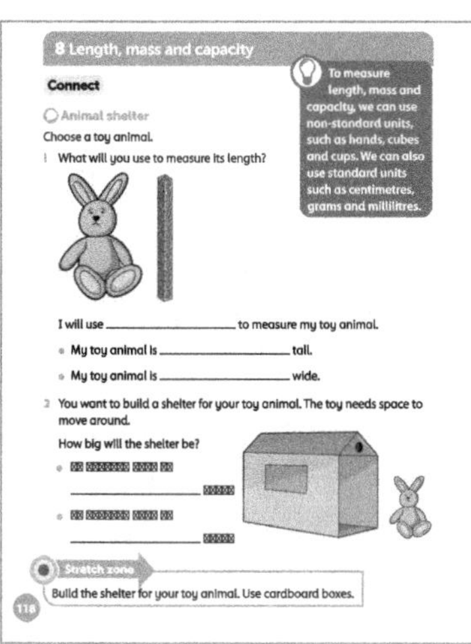

Differentiation

Supporting: Help students decide which shelter they will make, direct them towards choosing suitable dimensions and use direct comparisons.

Consolidating: Encourage students to explain the strategies they are using and use non-standard units.

Extending: Encourage students to measure some objects using standard units.

Stretch zone: *Build the shelter for your toy animal. Use cardboard boxes.*

Students should use the measurements in the activity and create the ideal home for their toy animal.

Student Book page 118

Observe students as they work and note who is measuring confidently, with a clear idea of how long or tall the shelter should be.

 **Reflection time**

Ask groups to describe the process they used to calculate the relative sizes of their animal shelter. Use the language of comparison throughout, particularly 'bigger than' and 'smaller than'.

You should also focus on the key vocabulary of length, e.g. length, width, long, short, tall, high, low, wide, narrow, deep, shallow, longer, shorter, taller, longest, shortest, tallest and highest.

Differentiated outcomes	
All students	should make direct comparisons.
Most students	will use the language of comparison with confidence.
Some students	may use standard units to support their animal shelter design.

Review   Student Book page 119 • Practice Book page 112

Global skills

- **Creative skills:** problem solving
- **Interpersonal skills:** communication
- **Self-development skills:** reflecting on learning

Student Book

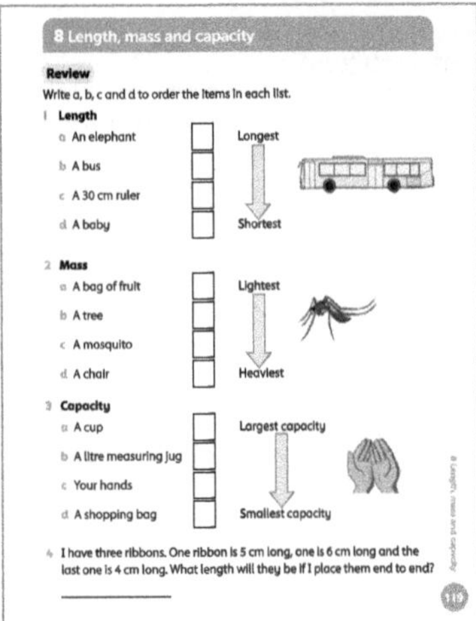

With young children, assessment activities are most effective when carried out as an everyday classroom activity. Students could have key vocabulary lists available so they can refer to these to support them.

Watch as students decide on the order of measurements of the separate lists for length, mass and capacity.

Answers

Student Book page 119

1 bus, elephant, baby, ruler

2 mosquito, bag of fruit, chair, tree

3 shopping bag, jug, hands, cup

4 15 cm

Practice Book

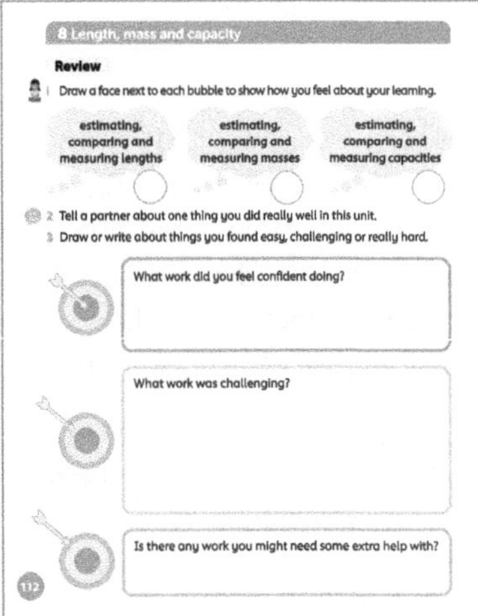

With young children it is appropriate to complete this as a whole-class discussion. You may choose to keep a record of the class discussion or a copy of the review page for your own records. Use the Student Book to briefly remind students of the areas of mathematics that they have worked on in this unit.

Give out the Student Book to support pairs of students as they discuss and answer the questions in the Practice Book.

Allow students plenty of time for discussion before asking them to share their responses with the rest of the class. If students complete this assessment at home, encourage them to discuss this with adults.

Make a note of areas that students still feel unsure about. You can build measures into everyday practice, estimating and measuring lengths, masses and capacities as opportunities arise.

Additional material

There are additional end-of-unit assessments available on the *Oxford Owl* website.

9 Money

Overview

Big Idea

To students, money often has no connection to measuring – it is used only for buying. They pay with notes and coins for what they want, giving the number of notes or coins for the price asked for, for example, £5.50, £3.25, 85p. The idea that money measures the exchange value of goods is a very advanced idea. Students only know that a price is attached to goods and they know what notes or coins to offer. This means we must treat money differently from the properties of length, mass and capacity which have been covered earlier.

Instead of choosing a unit which gives us a measure by repeating along a line, we must accept that a number of different notes and coins are already familiar to students and they must learn the relationship between the different denominations. In money, we assume the values of the coins as given, and learn the relationships between the different notes and coins.

Cashless shopping is reducing students' opportunities to observe money being handled, so it is crucially important that teachers introduce students to 'real' money in the classroom so that students can learn the currency.

The three key ideas needed for students to learn about and understand money are coin recognition, equivalence and practical situations.

Look out for

- **Students who do not understand that change needs to be given and might wonder why a shopkeeper is giving money back**. Role-play paying for items in a shop to help explain this.
- **Students who do not know how to exchange one coin for two coins that have the same total value,** for example, 10p = 5p + 5p.

Possible misconceptions

- **Students may not understand the fact that two coins have different values.**
- **Students may think, the bigger the coin, the greater the value.**
- **Students may think that more coins means a greater value,** for example, that five 10p coins have greater value than two £1 coins.

Key vocabulary

- coin, notes, currency, total, how much, less than, greater than, stamp, postage, value
- cent, dollar, 1-cent coin, 5-cent coin

Coverage in lessons

Learning focus	Learning outcomes (the ENC objectives)
Money amounts	Recognise and know the value of different denominations of coins and notes.
Notes and coins	Recognise and know the value of different denominations of coins and notes.

9 Money

Engage Student Book page 120

Big question

- What different notes and coins do we use?

Global skills

- **Creative skills:** exploring / investigating
- **Real-world skills:** research / financial literacy
- **Interpersonal skills:** communication / teamwork

Key vocabulary

- coin, note, currency, dollar, cents, value, price, cost

Resources

- notes and coins

Language support

Students should be shown a list of names and values of coins relating to their own currency system. For example, in the UK, they need the words 'pounds' and 'pence', while in the USA they need 'dollars', 'cents', 'nickels', 'dimes', 'quarters' and so on.

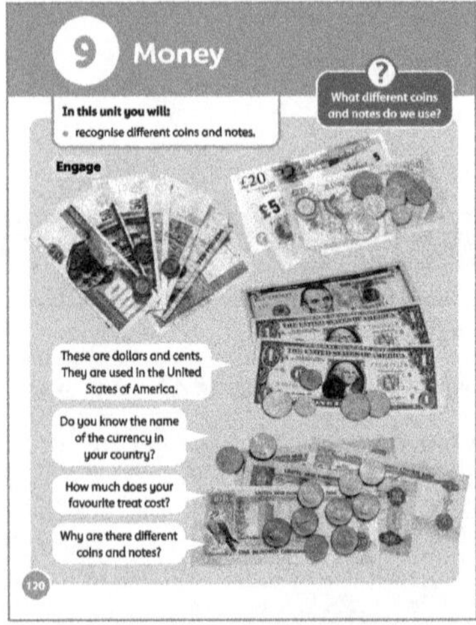

Introductory activity

Ask students what they can tell you about money. For example, they might tell you some of the **coins** they know, and what colour and size they are. Ask if they know what money is used for: for example, they might recognise that it is used to buy things in shops. *Do you know what the **notes** and coins are called in this country?* Mention that, in some countries, pieces of paper money are called bills, rather than notes. *Do you know any names of coins in other countries?* Write on the board a list of the names of the different denominations and have examples of the main coin and note values to pass round. Display page 120 in the Student Book on the IWB, if possible, to show and discuss what coins and notes look

like in various **currencies**. *What are the coins made of do you think? What colours are they?*

What shape are most of the coins? (round)

Can you see any that are not round? How many sides do they have?

What numbers can you see on these notes? What do you think they tell us?

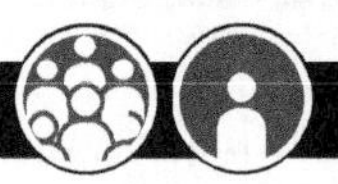 ### Main activity

Ask students to think about things they buy, such as food or drink, or comics and toys. Ask whether they know the **price** of any items, and then discuss which things might **cost** more or less than others. For example, which costs the most, a house or a cuddly toy? Next, think about actual prices in local currency for these items.

Return to the list of coins and notes that you produced for the introductory activity. Discuss the different **values** of the notes and coins. *Look at the numbers and symbols on the money. Which numbers are bigger or smaller?'*

Ask students to look at the face of a note and say what it shows. *Whose picture is on it? Where does it say the value of the note?* Ask students to design their own note. *What would you put on it?*

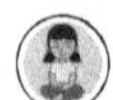 ### Reflection time

Look together at the notes that have been designed by students. *Have they all got the same features as the real notes?* Have a class poll to see which new notes are most popular and ask students why they like particular notes.

Discuss ideas about the value of money and start to explore the notion of exchange. Ask questions such as, *What would you rather have, lots of coins or notes? What can you buy more with? These big coins or these little ones? I want to swap ten of these coins for one of these? Is that a good swap?'*

9A Money amounts

Discover
Student Book page 121 • Practice Book page 113

Specific learning focus
- Recognise all coins and work out how to pay an exact sum using smaller coins.

Global skills
- **Creative skills:** problem solving / exploring / investigating
- **Real-world skills:** presenting information / financial literacy
- **Interpersonal skills:** communication / teamwork

Key vocabulary
- coin, cent, 1-cent coin, 5-cent coin, 50-cent coin, dollar (or currency vocabulary for your country), value

Resources
- selection of coins in local currency

Language support
Use student knowledge of counting in ones, twos, fives and tens. Make the connection that it is still the same concept but using money instead of objects or numbers.

 **Introductory activity**

Use a set of local coins. Look again at the values of each coin with students and list them on the board, showing how they are written. Ask students if they know any nicknames for coins, for example, nickels, dimes and quarters in the US. Students could look at the features on the coins and find out more about the faces used.

 **Main activity**

Write these amounts (or similar amounts in local currency) on the board:

6¢   21¢   25¢ $1.01 $1.05 $1.20

Give students a selection of coins (1¢, 5¢, 50¢, $1) to work with. Ask a student to pick an amount and see if they can make that amount using a combination of two coins. Ask them to say how they would make the amount, then write it down. Repeat with other students until all the amounts have been made.

Now ask a student to choose three of the coins. *What amount does it make?* Write it down. Repeat to make different amounts using three different coins.

Students can now complete the activities on page 121 of the Student Book. Look at the coins together and name each one in turn. Compare them to local currency, e.g. their their size and shape. Explain where the USA is in relation to your own country and show on a map.

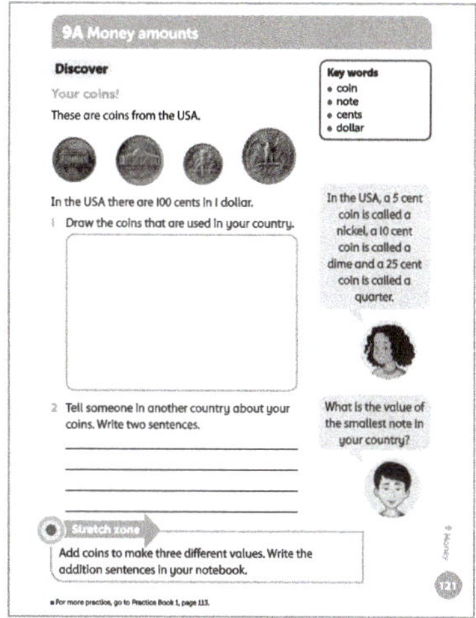

Differentiation

Supporting: Use **1-cent** and **5-cent** coins and model the exchange.

Consolidating: Use 1-cent, 5-cent and **10-cent** coins and model the exchange.

Extending: Use coins up to 1 **dollar**.

Stretch zone: *Add coins to make three different values. Write the addition sentences in your notebook.*

Check that the students have written the addition sentences correctly using money symbols.

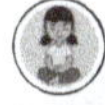 **Reflection time**

Ask students to share which amounts they made that were less than $1. *Which coins did they use? What about for amounts greater than $1? Which coins did they use?*

Who made the biggest amount less than $1? Who made the smallest amount greater than $1?

Practice Book: Students can complete Practice Book page 113. This can be done directly after the main activity, as homework, or as the focus of a separate mathematics session to help students consolidate their learning and build fluency. Adjust this activity to use local coins and prices, if appropriate. Some students may need additional support to choose items as they may be unfamiliar with prices and find it challenging to identify objects that could be purchased using coins only. However, they should be encouraged to choose items first (researching prices, if possible), rather than be given a list of items.

Differentiated outcomes	
All students	should make an amount using three coins.
Most students	will find different ways of using three coins to make an amount.
Some students	may make the same amount in more than one way using different numbers of coins.

Student Book page 121

Check that students have drawn appropriate coins for their currency and have written two sentences about them.

Stretch zone: Answers will vary as students choose the coins that they will use.

Check that the amounts that students have made are correct. For example:

1¢ coin and 50¢ coin = 51 cents

5¢ coin and $1 coin = $1 and 5 cents

Practice Book page 113

Check that the coins used to make the cost of an item total the actual price. For example, 50¢ coin and four 1¢ coins = 54 cents.

Stretch zone: Students should use the Practice Book activity to work out the fewest and most coins to pay for the tem.

9A Money amounts

Explore 1 Student Book page 122 · Practice Book page 114

Specific learning focus

- Recognise all coins and work out how to pay an exact sum using smaller coins.

Global skills

- **Creative skills:** problem solving / exploring / investigating
- **Real-world skills:** presenting information / financial literacy
- **Interpersonal skills:** communication / teamwork

Key vocabulary

- coin, cent, 1-cent coin, 5-cent coin, 50-cent coin, dollar (or currency vocabulary for your country), value

Resources

- Resource Sheet 5 (available on the Oxford Owl website) – one large version for display and one per pair
- pot of coins in local currency in values up to 1 dollar or the equivalent: more 1-cent coins than other values
- large 0–20 number line

Language support

When talking about the value of coins, use real money rather than plastic. Relate cents and dollars (or local currency) to real-life situations that students are familiar with, such as pocket money and shopping. Make, or ask students to make, a wall poster of money words, including pictures and labels.

 **Introductory activity**

Put a pot of mixed coins on each table. Ask a student from each group to take a coin from the pot and tell you the value. Ask one group at a time. Keep a record of the cumulative value.

Repeat until all students have chosen a coin. Link the addition of money to saving. *How much have we got? How much more do we need to make a dollar?*

Start with 100 cents (or similar amount in local currency) and reduce this by the amount of each coin that is chosen. Link the subtraction of money to spending. *How much have we spent? How much is left?*

 **Main activity**

Display Resource Sheet 5. Show a large pile of 1-cent coins. Explain that they are very heavy to carry in a bag or pocket. They must be exchanged for other coins.

Place five 1-cent coins along the top row of the exchange sheet. *There are 5 cents along the top.* (count them) *What can we exchange these for?* (one 5-cent coin)

Show how to take the 1-cent coins off and replace them with a 5-cent coin. Move the 5-cent coin to the row below. Continue filling in each row and exchanging for the next value coin. When each row is full, exchange it for the next value.

When all the 1-cent coins have gone, ask how much in total there is. Start adding from the highest value. Explain that each pair will carry on exchanging coins. Give each pair a pot of money and an exchange sheet to work with at their table and find the total amount in their pot. Some groups can have more or higher-value coins than others.

Students can now complete the activities on page 122 of the Student Book in pairs. Give students actual coins to work with to make combinations of coins for 20 cents. Alternatively, they could work with three colours of counters, assigning each colour a value, e.g. red = 1¢, yellow = 5¢, blue = 10¢.

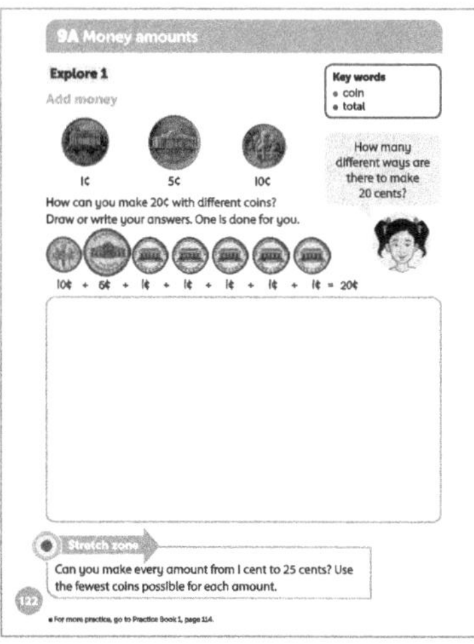

Differentiation

Supporting: Ask students to use 1-cent and 5-cent coins only, and model the exchange for students.

Consolidating: Ask students to use 1-cent, 5-cent and 10-cent coins, and model the exchange for students.

Extending: Ask students to use coins up to 1 dollar.

Stretch zone: *Can you make every amount from 1 cent to 25 cents? Use the fewest coins possible for each amount.*

Check that students have used the correct coins to make each amount with the fewest coins. For example, 25 cents can be made with just two coins (20 + 5). Choose an amount, e.g. 8 cents, and work through the possibilities systematically. *What is the biggest coin you can use to start to make 8 cents?* (5 cents / a nickel) *How many more cents to make 8 cents?* (3) *Which coins can you use to make 3 cents?* (three 1-cent coins) *That's four coins altogether to make 8 cents.*

 **Reflection time**

Share work done on page 122 of the Student Book. Ask students to share their different answers with a partner so they can check each other's answers. Ask them to choose one answer to share with the class. As you write the answers on the board, keep a cumulative total on a large 0–20 number line.

Practice Book: Students can complete Practice Book page 114. This can be done directly after the main activity, as homework, or as the focus of a separate mathematics session to help students consolidate their learning and build fluency. Encourage students to work with real coins or counters (assigning each a value, as with the Student Book activity). If students are doing this activity at home, tell them that they can work on this with an adult.

Differentiated outcomes	
All students	should use 1-cent and 5-cent coins.
Most students	will use 1-cent, 5-cent and 10-cent coins.
Some students	may use all the coins.

Answers

Student Book page 122

Answers will vary. Here are some example answers:

10 + 10

10 + 5 + 5

10 + 5 + 1 + 1 + 1 + 1 +1

Practice Book page 114

Check that students have made the amounts correctly each time, and that each way is different. For example:

22 = 20 + 1 + 1; 10 + 10 + 1 + 1; 5 + 5 + 10 + 1 + 1

Stretch zone: Yes, 10 + 10 + 5

9A Money amounts

Explore 2 Student Book page 123 • Practice Book page 115

Specific learning focus

- Recognise coin values and know to make different amounts

Global skills

- **Creative skills:** problem solving / exploring / investigating
- **Real-world skills:** research / presenting information / financial literacy
- **Interpersonal skills:** communication / teamwork

Key vocabulary

- coin, total, less than, greater than, value

Resources

- list of small classroom items and their prices, to display
- pot of coins in local currency in values up to 1 dollar or the equivalent (more smaller-value coins than other values)
- number rods, Resource Sheet 5 (available on the Oxford Owl website), one per pair (optional)

Language support

Create equivalence boards by tracing around or make a rubbing of local coins and their equivalences for students to refer back to as they are working.

Show students a set of four different coins, for example 1¢, 5¢, 20¢, $1 (or equivalent in local currency). Explain to students that $1 is the same as 100¢.

Ask a student to choose two coins and then say what the **total** amount is. Repeat with another student who chooses a different pair of coins. Explain that choosing two coins does not mean the amount is the same.

Tell students you want to go shopping for some things for the classroom, such as pencils, rulers and notebooks. Show students a list of prices for these items. Ask them to talk to their partner about which things they would buy and what the total cost would be.

Give time for discussion. Collect responses from students.

Initially, use amounts that can be made using just 1-cent coins. You may decide to ask some students to work on the Student Book activities on page 123 at this point using 1-cent coins only. For the rest of the students, use prices which can be made with 10-cent, 5-cent and 1-cent coins. Again, you may then ask students to work on the Student Book before moving on to more complex prices.

Focus on paying for the same item in more than one way, where possible, and recording as a number sentence. For example, a pencil costing 48¢ could be paid for by using four 10-cent coins, a 5-cent, 2-cent and 1-cent coin. It could also be paid for with four 10-cent coins and four 2-cent coins. Students can use their copies of the coin exchange sheet (Resource Sheet 5) from the previous lesson to help them.

This way students get to practise adding but also think about combinations that have the same monetary value, regardless of number or size of coins used.

For smaller amounts, you can use number rods to show the equivalences, for example:

10¢									
5¢					5¢				
1¢	1¢	1¢	1¢	1¢	1¢	1¢	1¢	1¢	1¢

Ask students to complete the activities on page 123 in the Student Book individually or in pairs.

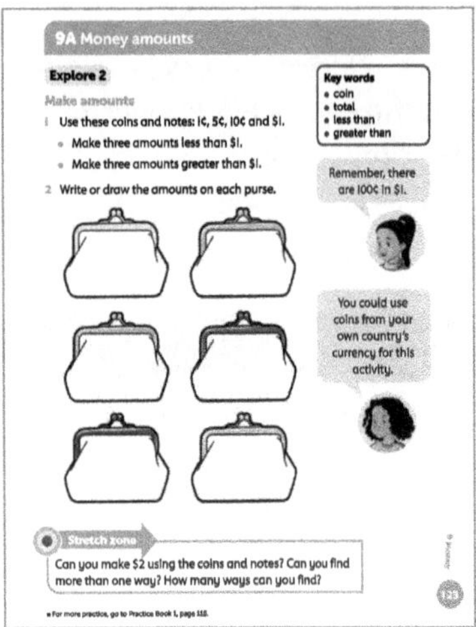

Differentiation

Supporting: Only use prices that can be made with 1-cent coins to start with, then swap simple amounts for different coins, e.g. 5-cent coin for five 1-cent coins.

Consolidating: Use prices that can be made with 1-cent and 5-cent coins up to 20 cents.

Extending: Use all the coins.

Stretch zone: *Can you make $2 using the coins and notes? Can you find more than one way? How many ways can you find?*

Check that students have made the amounts accurately.

Reflection time

Discuss the items students 'bought' and how much they cost. Record their responses on the board. Ask questions such as: *How much did your items cost altogether? Which coins did you use to make up that cost? Could you have made it using different coins?*

Practice Book: Students can now complete Practice Book page 115. This can be done directly after the main activity, as homework, or as the focus of a separate mathematics session to help students consolidate their learning and build fluency. Before beginning, take examples of amounts greater than and less than 50 to remind students of this vocabulary.

Differentiated outcomes	
All students	should understand amounts using just 1-cent coins.
Most students	will understand amounts using 1-cent and 5-cent coins and changing 5-cent coins for five 1-cent coins.
Some students	may be able to use all the coins correctly.

Answers

Student Book page 123

Check that students have made amounts less than or greater than $1 in different ways, and that they have written or drawn the amounts on the purses.

Practice Book page 115

Check that students have made the amounts according to the instructions, and written or drawn them correctly in the boxes.

Stretch zone: Students can make $1.25 in several ways, e.g. $1 \times \$1 + 5 \times 5$¢.

9B Notes and coins

Discover Student Book page 124 • Practice Book page 116

Specific learning focus

- Work out the notes or coins needed to buy items.

Global skills

- **Creative skills:** problem solving / exploring / investigating
- **Real-world skills:** research / presenting information / interpreting information / financial literacy
- **Interpersonal skills:** communication / teamwork

Key vocabulary

- coins, notes, value, price, cost, how much?

Resources

- list of 12 food items and local prices for Introductory activity
- local prices for the shopping items in the Student Book activity (bag of rice, fruit juice, bottle of water, box of cornflakes, melon), or internet access for students to research these prices
- 0–20 number lines, 100-squares

Language support

Practise reading money amounts with symbols. *What do the symbols mean? Point out cases where we write them differently from the way we say them, e.g. $2 and two dollars.*

 **Introductory activity**

Together, write a shopping list of 12 items and write down the typical price of each item. Choose some items that cost less than $1 and some that cost greater than $1 (or similar amounts in local currency). Talk with the students about which is the least expensive item and which is the most expensive.

 **Main activity**

Take the first two items from the shopping list you compiled in the introductory activity. Ask students to think how they might pay for that item. *If the amount is small, can you pay with just coins? If the amount is large, do you need notes?*

Help students to calculate how to use notes or coins to make the amounts. You can record the amount and how it might be paid. Students can use a 0–20 number line or 100-square to help them work out the total. They should then try to find how to pay for three more of the items from the list.

Look at page 124 in the Student Book together, on the IWB if possible. Discuss what the items are and ask students to think about how much they think these may cost. Conduct internet research to find out actual prices, if appropriate. Alternatively, provide students with prices. Students can now complete the activities on page 124 of the Student Book in pairs.

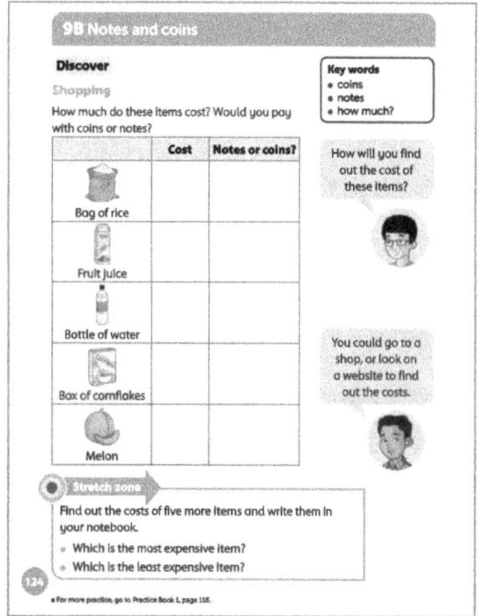

Differentiation

Supporting: Ask students to add notes and coins using a number line and 100-square for support.

Consolidating: Ask students to add amounts up to $1 using a number line.

Extending: Ask students to use amounts over $1 and add using a number line or 100-square.

Stretch zone: *Find out the costs of five more items and write them in your notebook. Which is the most expensive item? Which is the least expensive item?*

Check that students can identify the least and most expensive items.

 **Reflection time**

Ask students to describe how they worked out whether an item needed coins or notes. *Does it depend on anything other than the cost?*

Choose an item that you would pay for in coins. *How many could you buy before you need notes to pay for them?*

Practice Book: Students can complete Practice Book page 116. This can be done directly after the main activity, as homework, or as the focus of a separate mathematics session to help students consolidate their learning and build fluency. Ask students to estimate the price of each item before they ask an adult what the actual price is. Was their price close to the actual price or not?

Differentiated outcomes	
All students	should choose whether to pay in coins.
Most students	will choose whether notes are needed.
Some students	may work out the amount using the correct notes and coins.

Student Book page 124

Check that the amounts are accurate and that students have correctly identified which can be bought with coins and which with notes.

9B Notes and coins

Explore   Student Book page 125 • Practice Book page 117

Specific learning focus

- Work out the coins needed to buy stamps of different values.

Global skills

- **Creative skills:** problem solving / exploring / investigating
- **Real-world skills:** research / presenting information / interpreting information / financial literacy
- **Interpersonal skills:** communication / teamwork

Key vocabulary

- stamp, postage, value

Resources

- small denomination coins to pay for postage, real examples of stamps
- large 0–100 number line for front of class

Language support

Discuss with students about buying stamps so they can post parcels. Show them stamps and help them use the vocabulary of stamps and postage.

Introductory activity

Show students some examples of packages with **stamps** on. Ask if anyone knows what the stamps are for. Explain how a sender can pay for a stamp to show that they have paid the postal service the correct amount to have a parcel delivered. How much you pay depends on the size, mass, distance and speed you want the parcel delivered.

Students should look at the examples of stamps in current use, and notice the pictures and faces on the stamps and the different amounts. *Why are there stamps of different value?* Students might know that stamps of greater value are needed to send bigger items, or to send things over a greater distance.

Practice Book page 116

Check that the amounts are accurate and that students have correctly identified which can be bought with coins and which with notes. Check that the amounts have been made up correctly.

Stretch zone: Answers will depend on the items found in the activity.

Main activity

Tell students that they are going to buy some stamps to send some parcels. The stamps that are available have values of 5¢, 10¢, 20¢ and 50¢ (these can be adjusted to match the local currency).

The amounts of **postage** needed for sending the different parcels are:

| 25¢ | 40¢ | 45¢ | 60¢ | 75¢ |

Ask a student to come forward and choose stamps to make the correct value in postage. Help them by saying that they can buy more than one of the same stamp. Prompt them to start with the largest-value stamp first, and use a number line to build up to the total amount.

Repeat with other students until the amounts have all been made in stamps.

Students can now complete the activities on page 125 of the Student Book individually or in pairs. Before they begin, look at the stamps and ask questions such as *What is the value of each stamp? Which stamp has the greatest value? Which has the smallest value? Which stamps could I use that have the same value as one stamp with a tree on it?*

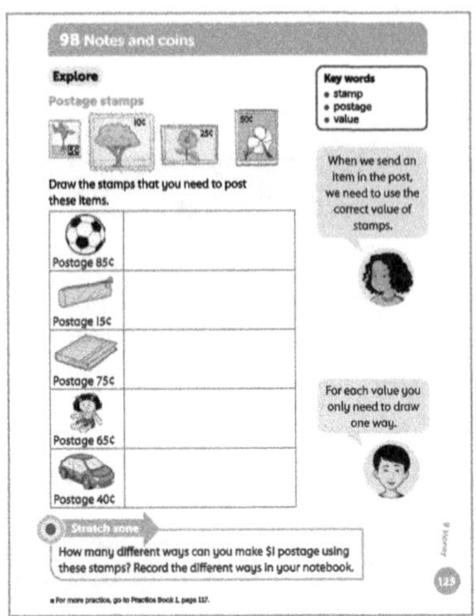

Differentiation

Supporting: Ask students to add stamp values from two stamps, for example, *how can you make 25¢ and 40¢?* Use this to help find, for example, 60¢ by adding onto 40¢.

Consolidating: Ask students to add three or more stamp values.

Extending: Ask students to look for ways to pay the postage using the most/fewest stamps.

Stretch zone: *How many different ways can you make $1 postage using these stamps? Record the different ways in your notebook.*

Check that students have made $1 correctly using the stamps.

 ## Reflection time

Discuss with students how they made the correct value for each postage amount. Could they have made some amounts in a different way? Show students how to represent this on the number line, modelling the values and the addition used. *How did you add the stamp values?*

Practice Book: Students can complete Practice Book page 117. This can be done directly after the main activity, as homework, or as the focus of a separate mathematics session to help students consolidate their learning and build fluency. If necessary, explain that 'p' stands for pence, and in the UK 100p makes £1 (1 pound).

Differentiated outcomes	
All students	should say the value of each stamp and add two together with support.
Most students	will add some stamp values together.
Some students	may work out how to make any amount in stamps.

Answers

Student Book page 125

85¢ = 50¢ + 25¢ + 10¢

15¢ = 10¢ + 5¢

75¢ = 50¢ + 25¢

65¢ = 50¢ + 10¢ + 5¢

40¢ = 25¢ + 10¢ + 5¢

Practice Book page 117

Check that students have shown eight ways of making 20p using the coins given. Examples include:

10p + 10p   10p + 5p + 5p   10p + 5p + 2p + 2p + 1p

20p   10p + 2p + 2p + 2p + 2p + 2p

Stretch zone: Any combination of up to five coins to make 20p.

9 Money

Connect   Student Book page 126

Big idea

- I know the different coins and notes in my country.

Global skills

- **Creative skills:** problem solving / exploring / investigating
- **Real-world skills:** research / presenting information / interpreting information / financial literacy
- **Interpersonal skills:** communication / teamwork

Key vocabulary

- notes, coins, value, currency, cost

Resources

- coins and notes from local currency, counters

Language support

Discuss with students the names of the coins and notes in their local currency. Have examples and images to show them and discuss the features on each.

 ## Introductory activity

Ask students which is worth more: 10p + 10p or 20p + 1p? This is to help them understand that, although both amounts are made of two coins, the values are different. This activity can also be done in the local currency. If you work with UK currency, you may choose to look at the coins on page 126 of the Student Book discussing and comparing their values.

Ask students to make up a similar question for a partner.

 ## Main activity

Students play a game called 'Clear the shelves' in pairs. The UK coins could be replaced by the local currency. Explain that the aim of the game is to buy fruit until there is none left on the shelves. You can adapt the names of fruit to reflect what is popular locally.

Give each pair 2p, 5p and 10p coins (five of each, in a bag or under a piece of material) and 25 counters (in five different colours to represent the different fruits in the shop: cherries, nuts, grapes, melon, apples). Cherries cost 1p, nuts cost 2p, grapes cost 3p, melon costs 4p and an apple costs 5p. Students may choose to label the 'fruits' and the prices (or you can help them to do this).

Students take turns to choose a coin at random from the bag. They then decide how to spend it on some fruit to that value – one or more pieces of fruit. Discuss with students how they are going to record what they have spent, and which fruits they have 'bought'. When they have bought some 'fruit', they remove these items from the 'shop'. Their partner checks that the items chosen match the amount of money the student has. If they do, then the student gets a point. If there are no 'fruits' left that can be bought with the student's coin then they do not score a point. Students return the coin to the bag for the next turn. The game ends when there is nothing left in the shop to buy (or when you end the game). The winner is the player with most points.

Students can then complete the activities on page 126 of the Student Book individually. You could remind students that, sometimes notes (bills) or coins are given a nickname instead of their real-value name. For example, in the UK, a £5 note is sometimes called a 'fiver', or in the US, a 10¢ coin is called a 'dime'.

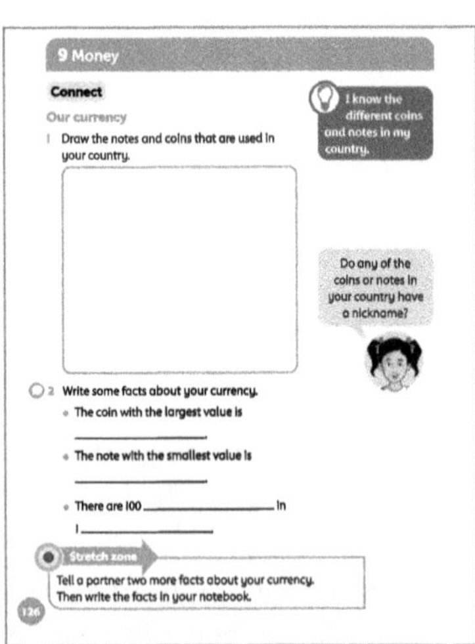

Differentiation

Supporting: Help students to choose fruit that they can spend their coin on.

Consolidating: Help students work out the best fruit(s) to buy, by asking questions such as, *How much do you have to spend? Are there any fruits that cost less than the value of your coin?*

Extending: Challenge students to spend up to a certain amount or to buy a certain number of pieces of fruit.

Stretch zone: *Tell a partner two more facts about your currency. Then write the facts in your notebook.*

This might require students to research information about the currency, or perhaps ask a parent or carer. Check that students have recorded accurate facts about the local currency.

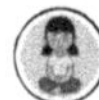 Reflection time

Ask students to talk about how they chose their fruit while playing the game. Did they go for the most expensive fruit or more of the cheaper fruits?

Differentiated outcomes	
All students	should say how much each coin is worth.
Most students	will choose to spend their coin on fruit.
Some students	may work out whether they can spend their coin when there is little fruit remaining.

9 Money

Global skills

- **Creative skills:** problem solving / exploring / investigating
- **Real-world skills:** presenting information / financial literacy
- **Interpersonal skills:** communication
- **Self-development skills:** reflecting on learning

Student Book

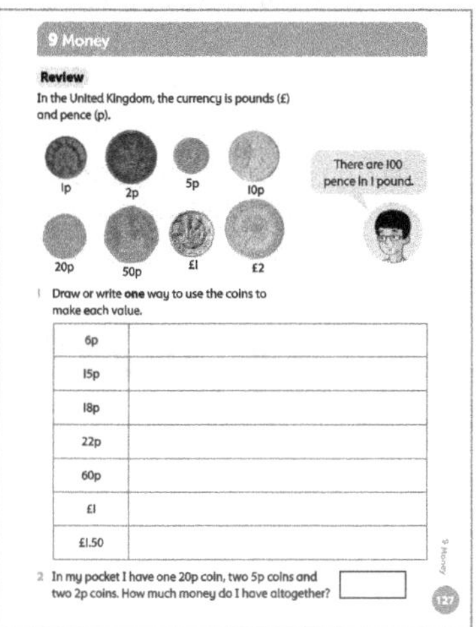

With young children, assessment activities are most effective when carried out as an everyday classroom activity. Students should have real coins and notes available to support them, if possible. Encourage students to use a number line to help them answer questions.

Watch as students complete the activities that require amounts to be made. Check that students understand the language of money with coins, notes and other vocabulary associated with the currency.

Extend by asking a range of similar questions or asking students to make their own problems. Encourage students to use a number line to help them.

Answers

Student Book page 127

1 6p = 5p + 1p

15p = 10p + 5p

18p = 10p + 5p + 2p + 1p

22p = 20p + 2p

60p = 50p + 10p

£1 = 50p + 50p

£1.50 = £1 + 50p

2 34p

Practice Book

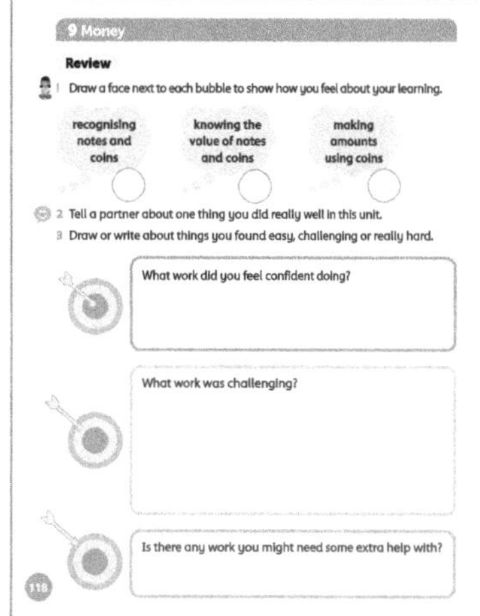

With young children it is appropriate to complete this as a whole-class discussion. You may choose to keep a record of the class discussion or a copy of the review page for your own records. Use the Student Book to briefly remind students of the areas of mathematics that they have worked on in this unit.

Give out the Student Book to support pairs of students as they discuss and answer the questions in the Practice Book.

Allow students plenty of time for discussion before asking them to share their responses with the rest of the class. If students complete this assessment at home, encourage them to discuss this with adults.

Make a note of areas that students still feel unsure about. You can build money into everyday practice. For example, talking about the cost of things, or collecting money for school items such as lunches.

Additional material

There are additional end-of-unit assessments available on the *Oxford Owl* website.

10 Time

Overview

Big Idea

Telling time is a skill as essential as addition and subtraction – especially in the age of digital clocks. It is important that students have a very strong grasp of numbers, at least to 12 for an analogue clock. Because there are 60 minutes in an hour, students must be able to read and order numbers to 60 before they can read time on the digital clock.

Teaching general references to time such as 'in the morning' and 'at night' is a good start. We use the word 'time' frequently throughout the day, often without realising it. What time is it? It's time to pack up. It's time for music. It's lunchtime.

Knowing the names of the days of the week makes this abstract concept of time more concrete for young students. A week is comprised of seven separate days, each with a morning, afternoon and evening.

An important part of learning about the days of the week is understanding the terms 'today', 'tomorrow' and 'yesterday'. These terms allow students to discuss their activities (and understand when other people are discussing activities) by properly referencing what they did the day before or what they plan to do the following day.

Look out for

- **Students who find it difficult to judge how long a time 'feels'.** You can develop students' understanding of the durations of given lengths of time by using sand timers for times up to 5 minutes, or asking students to do an activity for one minute.
- **Students who are confused by the fact that we say the time in many different ways.** This adds to the challenge of learning to tell the time. Include lots of examples showing that, e.g. for quarter past 4, we might also say four-fifteen or 15 minutes past 4.

Possible misconceptions

When using an analogue clock:

- **Students may confuse the minute and hour hands.** Making the hands of the clock different colours may help avoid this.
- **Students may assume that at 3.30, for example, the hour hand will point directly at the three.** They are not sure what time it is when the hour hand is pointing somewhere between two numbers on the clock face. Show students that the hour hand moves gradually from one number to the next and that, at the half hour, the hour hand will be half way between numbers on the clock face.

Key vocabulary

- time, before, after, next, first, last
- morning, afternoon, lunchtime, evening, night-time
- recent, a long time ago, older, oldest, earlier, later
- day, week, days of the week: Monday, Tuesday, …, Sunday
- month, year, calendar, months of the year: January, February, …, December
- yesterday, today, tomorrow
- clock, o'clock, hour, minute, hands, 1 o'clock, 2 o'clock, …, 12 o'clock, short hand, long hand, minute hand, hour hand
- second, minute, hour, how long, what time is it, half past

Coverage in lessons

Learning focus	Learning outcomes (the ENC objectives)
Ordering events	Sequence events in chronological order using language: for example, before and after, next, first, today, yesterday, tomorrow, morning, afternoon and evening.
Days of the week	Recognise and use language relating to dates, including days of the week, weeks, months and years.
Telling the time	Tell the time to the hour and half past the hour and draw the hands on a clock face to show these times.
Measuring time	Measure and begin to record time (hours, minutes, seconds).

10 Time

Engage Student Book page 128

Big question

- Why do we need to know the time?

Global skills

- **Creative skills:** exploring / investigating
- **Real-world skills:** interpreting information
- **Interpersonal skills:** communication / teamwork / leadership

Key vocabulary

- morning, lunchtime, afternoon, evening, night-time, time, clock

Resources

- photographs or pictures showing different common daily tasks, e.g. walking to school

Language support

Make a display using some of the students' pictures and link it to the key vocabulary.

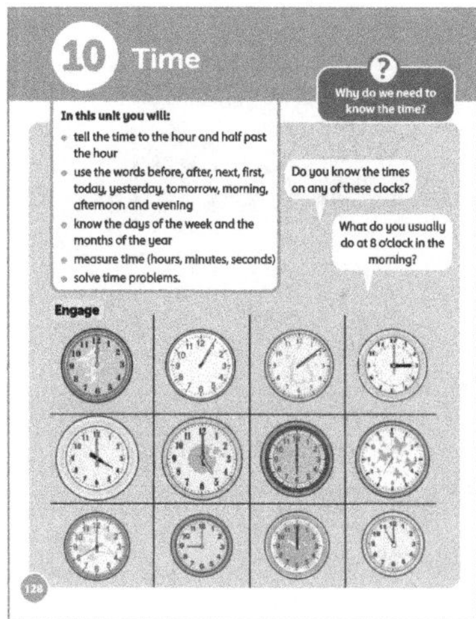

Introductory activity

Discuss the idea of **time**. Ask questions such as, *Why do we need to be able to tell the time when we are at home or at school? Imagine there are no **clocks**. We don't know the time. Is this OK? Why not? How would we know if it was **morning** or **night-time**? How would we know when to eat our breakfast or get up in time for school?* Display page128 in the Student Book on the IWB, if possible.

What can you see? (clocks) *What is the same? What is different?* Talk about the numbers students can see, noticing that they are in the same position on the different clocks, and the clock faces all have two hands.

What's the first thing you do in the morning? Ask different students to show, through mime, other things that they do before they come to school. (have breakfast, clean teeth, get dressed) *What do you do at 8 o'clock in the morning?* Extend beyond morning time so that they have more ideas for the game in the main activity.

 **Main activity**

Explain and then play the game 'What's the time Mr Wolf'.

Ask students to line up in a row. Choose one student to be 'Mr Wolf'. Mr Wolf stands at the opposite end of the classroom, hall or playground from the other players, facing away from the rest of the players. The object of the game is to get to the other end of the room without being caught by Mr Wolf.

All players, except for Mr Wolf, chant in unison, *What's the time, Mr Wolf?*, while slowly moving towards him. After each chant, Mr Wolf answers a time during the day (e.g. getting up time, bedtime, 9 o'clock, 5 o'clock) and turns around quickly. If he spots anyone moving, they are out of the game and must sit down.

Mr Wolf turns around again, away from the class, and the rest of the class move forward again and repeat the question, *What's the time, Mr Wolf?*

When he finally calls *'Dinnertime!'* all students stand still and the student closest to Mr Wolf takes his place and the game starts again.

Without upsetting the flow of the game, keep the focus on the key vocabulary – morning, **lunchtime**, **afternoon**, **evening**, night-time. Encourage a wide use of vocabulary, modelling as necessary. Mime times during the day to help students understand the new vocabulary, where suitable, e.g. lunchtime. You may want to show children images of common daily tasks alongside use of the vocabulary.

Differentiation

Supporting: Ask questions that encourage students to link activities to the time of the day.

Consolidating: Ask questions to make connections between periods of the day and times. For example, ask, *What time do you leave for school?*

Extending: Ask students to write times on their pictures.

 **Reflection time**

Each pair should draw things that they do during a day. Ask them to start in the morning and, using their pictures to help, tell the story of their day. Encourage students to ask other students 'before or after' questions, e.g. *What do you do after you eat supper?*

10A Ordering events

Discover Student Book page 129 • Practice Book page 119

Specific learning focus

- Begin to understand and use some units of time to order familiar events.

Global skills

- **Creative skills:** exploring / investigating
- **Interpersonal skills:** communication / teamwork

Key vocabulary

- before, after, first, next, last, morning, lunchtime, afternoon, evening, night-time, recent, long time ago, early, earlier

Resources

- photographs or pictures of familiar objects, family members or people from a magazine, from different periods of time
- picture/word cards for morning, afternoon, evening, night-time, cut out from Resource Sheet 6 (available on the Oxford Owl website)
- cards with daily activities (e.g. going to school, watching television) written on them, several for each of the main times: morning, afternoon, evening, night-time
- A4 paper

Language support

Observe to find out how well the students identify activities that occur at given times of the day. What knowledge of time do they demonstrate when sharing their experiences? Are they able to correctly use vocabulary and language associated with time? Use the display to model the correct vocabulary for students.

Introductory activity

Show the class photographs from different periods of time.

Choose one picture and ask, *Do you think this is from recent times or a long time ago? How do you know? What makes you think that?*

Show a second picture and ask whether the students think it is more recent or **older** than the **first** one. *Why do you think that?'*

Now ask students to order all the pictures from **oldest** to most recent, with oldest on the left and the most recent on the right. Can they give you examples of things they did recently and things they did longer ago?

Main activity

Place two chairs at the front of the class. Put the 'morning' card on one of them and the 'afternoon' card on the other.

Read the activity cards one at a time and ask students to sit in front of the chair where they think the activity belongs. For example, say, *Is 'getting out of bed' a morning or an afternoon activity?* Students will move to the picture for the morning activity.

Some activities will be neither a morning nor an afternoon activity, e.g. 'Go to bed'. Ask the class what you could do about such activities. Accept all answers. If students do not suggest adding a new chair, ask, *What about another chair? Where would the chair need to go? Would it go at the end or in the middle?*

Encourage the use of correct vocabulary and language such as **before**, **after**, **next**.

Continue to read out activities until there are four chairs at the front of the class: morning, afternoon, evening and night-time.

As a class, make lists on the board of things students do at different times of the day.

Students complete the Student Book page 129 individually. When they have completed the questions, ask them to look back at the pictures in question 1 and describe them to a partner using 'before', 'after' and 'next'.

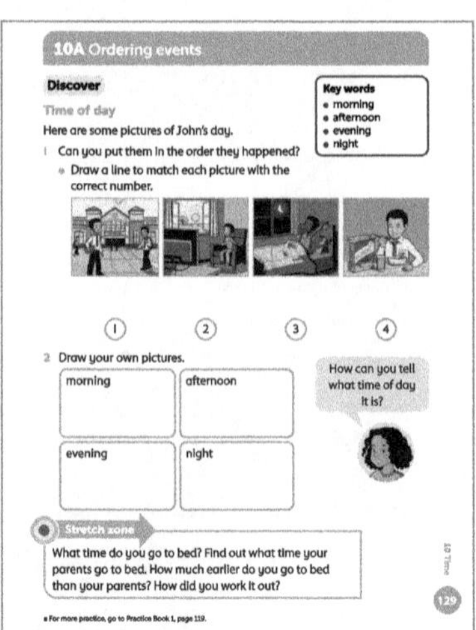

Differentiation

Supporting: Ask questions that encourage students to link activities to the time of the day, for example, *'Which meal do you have at midday?*

Consolidating: Ask questions to make connections between periods of the day and times. For example, *What time do you leave for school?*

Extending: Ask students to write times of the day on their pictures.

Stretch zone: *What time do you go to bed? Find out what time your parents go to bed. How much **earlier** do you go to bed than your parents? How did you work it out?*

Listen to students' explanations about how they work out the time difference.

Reflection time

Ask students to show their drawings from the Student Book activity to the class and choose one time of day and say what they do at that time. Use this as an opportunity to correct or reinforce the language used. Ensure that all times of the day are covered.

Practice Book: Students can complete Practice Book page 119. This can be done directly after the main activity, as homework, or as the focus of a separate mathematics session to help students consolidate their learning and build fluency. Students may need support linking the drawings to times of day and their own daily experiences. Encourage them to discuss with an adult to come up with and record their answers.

Differentiated outcomes	
All students	should order events in the day.
Most students	will know approximate times some of these events take place.
Some students	may know the times these events take place.

10A Ordering events

Explore Student Book page 130 • Practice Book page 120

Specific learning focus

- Begin to understand and use some units of time to order familiar events.

Global skills

- **Creative skills:** exploring
- **Real-world skills:** presenting information / interpreting information
- **Interpersonal skills:** communication / teamwork

Key vocabulary

- before, after, first, next, last, morning, lunchtime, afternoon, evening, night-time

Resources

- A4 paper, sticky tack
- stories from books or the internet that involve different times of the day/days of the week, e.g. *The Very Hungry Caterpillar* by Eric Carle, *The Tale of Peter Rabbit* by Beatrix Potter, *The Owl Who Was Afraid of the Dark* by Jill Tomlinson.

Answers

Student Book page 129

1 Breakfast   **2** Arrive at school

3 Watch TV   **4** Go to bed

If necessary, ask students to describe what is happening in each picture so that you can check that they have understood the vocabulary and the different times of day.

Practice Book page 119

Check that the students' pictures/writing are appropriate for the required time of day. Discuss with students what is happening in their pictures and why they do those activities at that time of day.

Stretch zone: Students' answers will vary but may relate to waking up, getting light, getting dark, going to bed.

Language support

Reflection time will give students opportunities to think about the cycle of the day and how they can use vocabulary and language to describe it. This can give insight into their thinking as well as their use of language.

Introductory activity

Read a story to introduce sequencing time. Ask students, *What was the very first thing that happened? Was it something that happened in the morning, afternoon, evening or was it during the night-time?* Ask them to tell you something else that happened. *Was that the next thing? Was there something before that? What happened after that? What was the **last** thing that happened?*

As students tell you parts of the story, draw simple pictures of what they say on sheets of A4 paper and stick the pictures on a board. *Are these pictures in the right order? Let's check them with the story book.*

Allow students to make changes to the picture story when necessary, so that it matches the original one.

Main activity

Tell the class that they are each going to plan their perfect day. If possible, display page 130 of the Student Book on the IWB.

In your Student Books, there are six blank rectangles. You will draw in each of them. The first picture of your day will be having breakfast. Then draw the best thing you like to do in the morning. This can be a day at school or a day at home. The next will be having a meal at lunchtime and then your favourite thing to do during the afternoon. After that, another meal. Your last picture should show you going to bed.

Model these instructions as you say them by displaying the key times as pictures (going to bed and meals) as fixed points for the students to pinpoint the other times during the day. Pairs should tell each other their perfect day and then draw the images on page 130 of the Student Book. Do they know the approximate or exact times these things happen?

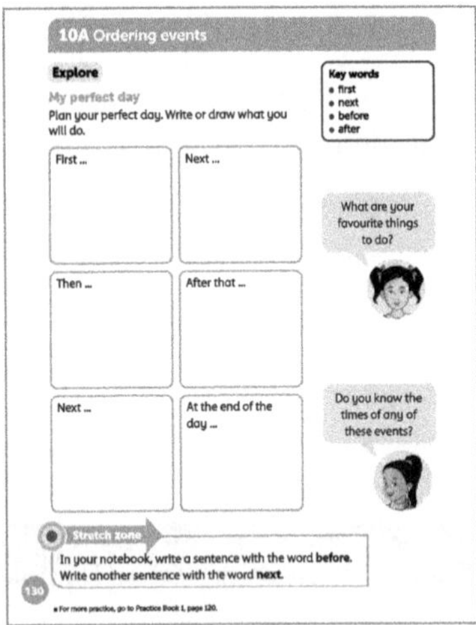

Differentiation

Supporting: Ask questions which help students with the vocabulary of ordering, or ask them them to draw the pictures and you help with the vocabulary.

Consolidating: Ask questions to make connections between periods of the day and times. For example, ask, *If you first arrive at school at 9 in the morning, at what time do you think you will do the activity for your next picture?*

Extending: Ask students to think about which hour of the day the pictures may happen.

Stretch zone: *In your notebook, write a sentence with the word **before**.*

*Write another sentence with the word **next**.*

Check that students have used the words correctly in their sentences.

 **Reflection time**

Ask some of the class to share their perfect day with the rest of the class. Use language in your questions such as 'before', 'after', 'next', e.g. *What did you do before you went to bed?*

Practice Book: Students can complete Practice Book page 120. This can be done directly after the main activity, as homework, or as the focus of a separate mathematics session to help students consolidate their learning and build fluency. Read each sentence together as a class or ask students to read them aloud to an adult. You could also ask them to describe each picture using the vocabulary 'first', 'next', 'after' and 'before'.

Differentiated outcomes	
All students	should order events in the day.
Most students	will know approximate times some of these events take place.
Some students	may know the times these events take place.

Answers

Student Book page 130

Observe who has correctly sequenced the events in their perfect day by time of day and ends with their going to sleep. As the class are working, walk around and ask questions such as, *Can you tell me what you have done so far? What is the next thing you will do? What was the last thing you have just done?*

Practice Book page 120

1 evening

2 night

3 morning

4 afternoon

Stretch zone: Students should add an explanation for each answer.

10B Days of the week

Discover
Student Book page 131 • Practice Book page 121

Specific learning focus

- Begin to understand and use some units of time such as days of the week.

Global skills

- **Creative skills:** problem solving / exploring / investigating
- **Real-world skills:** interpreting information /
- **Interpersonal skills:** communication / teamwork

Key vocabulary

- Monday, Tuesday, Wednesday, Thursday, Friday, Saturday, Sunday, week, day
- January, February, March, April, May, June, July, August, September, October, November, December, year, month
- calendar, yesterday, today, tomorrow

Resources

- large calendar
- large laminated cards of days of the week – school days in black, others in red (these could be from Resource Sheet 7, available on the Oxford Owl website, copied onto A3 sheets of coloured card and cut out)
- horizontal list of the days of the week
- laminated cards of the months of the year (from Resource Sheet 8, available on the Oxford Owl website, copied onto card and cut out)

Language support

Use more picture clues throughout the day, such as on the class weekly or daily timetable. Have calendars and days of the week vocabulary visible around the classroom.

Introductory activity

Show a large **calendar** to the class and ask them what they can tell you about it. Use the laminated cards to revise or introduce words such as the **days** of the **week** and the **months** of the **year**. Discuss what students do on different days of the week and in different months. *What did you do on **Monday**? In which month is your birthday?*

Find today's date on the calendar. Point to the day on the calendar and tell students what day it is. Ask them

which way they would move to find which day it will be **tomorrow**. *Is there anything that could help?* (Look at the sequence of numbers.) *If today has a six, what will the next number be? What is the number after six?* **Today** is *[Tuesday], so the next day will be [Wednesday].* Use the days-of-the-week line as a reference.

Main activity

Write on the board the words: **'Yesterday'**, 'Today', 'Tomorrow'. Explain that yesterday, today and tomorrow change as the day does. To reinforce this idea, say *Yesterday you were not at school.* (Change as necessary.) *Today you are at school. Tomorrow you will go swimming.* Ask students, *What did you do yesterday? What will you do tomorrow? What are we doing today?*

Ask the students to complete page 131 in the Student Book in pairs so that they can discuss the answers and discuss what they notice about the calendar.

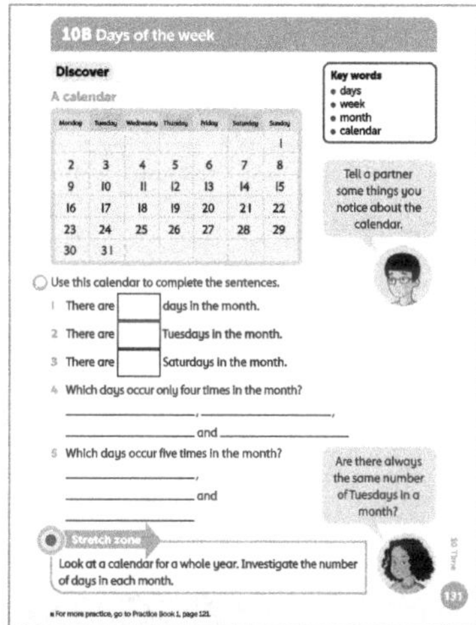

Differentiation

Supporting: Model the vocabulary carefully, repeating the days of the week in order, with students joining in when they feel confident.

Consolidating: Ask questions such as, *What day comes after **Thursday**? What is the day before **Friday**? What day was it yesterday? What day will it be tomorrow?*

Extending: Provide students with a calendar for them to find special dates.

Stretch zone: *Look at a calendar for a whole year. Investigate the number of days in each month.*

Which months have the most days? How many days do they have? Which month has the fewest days?

 Reflection time

Ask pairs to discuss their favourite day of the week. After 2 minutes, pick pairs to some to the front and select the correct day card and explain why this day is their favourite. Repeat for other days of the week and then repeat for months of the year. Ask students, for example:

What days is it two days before your favourite day? What month comes after your favourite month?

Practice Book: Students can complete Practice Book page 121. This can be done directly after the main activity, as homework, or as the focus of a separate mathematics session to help students consolidate their learning and build fluency. If time allows, look at the calendar on this page in the Practice Book and compare it with the one on page 131 of the Student Book. *What is the same? What is different?*

Differentiated outcomes	
All students	should recognise and order days of the week.
Most students	will recognise and order days of the week and months of the year.
Some students	may know special dates and find them on a calendar, and be able to put them in order across the year.

Student Book page 131

1 There are 31 days in the month.

2 There are 5 Tuesdays in the month.

3 There are 4 Saturdays in the month.

4 Wednesday, Thursday, Friday, Saturday

5 Monday, Tuesday, Sunday

Practice Book page 121

Monday	4
Tuesday	4
Wednesday	5
Thursday	5
Friday	5
Saturday	4
Sunday	4

Stretch zone: Saturday, Thursday, Friday

10B Days of the week

Explore   Student Book page 132 • Practice Book page 122

Specific learning focus

- Begin to understand and use some units of time such as days of the week.

Global skills

- **Creative skills:** problem solving / investigating
- **Real-world skills:** presenting information
- **Interpersonal skills:** communication / teamwork

Key vocabulary

- Monday, Tuesday, Wednesday, Thursday, Friday, Saturday, Sunday, week, day, calendar
- January, February, March, April, May, June, July, August, September, October, November, December, year, month

Resources

- large calendar
- large laminated cards of days of the week – school days in black, others in red (from Resource Sheet 7, copied onto coloured card and cut out)
- horizontal list of the days of the week, sticky tack
- large laminated cards of the months of the year (from Resource Sheet 8, available on the Oxford Owl website, copied onto card and cut out)

Language support

Make sure there is a large calendar in the classroom with days and dates clearly in view. Refer to it regularly, and maybe start each day by asking what date it is.

Introductory activity

Show the class the horizontal list of the days of the week. Read them out loud and ask the class to join in, pointing to each day as you say it. You could start on Monday and then repeat starting from a different day.

Place all the days-of-the-week cards in a large bag or box, counting them as they go in. Ask a student to take one of the cards out. If it is today's day, stick it on the horizontal line of days of the week so that it covers the same day. If it isn't, place it on a table and ask another student to choose a card until today's day is revealed. Point at one of the other days on the line and ask another student, *Can you find the card that looks like this? Is it before or after today?* Say the days in order until you reach the one that matches the card. Then stick the card on the line so that it covers the same day.

Continue until all of the cards have been placed in order. Read them through, asking the class to join in with you.

Ask the class, *How many days are in a week?* (7) *Can you name any of them?* Choose seven students to pick a day card at random and stand in a line. The rest of the class should sort the students into the correct order by giving instructions to move one student at a time. Encourage them to use the names of the days of the week, along with 'before', 'after'. Take this opportunity to model the language of the days of the week and the ordering process carefully.

Repeat with the month cards. When you have the month cards in the correct order ask students to stand in front of the month of their birthday.

Ask students to complete page 132 in the Student Book individually. They can refer to the calendar on page 131 of the Student Book to support them. When they have completed the questions, ask them to share them with a partner, reading each full sentence aloud to practise pronunciation of the names of the days of the week correctly.

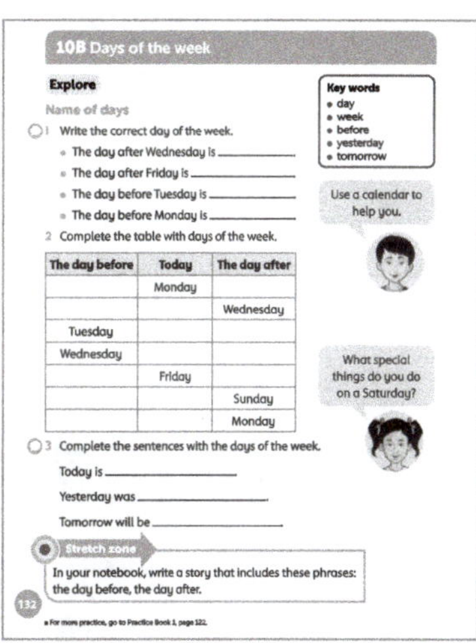

Differentiation

Supporting: Model the vocabulary carefully, repeating the days of the week in order, with students joining in when they feel confident.

Consolidating: Ask questions such as, *What day comes after Thursday? What is the day before Friday? What day was it yesterday? What day will it be tomorrow?*

Extending: Provide students with a calendar for them to find special dates.

Stretch zone: *In your notebook, write a story that includes these phrases: the day before; the day after.*

Check that students have used the phrases correctly in their story.

 **Reflection time**

Mix up the days-of-the-week cards and ask two students to come and put them in order as quickly as they can, but starting from a day you tell them. Repeat with different pairs.

Ask 12 students to come to the front and pick a month card. They should arrange themselves in the correct order, starting from the month you tell them.

Practice Book: Students can complete Practice Book page 122. This can be done directly after the main activity, as homework, or as the focus of a separate mathematics session to help students consolidate their learning and build fluency. Encourage students to think about something they do on these days that they have not yet described. Rather than drawing a picture for each day, they could find images from a magazine or printed from the internet and stick these in place, for example an image of a child playing the piano to show they have piano lessons each Tuesday.

Differentiated outcomes	
All students	should recognise and order days of the week.
Most students	will recognise and order days of the week and months of the year.
Some students	may know special dates and find them on a calendar.

Answers

Student Book page 132

1 Thursday
Saturday
Monday
Sunday

2

The day before	Today	The day after
Sunday	Monday	Tuesday
Monday	Tuesday	Wednesday
Tuesday	Wednesday	Thursday
Wednesday	Thursday	Friday
Thursday	Friday	Saturday
Friday	Saturday	Sunday
Saturday	Sunday	Monday

3 Students answers will vary.

Practice Book page 122

Ask students to tell you the day that they do each activity on so that you can check whether they have drawn it in the correct position in the table.

Stretch zone: False, False

10C Telling the time

Main activity

Discover Student Book page 133 • Practice Book page 123

Specific learning focus

- Begin to understand and use some units of time, hours and o'clock.

Global skills

- **Creative skills:** problem solving / exploring / investigating
- **Real-world skills:** presenting information
- **Interpersonal skills:** communication / teamwork

Key vocabulary

- clock, 12 o'clock, 1 o'clock, 2 o'clock, 3 o'clock, 4 o'clock,…, hands, short (hour) hand, long (minute) hand, What time is it?

Resources

- large geared clock for class use, a clock on the wall that all students in the class can see
- laminated time cards with each of the o'clock times on them (from Resource Sheet 9, available on the Oxford Owl website, copied onto card and cut out)

Language support

Use a clock during activities. As you say 'o'clock' ask students to make the time on the clock. Give students vocabulary cards of the time words used as a reference, including a clock picture of the time.

 Introductory activity

Ask pairs to discuss any times that they know, *What time do you wake up? What time does the school day start and finish?* Pairs who can say a sentence including a time should come to the front to share. Explain that, as there are two of the same **o'clock** times in every day, they will need to say 'in the morning' or 'in the afternoon'. (You may choose to teach the 24-hour clock at the same time.)

Pick out an o'clock time card (e.g. 7 o'clock) at random and make a statement about that time, e.g. *I went to bed at 7 o'clock in the morning.* Students decide whether it is true or false.

Repeat with other cards and select students to come to the front of the class to pick a card and make a true or false statement of their own.

Main activity

Show the class a geared **clock** set at **12 o'clock.**

*There are two **hands** on the clock: a **long hand** and a **short hand**. Can you see them both? Where is the short hand? Watch what happens when I move the long hand all the way round and back to 12.* Slowly move the long hand round so that the short hand moves to 1. *Where is the short hand now?*

Repeat the movement and tell students to watch the short hand. *Where is the short hand now?* Explain that every time we move the long hand from 12 back to 12, the short hand moves on to the next number. *The short hand is the **hour hand** because it tells us what hour it is; the long hand is the **minute hand** because it moves through all of the minutes to get back to 12.*

Set the clock to **1 o'clock**. Say, *It will always be o'clock when the long hand points to 12. But what o'clock is it? We look at the number the short hand is pointing to. The short hand is pointing to 1. So **what time is it?** It is 1 o'clock.*

Repeat for other o'clock times. Ask students to complete Student Book page 133 individually. They can then discuss their answers and the questions in the speech bubbles in pairs.

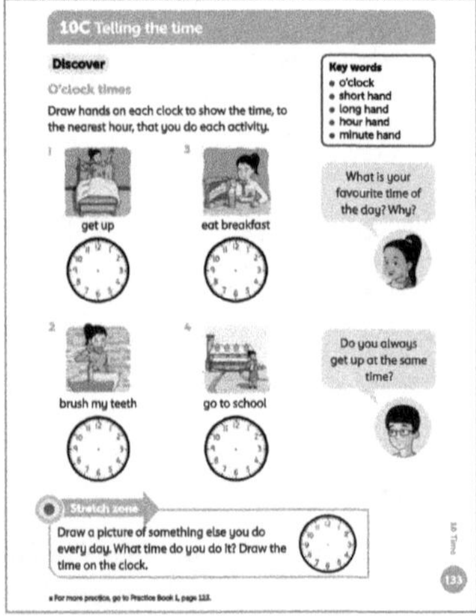

Differentiation

Supporting: Ask students to make times on clocks as you say them.

Consolidating: Ask students, *What is the hour before? What is the hour after?*

Extending: Ask if students know any times in between the hours.

Stretch zone: *Draw a picture of something else you do every day. What time do you do it? Draw the time on the clock.*

Check that students draw the time correctly to match their picture.

Draw or display a clock face on the board. Draw the hands to show 4 o'clock and ask the class what time it is. Write *4 o'clock*. Redraw the hands to a different time. *What time is it now?* Repeat two or three times, always writing the new time on the board.

Write a time on the board and ask a student to draw the hands, reminding them that one hand should be longer than the other one.

Practice Book: Students can complete Practice Book page 123. This can be done directly after the main activity, as homework, or as the focus of a separate mathematics session to help students consolidate their learning and build fluency.

Differentiated outcomes	
All students	should recognise o'clock times by looking at the position of the hour hand and checking the minute hand is on 12.
Most students	will have an understanding of how time progresses during the day and will know the hour before and the hour after.
Some students	may know times other than o'clock times.

10C Telling the time

Explore 1 Student Book page 134 • Practice Book page 124

Specific learning focus

- Begin to understand and use some units of time, hours and o'clock.

Global skills

- **Creative skills:** problem solving / exploring / investigating
- **Real-world skills:** research / presenting information
- **Interpersonal skills:** communication / teamwork / leadership

Key vocabulary

- 12 o'clock, 1 o'clock, 2 o'clock, 3 o'clock, 4 o'clock,…, What time is it? earlier, later

Resources

- large geared clock for class use, a clock on the wall that all students in the class can see
- large number cards 1–12 (laminated if possible)
- laminated cards with o'clock times (from Resource Sheet 9, available on the Oxford Owl website, copied onto card and cut out)
- clock face and two clock hands cut out from Resource Sheet 10 (available on the Oxford Owl website), copied onto thin card; split pin, paper plate, adhesive stick (one of each item per student)

Language support

Use a clock during activities. As you say *o'clock* ask students to make the time on the clock. Give students vocabulary cards of the time words used as a reference, including a clock picture of the time.

 **Introductory activity**

Place the large number cards 1–12 in a circle on the floor to look like a clock-face. Give 12 students (or 12 pairs) an o'clock time card each to keep face down.

The teacher and one other student stand in the centre of the circle of numbers. The teacher represents the minute hand and the student is the hour hand. Ask a student to turn over a time card and read it out. *Where should I point to? Why?* (Point to 12 because the time is o'clock) *Where should* [say the name] *point to? Why?*

Answers

Student Book page 133

Check that they have drawn each short hand correctly for the times that they say, and ask them to tell you the times that they do the activities they have drawn.

Practice Book page 123

Check that students have correctly drawn the hands on the faces for each time.

Stretch zone: Students should explain the position of both hands.

Students take it in turns to read their time and say where the teacher and student should point. Repeat until the entire class have had a turn to read a time.

 **Main activity**

Show the class how to make a paper plate clock face. Cut round the outside edge of the clock face from Resource Sheet 10, and stick it on to the paper plate. Cut round the two hands. Use the split pin to join the hands to the clock. Supervise as students make their own clock faces. You may prefer to cut out the clock faces and clock hands beforehand for students. You may also need to help students with fixing the split pin through the hands and clock face.

Move the hands so that your clock reads 4 o'clock. What do you need to remember when making an o'clock time? Check that all students have moved the long hand to 12 and the short hand to 4. Repeat using different o'clock times.

Ask students to complete Student Book page 134 individually making at least three of their own o'clock times using their clocks and then drawing the hands and recording the time in their books to match.

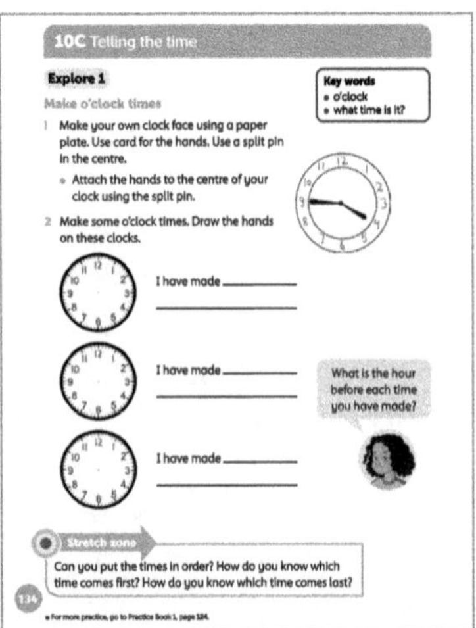

Then explain that students are going to play 'What's the time Mr Wolf?' again but with different rules. Ask students to set their clock to an o'clock time. Choose a Mr Wolf and then ask the rest of the class to stand up and ask, *What's the time Mr Wolf? Who is going to be safe from the Wolf?* If the Wolf calls out your time you can sit back down because you are safe. The last students standing are 'caught' by Mr Wolf.

Differentiation

Supporting: Ask students to make times on clocks as you say them.

Consolidating: Ask students, *What is the hour before? What is the hour after?*

Extending: Ask if students know any times in between the hours, or, for example, 2 hours before or 3 hours after.

Check by listening to students that they can order the times and explain how they did it.

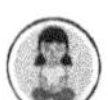 **Reflection time**

Ask students to show you 2 o'clock on their paper clocks. Check that they all show the correct time. *Show me the time 1 hour after 2 o'clock.* Give time for students to find the time and show you.

Continue with other times such as 1 hour earlier, 2 hours **later**, 2 hours earlier.

Practice Book: Students can complete Practice Book page 124. This can be done directly after the main activity, as homework, or as the focus of a separate mathematics session to help students consolidate their learning and build fluency. As this Practice Book page includes half-past times, you may prefer to complete this page after students have completed the Explore 2 activity.

Differentiated outcomes	
All students	should recognise o'clock times by looking at the hour hand and checking the minute hand is on 12.
Most students	will have an understanding of how time progresses during the day and will know the hour before and the hour after.
Some students	may know times other than o'clock times.

Answers

Student Book page 134

Check that students' times match the completed clock faces.

Practice Book page 124

1	5 o'clock	**5**	Half past 5
2	7 o'clock	**6**	6 o'clock
3	Half past 1	**7**	Half past 11
4	10 o'clock		

Stretch zone: The clock face should match the time given by the student.

Explore 2 Student Book page 135 • Practice Book page 125

Specific learning focus

- Begin to understand and use some units of time, hours, o'clock and half past.

Global skills

- **Creative skills:** problem solving / exploring / investigating
- **Real-world skills:** research / presenting information / interpreting information / financial literacy
- **Interpersonal skills:** communication / teamwork / leadership
- **Self-development skills:** reflecting on learning

Key vocabulary

- 12 o'clock, 1 o'clock, 2 o'clock, 3 o'clock, 4 o'clock', . . ., half past, What time is it?

Resources

- 2 large geared clocks for class use; a clock on the wall that all students in the class can see
- paper plate clock face with movable hands from pervious lesson (one per student)
- laminated cards with each of the o'clock times and half-past times on them (cut out from Resource Sheets 9 and 11, available on the Oxford Owl website), one set per pair
- laminated cards with daily activities (e.g. 'eat breakfast', 'go to school') written on them (one set per pair)

Language support

Use a clock during activities. Focus on the language of half past, for example, an hour can be split into halves (link to half of a circle). Half of the hour is finished and there is one more half until the next hour. We call this 'half past', meaning half-way past the hour.

 ## Introductory activity

Show 3 o'clock and **half past** 3 side by side on two clock faces. *What do you notice about these two times? What's the same? What's different?* If students do not suggest it, mention the term 'half past 3' and explain that we use this to mean half-way between 3 o'clock and 4 o'clock. Move the time on both clocks slowly ahead by half an hour. *What time is it on each clock now?* Get students to practise reading the time (to the hour and half past) and showing times on their small clocks. *How do you know it says 4 o'clock? Or half past 7?* Ask students to copy times

on their own clocks. *Where would the long hand and short hand have to be for 3 o' clock? For half past 10?*

 ## Main activity

Tell students they are going to be sorting some different events and putting them in order of the times they happen. Start with an example such as 8 o'clock for 'eat breakfast' and half past 3 for 'go home from school'. *Which one happens first? Which time is earlier/later? Can you make those times on your clocks?*

Give each pair of students a set of activity cards and a set of time cards (o'clock and half-past times). Ask them to match each activity to a time and then to put the activities and times in order from earliest to latest.

Then ask students to complete Student Book page 135 in pairs so that they can discuss and compare what they do at different times. Can they tell their partner about any special times they know?

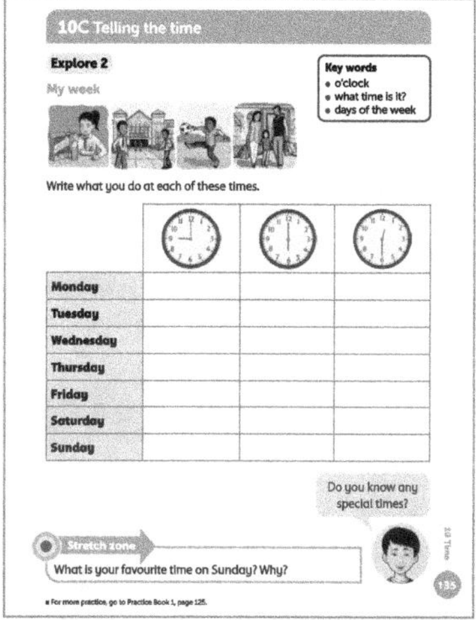

Differentiation

Supporting: Ask students to make times on clocks as you say them, and write out what the child says for each day and time.

Consolidating: Ask students, *What is the hour before? What is the hour after?*

Extending: Ask if students know any times in between the hours.

Stretch zone: *What is your favourite time on Sunday? Why?*

Listen as students explain their favourite time. Check that they are using the correct vocabulary and that they can make the time on a clock face.

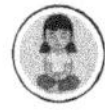 ### Reflection time

Talk through with students to see which order they put their events in the pairs activity. Check that they have the correct sequence.

Now give them a time that is not among the set they sorted and ask if someone can describe an activity they might do at that time. *Where would that activity go? What time and activity comes just before the new one?*

Practice Book: Students can complete Practice Book page 125. This can be done directly after the main activity, as homework, or as the focus of a separate mathematics session to help students consolidate their learning and build fluency. You may choose to ask students to think only about when they do these activities to the nearest half hour, or to record exact times, if they know them.

Differentiated outcomes	
All students	should recognise half-past times by looking for the minute hand on the 6.
Most students	will have an understanding of how time progresses during the day and will know the hour before and the hour after.
Some students	may be able to add on an hour past a half-past time.

10D Measuring time

Discover Student Book page 136 · Practice Book page 126

Specific learning focus

- Know the relationships between consecutive units of time.
- Measure activities using seconds and minutes.

Global skills

- **Creative skills:** exploring / investigating
- **Real-world skills:** research / presenting information / interpreting information
- **Interpersonal skills:** communication / teamwork

Key vocabulary

- second, minute, hour, how long?

Resources

- digital timer (large one for class use), stop watches, other timers (mobile phone, computer and so on)
- interlocking cubes, balls and other items for 1-minute challenges

Language support

Make vocabulary cards for some time words and display them with an illustration. This could be a drawing by one of the students or a photograph. Talk about other times when it is important to tell the time, for example, when cooking.

Answers

Student Book page 135

Check that students have written a suitable activity for each time in the table.

Practice Book page 125

Check that students have written a suitable time for each activity in the table.

Stretch zone: Make sure the time they write and the time they draw are the same.

 **Introductory activity**

Use a digital timer that the whole class can see. Tell them that you are setting it for one **minute**, and explain that it will count down from one minute to zero in **seconds**. *Let's count with it to find out how many seconds are in a minute.*

Set the timer to count down from one minute in seconds. Ask the whole class to count with the timer 1, 2, …. Discuss ways of counting seconds explaining that some people use the following rhyme to count seconds, e.g. *one elephant, two elephants, three elephants.* Agree that there are 60 seconds in one minute.

Set the timer again. This time ask students to count silently in their heads with their eyes closed. They should open them when they think one minute has passed. Talk through what they noticed. *Were most students too early or too late? Was anyone just right?*

 **Main activity**

Set some challenges for students to do in one minute. This could be with everyone doing the activity at the same time or with pairs of students timing each other. Challenges might include how many times a student can write their name, jump on the spot or click their fingers in a minute. Ask students to estimate how many times they do the activity in one minute before doing it. *How close where you?*

Alternatively, you could challenge students to stand on one leg for a minute or hold some other pose.

Challenge students to find out how many times they can clap in one second. If they can only clap once, does that mean they can clap 60 times in a minute? If they can clap twice in a second, will they be able to clap 120 times in a minute? Ask students to work in threes to test the ideas, one student clapping, another counting the claps and another timing the minute. Show students how to use a stop watch or other digital timer (e.g. mobile phone) to time the activity.

Show students Student Book page 136. Ask them to work on the activities in pairs, taking it in turns so that one carries out the activity while the other student does timing.

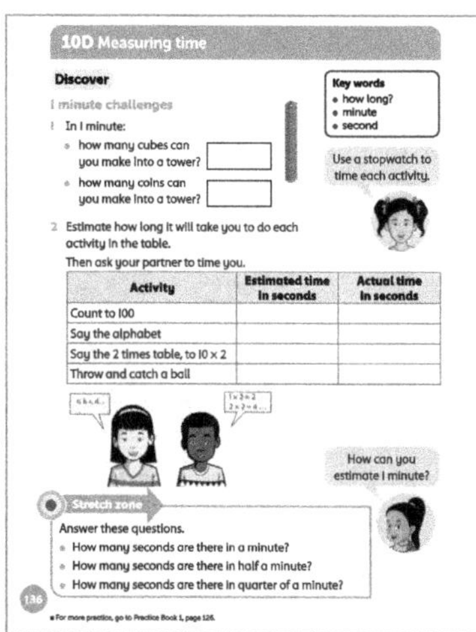

Differentiation

Supporting: Join in the minute activities either as timer or by asking students to time you.

Consolidating: Ask students how they are counting to estimate one minute.

Extending: Ask students if they know the relationships between units of time (seconds and minutes and **hours**).

Stretch zone: *Answer these questions.*

- *How many seconds are there in a minute?*
- *How many seconds are there in half a minute?*
- *How many seconds are there in quarter of a minute?*

Check that students know the equivalents between the units.

 ## Reflection time

Compare answers to the activity on page 136 in the Student Book. Ask pairs to feed back on how they estimated and the difference between their estimated time and the actual time. Ask for reasons for this difference.

Practice Book: Students can complete Practice Book page 126. This can be done directly after the main activity, as homework, or as the focus of a separate mathematics session to help students consolidate their learning and build fluency. Point out that they have done a few of the challenges in class. *Can you use your actual times to estimate how fast you will be this time? Do you think you will be faster or slower?*

Differentiated outcomes	
All students	should complete activities and time one minute.
Most students	will fairly accurately estimate one minute.
Some students	may understand the relationships between units of time (seconds and minutes and hours).

Answers

Student Book page 136

Observe students as they carry out the activities, checking to see if they are able to use the timers correctly. Ask students to explain how they worked out their estimates, and how close their estimates were to the actual times.

Stretch zone: 60, 30, 15

Practice Book page 126

Check whether the students' estimates were close to the actual times or very different. Do the 'actual times' seem reasonable or might the students have made mistakes when timing themselves?

Stretch zone: Students will have different ideas. Make sure these will all give an estimate of around a minute,

10D Measuring time

Specific learning focus

- Know the relationships between consecutive units of time.
- Measure activities using seconds and minutes.

Global skills

- **Creative skills:** exploring / investigating
- **Real-world skills:** research / presenting information
- **Interpersonal skills:** communication / teamwork

Key vocabulary

- second, minute, hour, how long?

Resources

- images of historical timing devices (e.g. sundials, water clocks, single-handed clocks)
- digital timers, sand timers, stop watches, other timers (mobile phone, computer, oven timer)
- small containers, sand, adhesive tape, paper circles (to make into cones for pouring sand)
- mini whiteboards and markers

Language support

Students should work in mixed-attainment groups so that less confident learners can hear the correct pronunciation of the vocabulary modelled by their peers.

 Introductory activity

Give each pair an image of a timing device used in the past, such as a single-handed clock, water clock or sundial. Ask students how they think the device worked and how accurate they think the timing device would be. For example, sundials will not work when there are no shadows, on cloudy days and at night. Take feedback from the pairs and explain how the devices worked. Pairs could write a sentence explaining how the device worked, or how accurate it was, and include it, alongside its image, to make a wall display.

 Main activity

Show students a 1-minute sand timer and explain how it works. Organise students in mixed-attainment groups and ask each group to make a 1-minute sand timer. Provide students with transparent containers (including small plastic bottles), sand, circular pieces of paper to make into cones and 1-minute sand timers or other timers to check with. When students are happy with their timers, groups could swap and test each other's timers. Have a class minute, when everyone's timer starts at the same time. *How many seconds were there between when the first and last timers finished?*

Students should complete Student Book page 137 in pairs. Ask them to discuss which category each activity should belong in. For some activities, they could use the 1-minute timers they made to check.

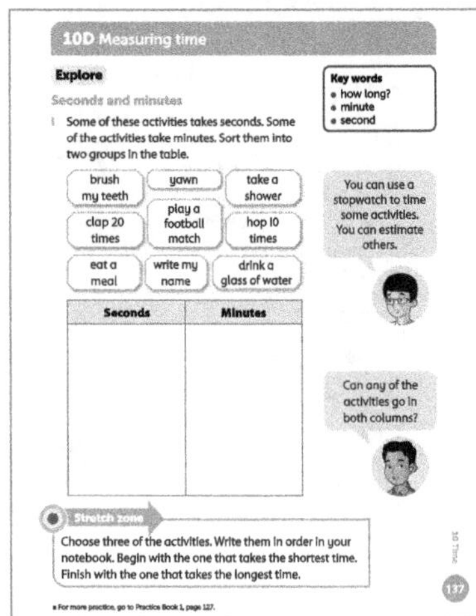

Differentiation

Supporting: Ask students if they think their timer will work and why.

Consolidating: Ask students what they think they need to adapt to make the timer work better.

Extending: Ask students questions such as, *How did you decide how your timer would work? How did you know how much sand to put in? How did you check if it was accurate?*

Stretch zone: *Choose three of the activities. Write them in order in your notebook. Begin with the one that takes the shortest time. Finish with the one that takes the longest time.*

Ask students to explain how they know and give estimates for how long each might take.

 Reflection time

Choose a piece of music around 2 minutes in length. Tell students they are going to listen to a short piece of music and they need to think about **how long** it is. Play the piece of music and then ask students to record their estimates on mini whiteboards.

Ask students to share their estimates and record them on the board. Find out the number of seconds between the shortest and longest estimates. *Did anyone estimate the exact time? Which estimate was the highest? Which was the shortest? What is the difference in time between them and the actual length of the song?* Ask students to talk about how they estimated. Did they count or do something else?

Practice Book: Students can complete Practice Book page 127. This can be done directly after the main activity, as homework, or as the focus of a separate mathematics session to help students consolidate their learning and build fluency. Once they have chosen an activity for each time, they could ask an adult to check using a stopwatch.

Differentiated outcomes	
All students	should share ideas to make the sand timer.
Most students	will adapt ideas to make an effective sand timer.
Some students	may explain how their sand timer works and how they checked its effectiveness.

10 Time

Connect   Student Book page 138

Big idea

- We need to know what day it is, what month it is and what time it is, so we know when to do things.

Global skills

- **Creative skills:** problem solving
- **Real-world skills:** presenting information /
- **Interpersonal skills:** communication / teamwork

Key vocabulary

- 12 o'clock, 1 o'clock, 2 o'clock, 3 o'clock, 4 o'clock, …, What time is it?

Resources

- picture book, such as *The Bad-Tempered Ladybird (The Grouchy Ladybug)* by Eric Carle
- two sets of large day-of-the-week cards (cut out from Resource Sheet 7)
- several sets of a morning card, an afternoon card and a night-time card (word and picture), cut out from Resource Sheet 6

Language support

Working in mixed-attainment groups will support less confident language learners and will allow them to hear peer modelling of the vocabulary.

Answers

Student Book page 137

Seconds	Minutes
yawn	brush my teeth
clap 20 times	take a shower
hop 10 times	play a football match
write my name	eat a meal
	drink a glass of water

Practice Book page 127

Check whether the students' suggestions for things that take each time interval seem reasonable. Discuss with them how they decided on each activity.

Stretch zone: Students answer will vary. Make sure their responses are reasonable.

 **Introductory activity**

Read a story from a book or show a video from the internet that includes times of the day, e.g. *The Bad-Tempered Ladybird* by Eric Carle. Ask questions about what happened at specific times of the day. Change what happened and at what times as appropriate, to match your story.

Who did he meet at 11 o'clock?

What did he do at 2 o'clock?

Go through the story again, pausing at each time. Ask the class to fill in the correct time to say what happened. This can be an individual or pair activity.

Encourage students to ask questions relating to the times in the book.

 **Main activity**

Split the class into seven groups and give each group a day-of-the-week card. Using a matching set of cards, hold one up at a time and read the name. Ask students in that group to put their hands up. Practise two or three times.

Then say, *I am going to give you a clue and if you think you are in that group, put your hand up but talk to the rest of the group first. Every time you get it right you will score a point.*

Give clues such as, *You come after Thursday. You are between Wednesday and Friday.* Include clues that relate to the students' day, for example, *You are a school day* (lots of groups get a point). *You are the day we have PE.* Give each group the chance to score two points.

Play again but give each group either a morning, afternoon or night-time card as well as their day-of-the-week card. *You are a morning group.* At this point, more than one group may hold their hands up so you need to give another clue. *You are a morning group and the day after Saturday.* At this point, only one group will fit the clues (Sunday morning). Repeat several times so that each group has an opportunity to score two points.

To increase the challenge, include time cards, giving each group a card with an o'clock time, to build up clues to, for example, *You are an afternoon group and before 3 o'clock and the day before Friday.*

Students should then complete Student Book page 138 individually. Review how to draw half-past times on a clock face before they begin, as necessary. Explain that it is fine to ask other students or yourself what time these activities happen if they are unsure, but they should try to draw these times independently.

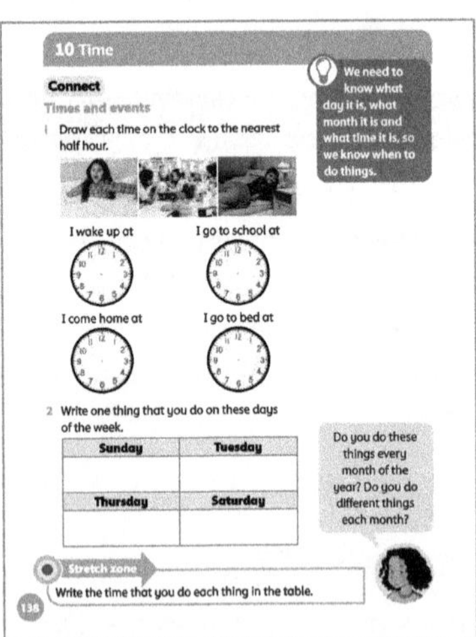

Differentiation

Supporting: Ask students to make times on clocks as you say them.

Consolidating: Ask students, *What is the hour before? What is the hour after?*

Extending: Ask if students know any times in between the hours.

Stretch zone: *Write the time that you do each thing in the table.*

Check that students have written a suitable time for each activity.

 Reflection time

Ask a student to read out their activities for each day. Other students should decide on a time that they do these activities. The first student says whether these times are correct or incorrect. Repeat for other students.

Do you do these things every month of the year? Do you do different things each month?

Differentiated outcomes	
All students	should order times and days of the week.
Most students	will use the vocabulary of time with confidence.
Some students	may use a more complex vocabulary of time.

10 Time

Global skills

- **Creative skills:** problem solving
- **Real-world skills:** research / presenting information / interpreting information
- **Interpersonal skills:** communication
- **Self-development skills:** reflecting on learning

Student Book

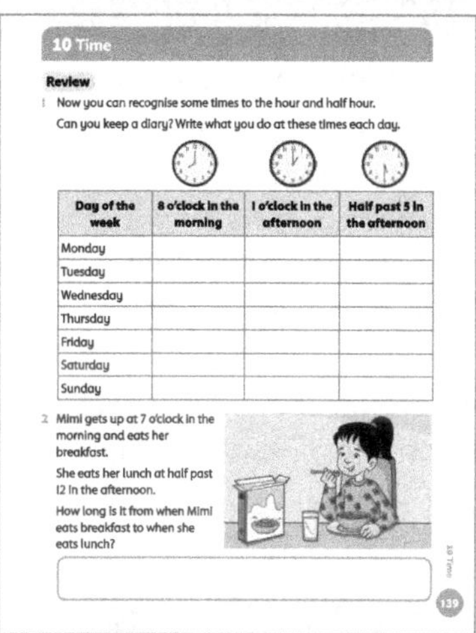

With young children, assessment activities are most effective when carried out as an everyday classroom activity. Students should have clock faces available so that they can refer to these to support them.

Watch as students complete the diary of different activities for each time. Listen to check that they understand the time and what things they might do at that time.

Encourage students to use the clock face for question 2. Extend by asking a range of similar questions, for example, *Mimi has her dinner 6 hours after her lunch. What time does she have dinner? It takes Mimi half an hour to eat lunch so what time does she finish her lunch?* You could also ask students to make their own time problems.

Answers

Student Book page 139

While students are working, ask them to tell you the time shown on each clock in the table (8 o'clock, half past 12, half past 5).

It is 5½ hours from breakfast to lunch for Mimi.

Practice Book

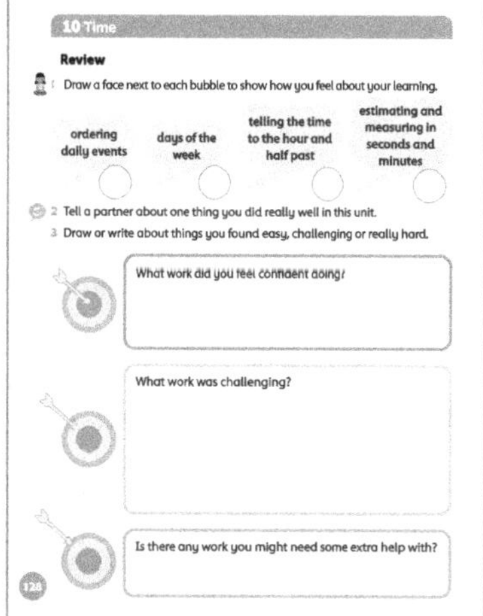

With young children it is appropriate to complete this as a whole-class discussion. You may choose to keep a record of the class discussion or a copy of the review page for your own records. Use the Student Book to briefly remind students of the areas of mathematics that they have worked on in this unit.

Give out the Student Book to support pairs of students as they discuss and answer the questions in the Practice Book.

Allow students plenty of time for discussion before asking them to share their responses with the rest of the class. If students complete this assessment at home, encourage them to discuss this with adults.

Make a note of areas that students still feel unsure about. You can build time into everyday practice. For example, *how long is it until breaktime*, or *how long does it take to walk home from school.*

Additional material

There are additional end-of-unit assessments available on the *Oxford Owl* website.

11 Geometry

Big Idea

The Big idea for this section is that shapes are all around us, not just in a book or a classroom. Many young students encounter the world of shapes through play, but every student, however young or old, can be challenged to think about and describe what they are doing and seeing, and to explore new possibilities and ideas.

A shape is the appearance of something, especially its outline. It is not dependent on size, position or orientation. It need not be two-dimensional (2D).

A regular shape is one where all of the sides and angles are equal.

 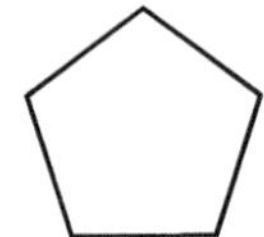  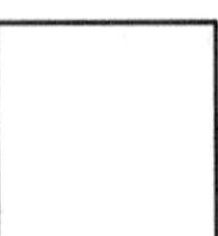

It is important that plenty of practical work with a range of materials is used during exploratory work. Discussion, mental work and seeing 'what would happen if' are included in the following activities. Students can classify (sort) shapes in any way that makes sense to them so that they begin to notice the properties of shapes.

Look out for

- **Students who do not recognise common shapes unless they are upright or in their usual orientation.** For example, students may not realise that this is a triangle:

- **Students who have limited understanding of properties of shapes.** A three-dimensional (3D) shape may be a solid but may not be. It may be open or closed; it may be regular or irregular.

- **Students who think that a line that divides a shape into two congruent parts is always a line of symmetry.** A line of symmetry always divides a shape into two congruent parts, but other lines may also divide a shape into congruent parts, and not be lines of symmetry. For example, this is not a line of symmetry:

Possible misconceptions

- **Students may think a square is not a rectangle.** Asking them to check the properties of both will help them to see that it is.

- **Students may be unfamiliar with the difference between 2D and 3D shape names.** For example, they may call a cube a square and a sphere a circle.

- **Students may refer to the vertices of a 3D shape as corners.** Be sure to model correct vocabulary when you are describing 2D and 3D shapes so that students have many opportunities to hear the vocabulary used correctly.

Key vocabulary

- shape, curved, straight, round, corner, side
- circle, square, rectangle, triangle
- cube, cuboid, pyramid, sphere, cone, cylinder, edge, flat, face, length, vertex
- two-dimensional (2D), three-dimensional (3D)
- symmetrical, pattern, line of symmetry, reflection, mirror line, horizontal, vertical, diagonal
- front, back, in front of, at the back of, behind, next to, between, underneath, first, last, on top of
- full turn, half turn, quarter turn, three-quarter turn, clockwise, anti-clockwise

Coverage in lessons

Learning focus	Learning outcomes (the ENC objectives)
2D shapes	Recognise and name common 2D shapes (e.g. circles, squares, rectangles and triangles).
3D shapes	Recognise and name common 3D shapes (e.g. cube, cuboids, cylinder, cone and sphere).
Symmetry	Recognise basic line symmetry [optional non-ENC objective]
Position and movement	Describe position, direction and movement.
Turns	Describe position, direction and movement, including whole, half, quarter and three-quarter turns.

11 Geometry

Engage Student Book page 140

Big question

- How can you describe the properties of the shapes you can see around you?

Global skills

- **Creative skills:** exploring / investigating
- **Interpersonal skills:** communication / teamwork

Key vocabulary

- curved, straight, round, corner, edge, circle, square, rectangle, triangle, cube, cuboid, pyramid, sphere, cone, cylinder, side, flat, face

Resources

- none needed

Language support

Support the vocabulary of shape with pictures, diagrams or whole-class participation in a physical activity to demonstrate the meanings of the words. For example, drawing a circle in the air.

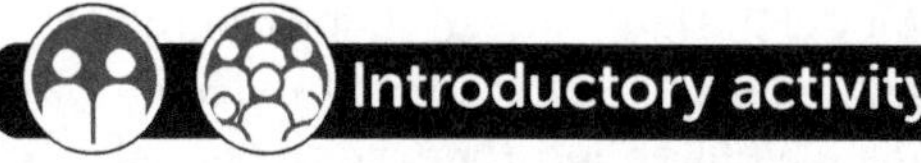 Introductory activity

Ask students to talk about the picture in the Student Book on page 140 and to relate it to any experiences they may have had in a playground or in school. If you have access to an IWB, you could display the Student Book Engage page. They should focus on looking for shapes that they already know the name of.

Ask students to share some of their ideas from their pairs talk with the rest of the class. To encourage discussion, use questions such as, *What can you see? What is the same/different about...?*

 Main activity

Choose a shape from those on page 140 in the Student Book. Describe its position to your partner. Can they work out which shape you chose? Encourage students to use positional language, for example 'my shape is in the centre of the climbing frame and has 4 straight sides'. *Choose a different shape. Describe its position to your partner. Can they work out which shape you chose?* Help students understand that the position and shape is important but the colour is not.

If you were going to climb over a climbing frame like the one in the picture, which ways could you go? How many different ways do you think there are?

Encourage correct mathematical language and vocabulary during the discussions. For example, rather than 'I see a football that is a circle', substitute sphere for circle.

Differentiation

Supporting: Model the language of shape carefully by pointing to the shapes in the image.

Consolidating: Ask students to describe the properties of shapes they see.

Extending: Develop the language of properties of shapes by comparing 'similar' shapes: for example, different-sized rectangles around the room or circles in the picture. *What is the same/different?*

 Reflection time

Ask pairs to look around the class for all the different shapes they can find. Ask them to give clues to their partner about what shape they are looking at and where it is. The partner uses these clues to guess the shape, and then shares with the class afterwards.

11A 2D shapes

Specific learning focus

- Name and sort common 2D shapes using features such as number of sides, curved or straight sides. Use them to make patterns and models.

Global skills

- **Creative skills:** problem solving / exploring / investigating
- **Interpersonal skills:** communication / teamwork

Key vocabulary

- corner, square, rectangle, triangle, side, straight, length

Resources

- elastic bands, geoboards (5 × 5), square dotted paper for recording, dictionary of shapes

Language support

Make a class poster showing all the different shapes labelled correctly, for students to refer to the names of the shapes at all times.

Introductory activity

Introduce the class to geoboards and demonstrate ways of using the elastic bands to make different shapes.

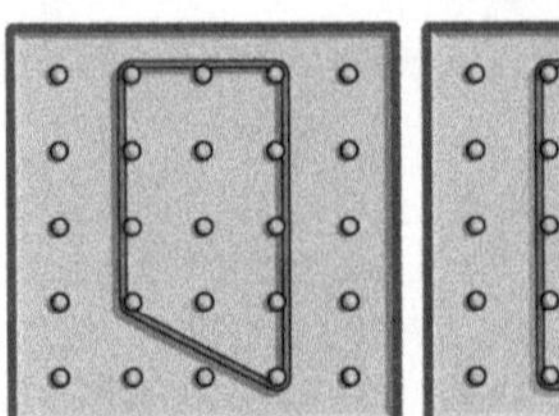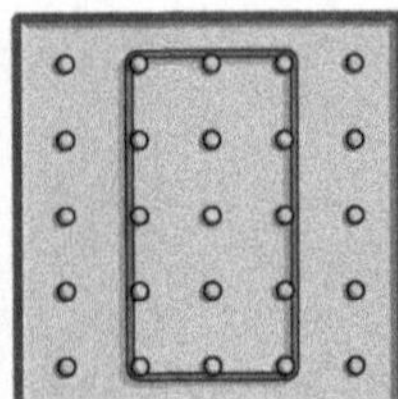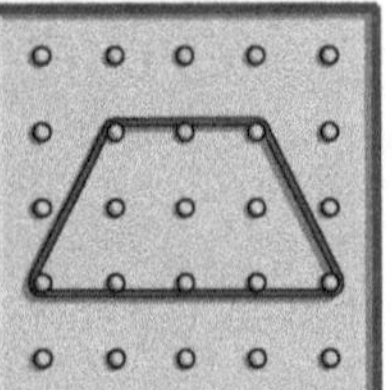

Ask students to work individually at first until they find out what shapes they can make and then share what they have found out with a partner.

Tell students to explore what they can do with a geoboard. Ask, for example, *What's the biggest/smallest shape you can make? Can you make a shape with* **sides** *that are not the same* **length***? Can you make a shape with 8 sides?*

Ask the class, *Can you make a* **square***? Can you make a* **triangle***? How many different triangles can you make?* Show them an example of each, showing acute, equilateral, obtuse, right-angled triangles but without naming them. *Are there any shapes you can't make?* (Circle) *Why not?* After each shape is made, students can hold them up to show you. Ask students to explain how they know they have made the right shape.

Main activity

Draw a square on the board and write 'square' next to it. *This is a square.* Next, draw a **rectangle** and write 'rectangle' next to it. *This is a rectangle.* Ask what is the same and different about the two shapes. Offering help with any new vocabulary, discuss the properties of each and agree that both have four **straight** sides and four **corners** but the rectangle's sides are not all the same length. Write what the properties tell you about the shapes, and demonstrate it with your drawing, for example, 'it has four sides'. Point to each side in turn as you describe the property. Do the same with other shapes:

Square – four sides (all the same length); four corners all the same.
Triangle – three sides, may or may not be equal in length, three corners.
Circle – no straight sides
Rectangle – four sides (two shorter and two longer), four corners all the same.

Ask students to work on the geoboards with a partner to find the biggest and smallest squares on the geoboard. Give students square dotted paper to record. Then ask them to find the biggest and smallest triangle they can and record these shapes.

Students can now work individually on the activities on page 141 in the Student Book, continuing to work with their geoboards to make shapes.

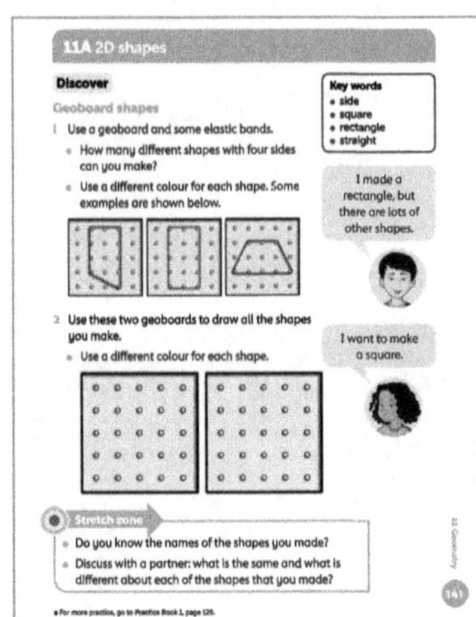

Differentiation

Supporting: Model the language of shape carefully, pointing to the shapes on the geoboards and any properties you are highlighting, for example, equal-length sides.

Consolidating: Ask students to describe the properties of shapes they see.

Extending: Encourage students to ddevelop the language of properties of shapes by comparing 'similar' shapes: for example, different-sized shapes they make on the geoboard. *What is the same? What is different?*

Stretch zone: *Do you know the names of the shapes that you made? Discuss with a partner: what is the same and what is different about each of the shapes you made*

Check that students' vocabulary when talking about shapes is accurate. If you have access to a shape dictionary, students could use this to find out the names of shapes if they don't know them.

 **Reflection time**

Ask some of the students to show what they did and what they found out. Raise the idea of a tilted square.

Cut a square from paper and ask what shape it is. Slowly rotate the shape, all the time asking what shape it is. Explain that it is still a square, whichever way up it is. Repeat with a triangle and a rectangle.

Do the same with a circle. What do you notice? *Does the same thing happen as with the other shapes? What is special about the circle?*

Practice Book: Students can complete Practice Book page 129. This can be done directly after the main activity, as homework, or as the focus of a separate mathematics session to help students consolidate their learning and build fluency. Encourage students to use a straight edge or a ruler to make their shapes as accurate as possible.

Differentiated outcomes	
All students	should make and name squares, rectangles and triangles.
Most students	will name squares, rectangles, triangles and circles confidently.
Some students	may begin to recognise that there are different types of triangle (without naming them as scalene, isosceles, right-angled, equilateral).

Answers

Student Book page 141

Check that students have drawn shapes with four sides. Ask them whether they think that any of their four-sided shapes are squares or rectangles. How do they know?

Practice Book page 129

Check that students have drawn different shapes on the geoboards. Ask students to explain how the shapes they have drawn differ from one another.

Stretch zone: Students can describe their shape any way they please. Check that the description is accurate..

11A 2D shapes

Explore 1   Student Book page 142 • Practice Book page 130

Specific learning focus

- Name and sort common 2D shapes using features such as number of sides, curved or straight. Use these to make patterns and models.

Global skills

- **Creative skills:** problem solving / exploring / investigating
- **Interpersonal skills:** communication / teamwork

Key vocabulary

- corner, square, rectangle, triangle, side, vertex, two-dimensional (2D) shape

Resources

- 2D shapes made from card, including a range of triangles and different sizes of squares and rectangles
- mini whiteboards or paper for recording, scissors, squares of paper

Language support

Make a display showing the names and properties of 2D shapes. Students should use this language of shape to help them discuss the properties of shapes in all the activities.

 **Introductory activity**

Show students the cardboard shapes. Ask them to sort through the shapes and, working with a partner, discuss what is the same and what is different about each shape. Ask questions such as, *How many sides do they have? How many corners do they have?* Introduce the word **vertex** and explain that it is a special word for the corner of a shape.

Explain that all of these shapes are flat – they have height and width, but no depth. Compare a square with a box to demonstrate this (a box has depth as well). Explain that we call these flat shapes **two-dimensional (2D)**.

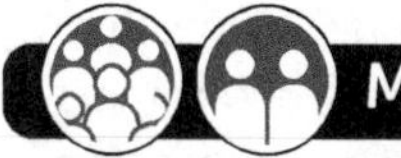 **Main activity**

Ask students, in their pairs, to use the shapes to make pictures or patterns.

Ask what they found out about the shapes. Record on the board the name of each 2D shape and the things that students discovered, including the properties of the shapes that students tell you. For example, they may discover that they can put two squares together to make a rectangle, or two or three triangles make a square.

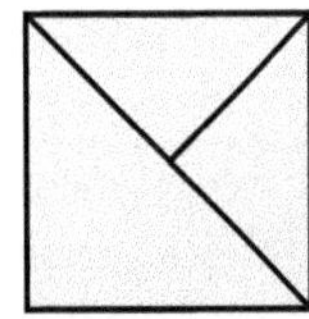

In pairs, students cut up a square of paper as described on page 142 of the Student Book, and rearrange the pieces to make the shape shown in question 5 of the Student Book. They should then make as many different shapes as they can. Repeat this for four-sided and five-sided shapes.

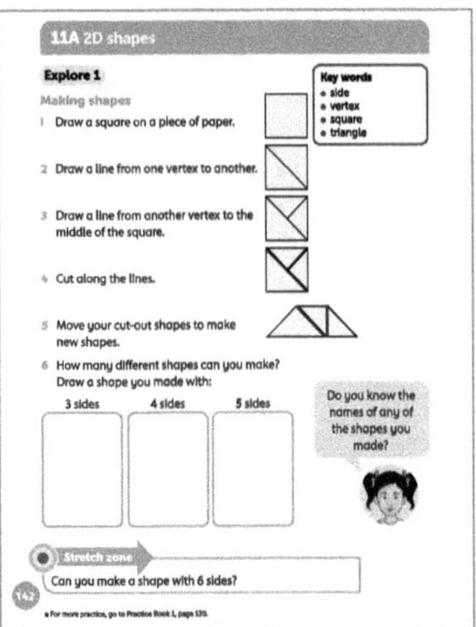

Differentiation

Supporting: Model the language of shape carefully, pointing to the sides and corners of the cardboard shapes and the shapes in the Student Book.

Consolidating: Ask students to describe the properties of shapes they make.

Extending: Encourage students to develop the language of properties of shapes by comparing 'similar' shapes. For example, for different shapes they make, ask, *What is the same? What is different?*

Stretch zone: *Can you make a shape with 6 sides?*

Check that students can make a six-sided shape using the pieces.

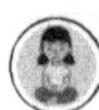 **Reflection time**

Students should come to the front of the class and show the different shapes they have made. Use the correct names for the different shapes, reminding students that the names are linked to the number of sides and corners the shapes have. For example, a shape with five sides is a pentagon. Ask students to show different examples of the same shape, for example, hexagons. However, students are not expected to know these names at this stage.

Practice Book: Students can complete Practice Book page 130. This can be done directly after the main activity, as homework, or as the focus of a separate mathematics session to help students consolidate their learning and build fluency. Encourage students to use a straight edge or a ruler to make their shapes as accurate as possible. Challenge them to follow the instructions to make two different shapes, if possible, rather than one.

Differentiated outcomes	
All students	should recognise and name squares, rectangles and triangles.
Most students	will name squares, rectangles, triangles and circles confidently and use appropriate vocabulary to describe the properties of shapes.
Some students	may begin to name different types of triangle (scalene, isosceles, right-angled, equilateral) and use the language of properties confidently and appropriately.

Answers

Student Book page 142

Check that students are following the instructions carefully, and that their shapes have the right number of sides. While students are working, ask them to name the different shapes that they make.

Practice Book page 130

Check that students have followed the instructions and that each shape matches its description. Students could use rulers to help them to draw straight lines.

Stretch zone: Students should name square, triangle and rectangle but may need help with the other shapes.

11A 2D shapes

Explore 2 Student Book page 143 • Practice Book page 131

Specific learning focus

- Recognise and name simple 2D shapes around the classroom.

Global skills

- **Creative skills:** problem solving / exploring / investigating
- **Real-world skills:** research / presenting information
- **Interpersonal skills:** communication / teamwork

Key vocabulary

- rectangle, face, circle, square, triangle

Resources

- set of simple 2D shapes

Language support

Use the display showing the names and properties of 2D shapes to help students use the language of shape when looking for shapes around the classroom.

Introductory activity

Hold up a rectangle. Ask students to count how many sides it has. (4) *Can you find an object that has a **face** that is a rectangle?* Show them an example to demonstrate what you mean by the word 'face', e.g. a face of a tissue box. Students may suggest, for example, a book, a desktop, or a ruler.

Draw two different rectangles on the board. Ask students to describe what is the same about the two rectangles, and what is different about them.

Main activity

Have a selection of 2D shapes for students to pick up and examine. Choose a student to come forward and pick a shape. Ask them if they can describe its properties, the number of sides and types of sides. Can they name something in the class that matches the shape?

Repeat for another shape and another student. Remind students to focus on the properties, such as how many sides, and not the size of the shape, since a small square and large square are still both squares.

Students can complete the activities on page 143 of the Student Book individually. Remind them of how they looked for objects that matched the shapes they already examined. Can they use this information to help them complete the table?

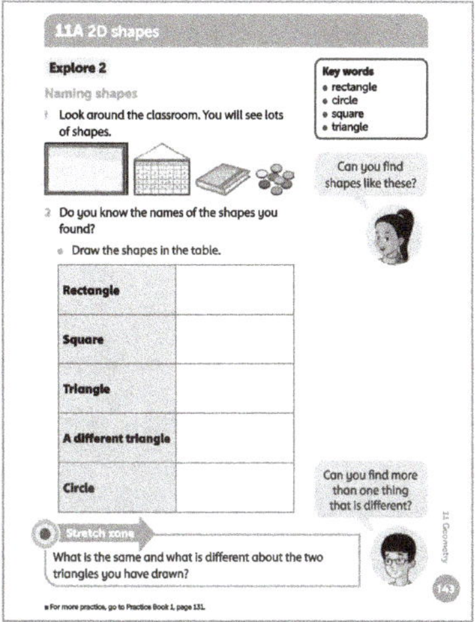

Differentiation

Supporting: Model the language of shape carefully, pointing to the sides and corners of the cardboard shapes and the shapes in the Student Book.

Consolidating: Ask students to describe the properties of shapes they see. Encourage them to use precise language.

Extending: Encourage students to develop the language of properties of shapes by comparing 'similar' shapes. For example, for different shapes they found, ask, *What is the same? What is different?*

Stretch zone: *What is the same and what is different about the two triangles you have drawn?*

Check that students can describe the features that are the same and different on their triangles.

Reflection time

Ask students to compare two shapes they have seen, for example, a triangle and a rectangle. *How are they the same? How are they different?*

Show students three shapes, for example, two different-sized squares and a triangle. *Can they name a shape that is the odd one out?* Explain that 'odd one out' means that there is one shape that is different from the others. *What makes it the odd one out?*

Practice Book: Students can then complete Practice Book page 131. This can be done directly after the main activity, as homework, or as the focus of a separate mathematics session to help students consolidate their learning and build fluency. Students should ask an adult for help in naming any objects they find around the house, using a dictionary to identify the names if necessary.

<table>
<tr><td colspan="2">Differentiated outcomes</td></tr>
<tr><td>All students</td><td>should recognise a rectangle, square, circle and triangle.</td></tr>
<tr><td>Most students</td><td>will recognise and name each of the shapes.</td></tr>
<tr><td>Some students</td><td>may draw different versions of some shapes, such as triangles and rectangles.</td></tr>
</table>

11B 3D shapes

Discover Student Book page 144 • Practice Book page 132

Specific learning focus

- Name and sort common 3D shapes (e.g. cube, cuboid, cylinder, cone and sphere) using features such as number of faces, flat or curved faces. Use these to make patterns and models.

Global skills

- **Creative skills:** problem solving / exploring / investigating
- **Real-world skills:** research / presenting information / interpreting information / financial literacy
- **Interpersonal skills:** communication / teamwork / leadership
- **Self-development skills:** reflecting on learning

Key vocabulary

- cube, cuboids, cylinder, cone, sphere, edges, faces, flat, curved

Resources

- empty containers (e.g. cereal boxes, cardboard cylinders, tin cans), wooden blocks of different 3D shapes

Language support

Create a poster with names and properties of 3D shapes (including face and edge), which you can refer to throughout the unit. *This shape rolls in one direction* (cylinder), *this shape rolls in any direction* (sphere).

Answers

Student Book page 143

Check that students have drawn the correct shapes in the table.

Practice Book page 131

Check that students have drawn the correct shapes and written the name of an object that matches the name of the shape.

Stretch zone: Check that students can describe the features that are the same and different on their triangles.

 **Introductory activity**

Show students the selection of empty containers of different shapes and sizes. In pairs, students should select one of the containers. Ask each pair to explore and discuss their container for two minutes. Invite each pair to share what they noticed about their shape. Discuss and describe the properties of each shape with students using correct vocabulary and share the name of each 3D shape. Encourage correct use of mathematical vocabulary and language relating to 3D shapes, e.g. rather than, *The box has writing on it and the wooden block doesn't,* say, *The **cube** has writing on it but the **cuboid** doesn't.* You can also model the correct language of the properties of the shapes.

Which of these containers can we roll? Can you find a shape from the wooden blocks that we can roll? **(cylinder, sphere)** *Are they the same shape as the container? Are they the same shape as each other? What is the same and what is different?*

 **Main activity**

Give each student or pair some empty containers of different shapes and a selection of wooden building blocks. Ask students to look at page 144 of the Student Book, and to stack the containers and blocks to make a tower. *Which blocks should go at the bottom? Which blocks will go on the top?* Demonstrate with the shapes when you are describing some of the properties. *Would it be a good idea to put a sphere or a cylinder at the bottom of your building? Why not?*

What do you think would make the tallest tower? What do you think would be the most containers or blocks you could use? Try it and see what happens. How will you make your tower with cylinders? Why?

Encourage students to feel the surfaces of the containers and blocks and make decisions about using the **flat** surfaces in their tower.

Ask students to record information about their towers on page 144 of their Student Books.

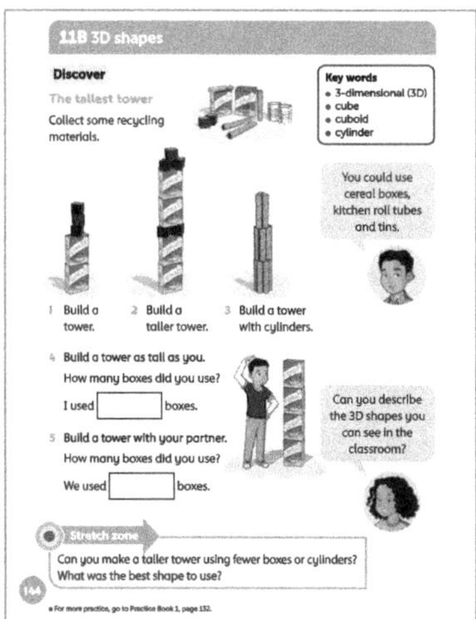

Differentiation

Supporting: Model the language carefully, pointing to the shapes as you use the vocabulary.

Consolidating: Ask students to describe the properties of shapes they see and use.

Extending: Encourage students to develop the language of properties of shapes by comparing 'similar' shapes: for example, different blocks which are 'similar'. *What is the same? What is different?*

Stretch zone: *Can you make a taller tower using fewer boxes or cylinders?*

What was the best shape to use?

Listen to students' reasoning and check they can use the correct vocabulary in describing their towers.

 **Reflection time**

Choose some of the pairs/groups to feed back to the rest of the class how they built their tower. Use sentence frames on the board to support students, for example,

We used a _____ at the bottom of our tower. We used a _____ at the top. We also used a _____.

Ask questions such as, *What was the tallest tower you made? How can you make a taller tower with fewer containers or blocks? Why did you choose this shape?*

Practice Book: Students can complete Practice Book page 132. This can be done directly after the main activity, as homework, or as the focus of a separate mathematics session to help students consolidate their learning and build fluency.

Differentiated outcomes	
All students	should name cubes, cuboids, cylinders and spheres and begin to see the differences between them.
Most students	will name cubes, cuboids, cylinders and spheres and begin to describe their properties.
Some students	may describe some 3D shapes and their properties with confidence.

Answers

Student Book page 144

Observe students as they work through the different activities and note who recognises how to build stable towers. Ask them to describe the faces of the shapes that they can lay on top of one another. (They're flat.)

Practice Book page 132

Check that students appreciate that they should put flat faces together when they build the towers. Ask them to name each shape that they use in their towers, and to explain how they know what a shape's name is (e.g. It has a circle at either end and a curved side, so it is a cylinder).

Stretch zone: Answers will vary depending on the towers students have built.

11B 3D shapes

Explore 1   Student Book page 145 • Practice Book page 133

Specific learning focus

- Name and sort common 3D shapes (e.g. cube, cuboid, cylinder, cone and sphere) using features such as number of faces, flat or curved faces. Use them to make patterns and models.

Global skills

- **Creative skills:** exploring / investigating
- **Real-world skills:** research / presenting information
- **Interpersonal skills:** communication / teamwork

Key vocabulary

- three-dimensional (3D), cube, cuboid, cylinder, cone, sphere, edges, faces, flat, curved, vertices

Resources

- set of 3D shapes (cubes, cuboids, cylinders) for each small group

Language support

Encourage the use of correct mathematical language throughout the activities. Students will learn from talking, and listening to others talking, and show their understanding to you.

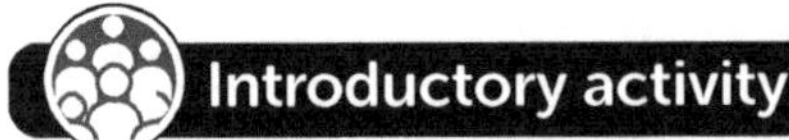

Introductory activity

Tell students they are going to learn more about **three-dimensional (3D) shapes**. Show them a cube, cuboid and cylinder. *What is different about these shapes compared to two-dimensional shapes such as squares or triangles?* (They have depth – we call them solid shapes.) Hide some 3D shapes in a bag or behind a screen. Place a matching set on the table at the front of the classroom so that all students can see them. Leave this display of shapes in view throughout the lesson.

Choose a student to pick one of the hidden shapes from the bag and, without looking at it, tell the others what they can feel. Encourage students to use correct mathematical vocabulary and language to describe the properties of the shapes, such as, it has no corners, it won't roll, it will roll, it has faces. Ask the class if they can see a matching shape on the table. If they can, choose a student to pick up the one that matches. Then ask the first student to reveal the hidden shape to see if it matches. Repeat with different students.

Main activity

Give each small group a set of 3D shapes. Tell students to look at the shapes and discuss their properties, referring to the set of 3D shapes on the table at the front of the classroom to help them.

Groups should then sort the shapes into two different groups and tell you the criteria they have used to sort them, for example shapes with more than four faces, shapes with more than eight **edges**, shapes that have a **curved** surface and can roll.

Students should record the activity in the Student Book page 145. They can continue to work in their small groups to complete question 3 or break off into pairs.

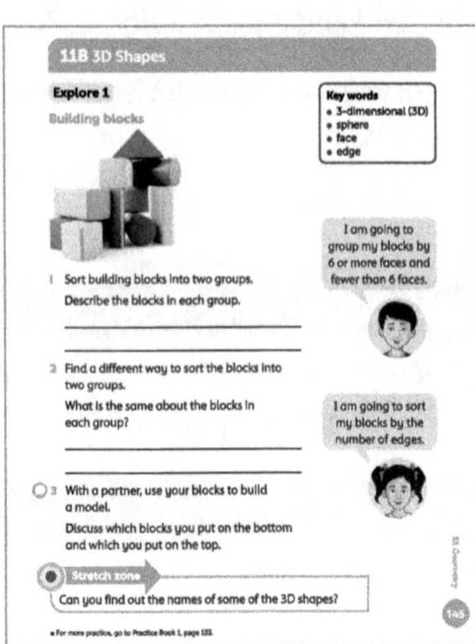

Once students have sorted their shapes and told you their criteria, ask them to sort them in another way. They should be able to tell you what criteria they are using again. Repeat a third time. Take this opportunity to model the use of the key vocabulary.

Differentiation

Supporting: Model the language carefully, pointing to the shapes and its properties as you use the vocabulary.

Consolidating: Ask students to describe the properties of shape they see and use.

Extending: Encourage students to develop the language of properties of shapes by comparing 'similar' shapes: for example, different blocks which are 'similar'. *What is the same? What is different?*

Stretch zone: *Can you find out the names of some of the 3D shapes?*

Students should be able to find out the names of the 3D shapes in the activity.

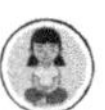 ## Reflection time

Give each student a 3D shape. Tell them that you are going to say something about a shape and if they think the description fits their shape, they must hold it in the air.

Choose some properties: for example, has 8 corners, has no corners, has 6 faces. When a student holds up the shape, ask them to show you why their shape has that property by demonstrating on the shape.

Practice Book: Students can complete Practice Book page 133. This can be done directly after the main activity, as homework, or as the focus of a separate mathematics session to help students consolidate their learning and build fluency. Remind students of what edge, face and vertex mean, using a 3D shape to help them visualise these features.

Differentiated outcomes	
All students	should name cubes, cuboids, cylinders and spheres.
Most students	will name a wider range of 3D shapes and begin to describe their properties, and use these to sort the shapes.
Some students	may name a wide range of 3D shapes and describe their properties with confidence, and sort the shapes according to two criteria, for example, 'all flat faces' and more than '3 faces'.

Answers

Student Book page 145

1 Some blocks can roll, others cannot – this is one possible answer.

2 Check that students can identify what is the same about all the blocks in their groups.

3 Check that students can use accurate vocabulary to describe their model.

Practice Book page 133

Ask students to explain how they know that the object they have found is, for example, a pyramid. They should count the number of faces, edges and vertices it has.

Stretch zone: Students should be able to name the shapes they have drawn.

Explore 2 Student Book page 146 • Practice Book page 134

Specific learning focus

- Name and sort common 3D shapes (e.g. cube, cuboid, cylinder, cone and sphere) using properties such as number of faces, flat or curved faces. Use them to make patterns and models.

Global skills

- **Creative skills:** problem solving / exploring / investigating
- **Interpersonal skills:** communication / teamwork

Key vocabulary

- three-dimensional (3D), straight edge, curved edge, face

Resources

- set of different 3D shapes – one per group or pair

Language support

Encourage the use of correct mathematical language throughout the activities. Students will learn from talking, and listening to others talking, and show their understanding to you. You could also label the actual shapes with stickers or sticky notes to help with counting faces.

 **Introductory activity**

Show students a cube. Ask them to tell you what properties it has. Can they identify the faces and edges? *How many faces does it have? How many edges?* They should be able to count the 6 faces and the 12 edges.

Ask them to describe the faces. Can they identify that all the faces are squares? How do they know?

 **Main activity**

Arrange students into small groups or pairs. Give them a selection of 3D shapes with straight edges, e.g. a cuboid, triangular-based pyramid and square-based pyramid.

For each shape, ask them to count the number of faces and edges and try to describe the shape of the faces. Ask students what they notice about the faces on the 3D shapes: for example, some 3D shapes have faces that are all the same shape, but some have different-shaped faces. Encourage them to describe the properties of the faces and use their correct names, if they are familiar with them, e.g. square, triangle and rectangle.

Now give each group a selection of 3D shapes with curved edges: different-shaped cylinders. For each shape, ask students to count the number of faces and edges and to try to describe them. Ask students what they notice. Draw out that these shapes have no straight edges, but they have a curved edge. Students may also notice that they have two flat faces and a curved surface.

Students can work on the activities on page 146 of the Student Book individually. When they have completed the questions, they should decide on another way to sort the shapes.

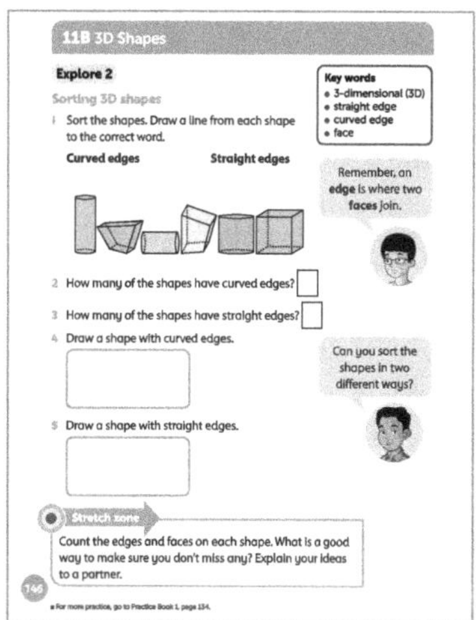

Differentiation

Supporting: Model the language carefully, pointing to the shapes as you use the vocabulary. When counting the faces on a 3D shape, you can mark the faces to help with counting.

Consolidating: Ask students to describe the properties of shapes they see and use.

Extending: Encourage students to develop the language of properties of shapes by comparing 'similar' shapes: for example, different blocks which are 'similar'. *What is the same? What is different?*

Stretch zone: *Count the edges and faces on each shape. What is a good way to make sure you don't miss any? Explain your ideas to a partner. Test it to see if it works.*

Check that students have counted edges and faces correctly and ask them how they tested their ideas to see if they worked.

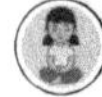 **Reflection time**

Show students a sphere and a cylinder. Ask them to describe to a partner what is the same and what is different about the two shapes. *Can you describe the faces?* (A sphere has no flat faces but one curved surface, a cylinder has two flat faces and a curved surface.) *Can you describe the edges?* (A sphere has no edges, a cylinder has two curved edges.)

Practice Book: Students can complete Practice Book page 134. This can be done directly after the main activity, as homework, or as the focus of a separate mathematics session to help students consolidate their learning and build fluency. Explain that they can use whatever they can find around their home to make their model if they don't have blocks. Encourage them to think about the properties of each of their blocks as they make their model.

Differentiated outcomes	
All students	should count faces on a 3D shape.
Most students	will count faces and edges on a 3D shape.
Some students	may describe all the faces and edges on a 3D shape.

Student Book page 146

1   Curved edges – shapes 1, 3 and 5. Straight edges – shapes 2, 4 and 6

2   Three shapes have curved edges.

3   Three shapes have straight edges.

4   Check that students have drawn a shape with curved edges.

5   Check that students have drawn a shape with straight edges.

Practice Book page 134

Check that students have made a model and drawn it carefully.

Stretch zone: Check that students have written a correct sentence using the shape vocabulary to describe their model.

11B 3D shapes

Explore 3   Student Book page 147 • Practice Book page 135

Specific learning focus

- Name and sort common 3D shapes (e.g. cube, cuboid, cylinder, cone and sphere) using properties such as number of faces, flat or curved faces. Use them to make patterns and models.

Global skills

- **Creative skills:** exploring / investigating
- **Real-world skills:** research
- **Interpersonal skills:** communication / teamwork

Key vocabulary

- three-dimensional (3D), straight edges, curved edges, faces

Resources

- set of different 3D shapes
- collection of empty tins and packets of different shapes (one per group)

Language support

Help students to understand that some mathematical words are used in everyday context without always having the same meaning, for example 'face'.

 **Introductory activity**

Show students a set of 3D shapes including a cube, cuboid, cylinder and sphere. Also have a set of everyday objects of different shapes, such as a cereal box, tin of beans and a tennis ball.

Let students look at the two sets of objects and then ask if they can match up each 3D shape with an everyday object. Ask them to explain their reasoning as they compare and match the shapes. Ask, for example, *What did you notice first about the cereal box? How is it the same as the cuboid? Why did you not match it with the cube? Can you think of another example of each shape?*

 **Main activity**

Give each group of students a collection of empty packets, tins and containers. Ask them to sort their items into two groups according to a property that you give, for example, whether they have a curved surface and can roll or not. Give them time to sort their items then ask each group in turn to explain how they sorted their shapes and why.

Now ask students to choose their own sorting criteria and let them have some time to re-sort their items. When they have finished, ask each group to explain to the rest of the class what criteria they used to sort their shapes and how they sorted them. You could ask other students to guess what the sorting criteria were for each group before they explain, by looking at how each group sorted their shapes.

Students can then complete page 147 in the Student Book individually. Drawing 3D shapes is very challenging. Explain this to students to manage their expectations and suggest that, if they prefer, they can draw each face of the shape instead.

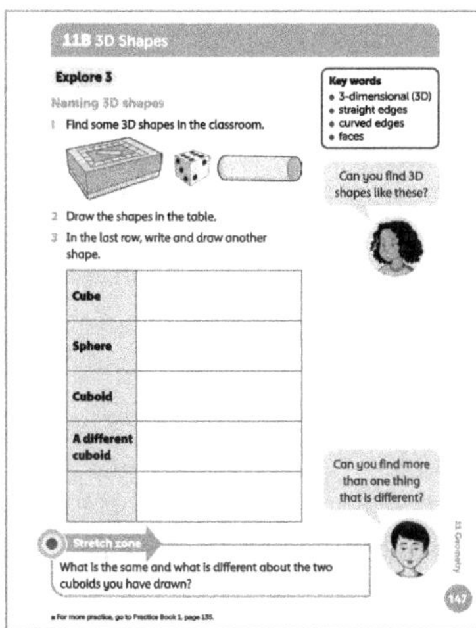

Differentiation

Supporting: Give students 2D shapes to trace around or let them draw their shapes on a separate sheet of paper so they can erase when necessary and make the drawing bigger than the boxes in the Student Book will allow.

Consolidating: Ask students to describe the properties of shapes they see and use.

Extending: Ask students to draw each face of their shapes.

Stretch zone: *What is the same and what is different about the two cuboids you have drawn? Write your answer in your notebook.*

Check that students have identified things that are the same and different about the two cuboids.

 Reflection time

Review the properties of 3D shapes with the students. What can they tell you about 3D shapes? Can they recognise, count and describe the faces? Edges? Can they relate the shapes to everyday items like boxes and tins? Encourage them to look for examples from around the classroom.

Practice Book: Students can then complete Practice Book page 135. This can be done directly after the main activity, as homework, or as the focus of a separate mathematics session to help students consolidate their learning and build fluency. Ask students to challenge themselves to describe the properties of each of the shapes and share them with an adult.

Differentiated outcomes	
All students	should identify some 3D shapes.
Most students	will think of everyday objects to match the 3D shapes.
Some students	may describe the properties of a 3D shape using its faces and edges.

Answers

Student Book page 147

Check that students have drawn and listed the correct shapes.

Practice Book page 135

1 Cuboid

2 Sphere

3 Square-based pyramid

4 Cuboid

5 Triangular-based pyramid

6 Cube

Stretch zone: They are both pyramids with triangular faces. Their bases are different – one is square, one is triangular.

11C Symmetry

Specific learning focus

- Recognise basic line symmetry.

Global skills

- **Creative skills:** problem solving / exploring / investigating
- **Real-world skills:** research / presenting information / interpreting information / financial literacy
- **Interpersonal skills:** communication / teamwork / leadership
- **Self-development skills:** reflecting on learning

Key vocabulary

- symmetrical, reflect, reflection, line of symmetry, mirror line, pattern

Resources

- pattern blocks or cardboard 2D shapes, mirrors

Language support

Encourage the use of correct mathematical language throughout the activities. Students will learn from talking, and listening to others talking, and show their understanding to you.

 **Introductory activity**

Show students some 2D shapes that they have seen previously, for example, square, rectangle, triangle, circle, pentagon, hexagon. Review their names and properties using questions such as, *How many sides does this shape have? How many corners does it have?* Place shapes in different arrangements and ask questions to practise the language of position. For example, *Look at these two shapes. Can you describe their positions?* (e.g. The triangle is below the square. The square is next to the rectangle.)

 **Main activity**

Place two squares so that they touch edge to edge on the table in front of the class where all students can see them. *What can you see?* (Two squares next to each other) Then slowly slide one of the squares away and replace it with a mirror held vertically next to the first square. *I put the mirror here. What can you see now? How many squares can you see?* (Two squares next to each other) *We have one square here and now we can see a second square, which is a **reflection** in the mirror.*

Draw a diagram on the board to show the two squares, with a broken line to represent the mirror.

*This line shows where I put the mirror. We call it a **mirror line**. Can you describe where the reflection of the square is?* Agree that it is in a similar position but on the other side of the mirror line.

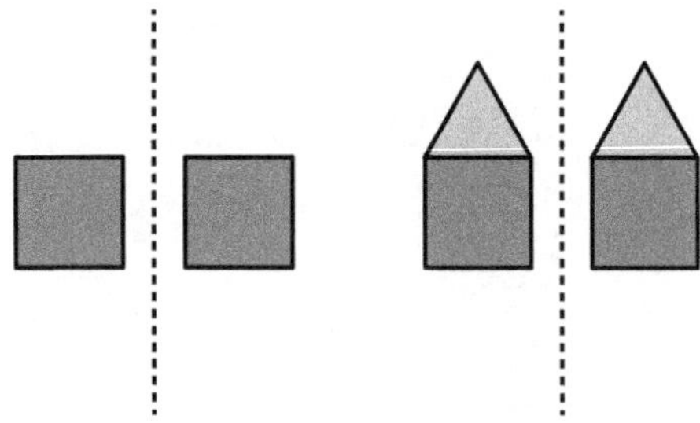

Now go back to the shapes on the table, and show the two squares as originally. Add a triangle to the first square. Explain that you want the pattern to be the same on both sides of the mirror. Agree where to add a triangle to the second square. *Is the **pattern** the same?* After listening to responses, say, *Let's check by using a mirror.*

Move one of the squares and triangles away and replace it with the mirror as before, showing that the reflection of the shape matches the original.

The square and triangle are reflected in the mirror. Add a triangle to each of the squares on the diagram on the board (as shown above) and point to the mirror line. *We can call the mirror line a **line of symmetry**. The pattern is the same on both sides of the line. We say it is **symmetrical**.*

Give each pair of students a mirror and shape blocks. Ask them to make a pattern using one, two and three blocks and reflect it in the mirror. They should explain the pattern they made on one side of the mirror and then describe the reflection.

Ask students to complete Student Book page 148 individually, which involves them colouring in grids to make symmetrical patterns. Give students mirrors to help them check whether the patterns are symmetrical.

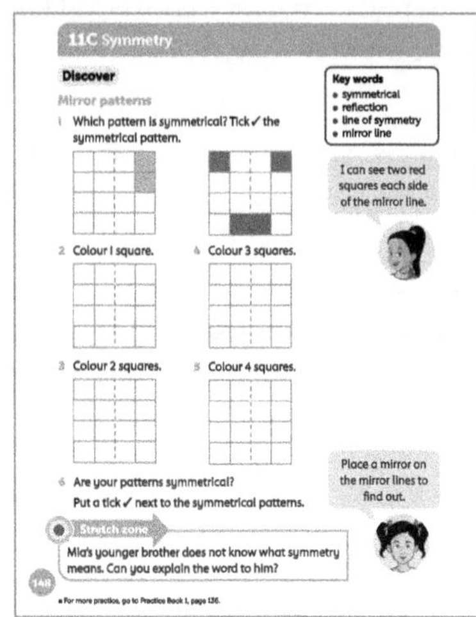

Differentiation

Supporting: Make a pattern with students. Allow them to colour a square in and then you colour a square to make it symmetrical. Gradually shift so that they are finding the symmetrical square.

Consolidating: Ask students to show you where the lines of symmetry are on the patterns that they are making.

Extending: Ask students to make more complex patterns with a variety of 2D shapes.

Stretch zone: *Mia's younger brother does not know what symmetry means. Can you explain the word to him?*

Check that students can describe symmetry reasonably, with examples if possible.

 **Reflection time**

Ask pairs to look around the classroom for examples of shapes that make a symmetric pattern. Then ask them to share the shapes that they have spotted with the whole class and explain where the line of symmetry is, and how they know.

Practice Book: Students can complete Practice Book page 136. This can be done directly after the main activity, as homework, or as the focus of a separate mathematics session to help students consolidate their learning and build fluency. Explain that this is a game and they will need to play with a partner, who can be an adult or another child. Read through the rules together so that students understand how to play the game, using the sample game artwork to support your explanation.

Differentiated outcomes	
All students	should recognise symmetrical shapes and make simple symmetrical patterns with support.
Most students	will make simple symmetrical patterns.
Some students	may make more complex symmetrical patterns and notice symmetry in their environment.

Answers

Student Book page 148

While students colour in squares on each grid, ask them to predict whether their patterns will be symmetrical or not. When they have used a mirror to check which patterns are symmetrical, ask them to explain what the symmetrical patterns have in common. (Coloured squares on opposite sides of the pattern are equal distances from the centre lines of the grid.)

Practice Book page 136

Observe students and note who makes the pattern symmetrical confidently by counting squares from the mirror line.

Stretch zone: Check that students have drawn two symmetrical patterns.

11C Symmetry

Explore 1   Student Book page 149 • Practice Book page 137

Specific learning focus

- Recognise basic line symmetry.

Global skills

- **Creative skills:** problem solving / exploring / investigating
- **Interpersonal skills:** communication / teamwork

Key vocabulary

- line of symmetry, symmetrical, reflection, mirror line

Resources

- pictures showing line symmetry in the environment (e.g. butterfly, starfish, leaf, flower), to display
- objects that have symmetry (e.g. an orange or certain 3D shapes), mirrors

Language support

During the Main activity, observe and listen as the class is working. Notice pairs who are having difficulty with the language of symmetry and model the correct key words and phrases.

 **Introductory activity**

Show students a picture of a butterfly and draw a vertical line through the image where the line of symmetry is. *What do you notice about the butterfly?* Students should be able to describe the two halves as having the same shape and pattern and may say that the butterfly is symmetrical. Show another picture of a symmetrical object and repeat the drawing of a vertical line of symmetry and the question. Discuss the fact that it is not only patterns that can be symmetrical; objects can be symmetrical too. Then show students a fruit, for example an orange. Ask if they think this is symmetrical. Cut it in half vertically. *Does it look symmetrical from the outside? What about on the inside?*

Draw a 3 × 3 square grid on the board and add a vertical line of symmetry, then shade two squares to form a symmetrical pattern, e.g.

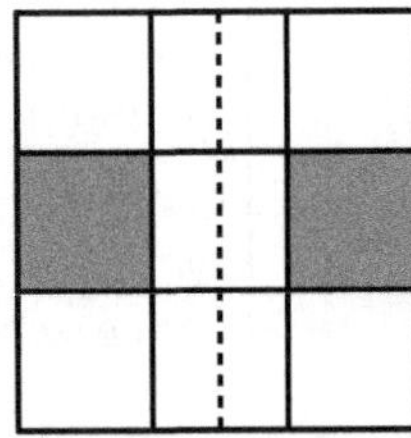

Explain that the dashed line is a vertical line of symmetry; it runs from top to bottom. *What do you notice?* Students may notice that there are two squares shaded to make a pattern and the pattern is symmetrical. *How do you know it is symmetrical?* (Because the shaded square on one side is a reflection of the pattern on the other side.)

Shade another square in the pattern, e.g. the top left-hand corner. *Is the pattern still symmetrical?* (No) *What square could we shade to make the pattern symmetrical?* Ask a student to come to the front of the class to show you by, e.g., shading the top right square. Shade further squares on both sides of the mirror line to make new symmetrical patterns. Include one with the mirror line running through a shaded square, e.g.

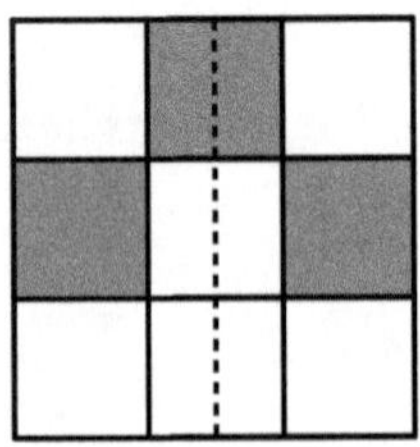

Ask students to draw and check symmetrical patterns on the 3 × 3 grids on page 149 of the Student Book. They should work in pairs and use a mirror to check that patterns are symmetrical. Before beginning, remind them that they need to draw a vertical line of symmetry on their grids. Give time for exploration. Show how a mirror line can be placed halfway through the shaded square in any direction, including the diagonals.

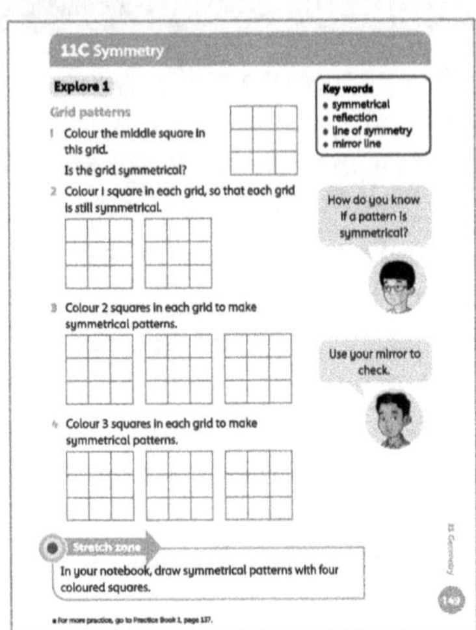

Differentiation

Supporting: Before starting work on the Student Book activity, draw or help students draw a line of symmetry on each grid.

Consolidating: Ask students to show you where the lines of symmetry are on the shapes that they are making, and encourage them to use horizontal as well as vertical lines of symmetry.

Extending: Ask students to use images from the Internet to create a poster to show symmetry. They should draw the lines of symmetry on the images.

Stretch zone: *In your notebook, draw symmetrical patterns with four coloured squares.*

Check that students have drawn symmetrical patterns, and use these as the focus of the reflection time.

 Reflection time

Choose pairs of students who have drawn more complex symmetrical patterns to come up to the board and draw them a square at a time on a 3 × 3 grid you have prepared. Other students should try to predict where the next square is to be coloured in.

When the pattern is complete, pairs should agree where the mirror line, or line of symmetry, will be. Some patterns may have more than one line of symmetry. Point this out to students.

Practice Book: Students can complete Practice Book page 137. This can be done directly after the main activity, as homework, or as the focus of a separate mathematics session to help students consolidate their learning and build fluency. Look together at the photograph of the snowflake and explain that it has a number of lines of symmetry. Can they find some of them? Say that when they have found pictures that have a line of symmetry and they have stuck them in their books they may also want to draw the line of symmetry on their pictures.

Differentiated outcomes	
All students	should recognise symmetrical shapes and make simple symmetrical patterns with support.
Most students	will make simple symmetrical patterns.
Some students	may make more complex symmetrical patterns and notice symmetry in their environment.

Answers

Student Book page 149

While students are working, ask them to show you how they can check that they have drawn symmetrical patterns. Ensure that the mirror is placed in the middle of each grid, either horizontally or vertically, or diagonally from corner to corner.

Practice Book page 137

Observe students and check that they have found examples of images that show symmetry.

Stretch zone: Check that the student's explanation makes it clear they understand symmetry.

11C Symmetry

Explore 2   Student Book page 150 • Practice Book page 138

Specific learning focus

- Recognise basic line symmetry and form symmetrical patterns.

Global skills

- **Creative skills:** problem solving / exploring / investigating
- **Interpersonal skills:** communication / teamwork

Key vocabulary

- symmetrical, reflection, line of symmetry, horizontal, vertical, diagonal

Resources

- pattern blocks or cardboard 2D shapes, mirrors, images that show line symmetry

Language support

Ask students questions about their pattern to reinforce the language of position and shape, for example *Show me which block is symmetrical to this one.* (Point to one colour block.) *Where does this block need to go to be symmetrical?*

 **Introductory activity**

Show students examples of patterns with different lines of symmetry, e.g. diagonal and horizontal. *The patterns we looked at in the last lesson had **vertical** lines of symmetry. This pattern is symmetrical but it has a **[horizontal/diagonal]** line of symmetry.* Ask how students can tell if the pattern is symmetrical. *Can you point out where the line of symmetry is? Which shapes have I used? How many shapes have been used? Can you describe where [the red rectangle] is?*

Main activity

Give each pair of students a set of pattern blocks. Ask each pair to take two of the same block and place them together along an edge, as below.

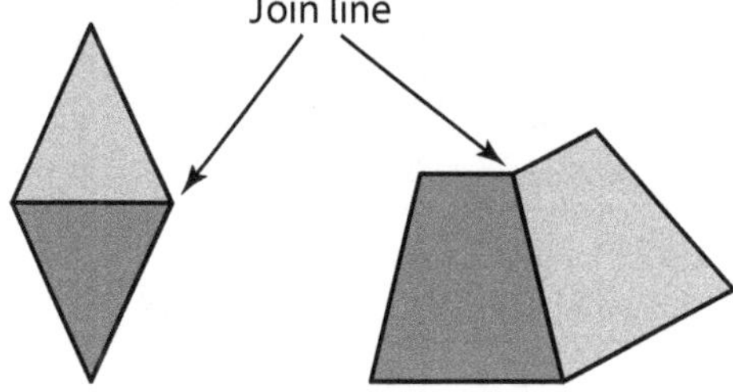

Have you made a symmetrical shape? Is the join line a line of symmetry? If not, can you join them another way to make a symmetrical shape?

Now ask students to use any three pattern blocks. Can they make a symmetrical shape using them? What about four blocks? *Can you make a symmetrical pattern in more than one way using the same blocks?*

Students can now complete the activities on page 150 of the Student Book. Explain that both the shapes and colours should be symmetrical, e.g. the reflection of a green triangle on one side of the line of symmetry should also be a green triangle, not blue for example.

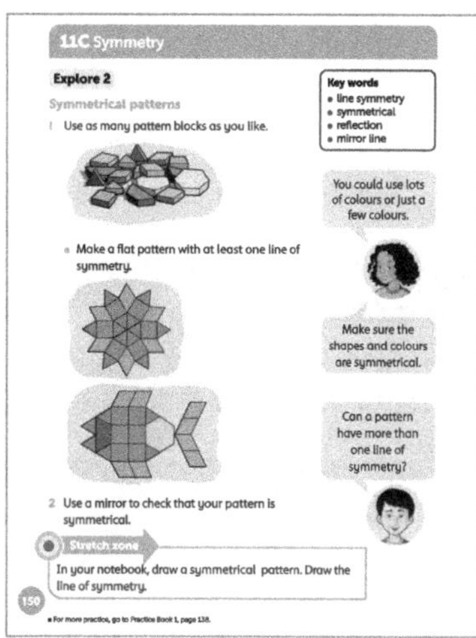

Differentiation

Supporting: Make shapes with students. Allow them to place a block and then you place one to make the pattern symmetrical. Include some intentional errors. Check with students if they agree with you each time you place a block.

Consolidating: Ask students to show you where the lines of symmetry are on the shapes that they are making.

Extending: Ask students to use images from the Internet to create a poster to show symmetrical patterns. They should draw the lines of symmetry on the patterns.

Stretch zone: *In your notebook, draw a symmetrical pattern. Draw the line of symmetry.*

Check that students have drawn a symmetrical pattern and marked its line of symmetry correctly.

 **Reflection time**

Each pair should be given the opportunity to describe their symmetric pattern to the rest of the class. Can they explain where the line of symmetry is?

Practice Book: Students can then complete Practice Book page 138. This can be done directly after the main activity, as homework, or as the focus of a separate mathematics session to help students consolidate their learning and build fluency. Tell students they can make a symmetrical pattern by colouring in complete squares but they can also use the grid to help them draw a symmetrical pattern using other shapes.

<table>
<tr><td colspan="2">Differentiated outcomes</td></tr>
<tr><td>All students</td><td>should recognise a pattern with a line of symmetry.</td></tr>
<tr><td>Most students</td><td>will make their own symmetrical pattern.</td></tr>
<tr><td>Some students</td><td>may make and describe a more complex symmetrical pattern.</td></tr>
</table>

Student Book page 150

Check that students have made a symmetrical pattern.

Practice Book page 138

Check that students have made a symmetrical pattern on the squared grid.

Stretch zone: Check that students have made a non-symmetrical pattern.

11D Position and movement

Discover Student Book page 151 • Practice Book page 139

Specific learning focus

- Use everyday language of direction and distance to describe movement of objects.

Global skills

- **Creative skills:** problem solving / exploring / investigating
- **Real-world skills:** research / presenting information / interpreting information / financial literacy
- **Interpersonal skills:** communication / teamwork / leadership
- **Self-development skills:** reflecting on learning

Key vocabulary

- front, back, in front of, at the back of, behind, next to, between

Resources

- toy cars and trucks (or similar)
- cubes, beads

Language support

Use an image of a traffic jam and write the key vocabulary such as 'in front' and 'behind' to create a poster to refer to. Include sentences which describe the relative positions of different cars.

 Introductory activity

Ask all students to come to the front of the classroom. They should sit in a circle around a selection of toy cars and trucks. Tell them that they are going to make traffic jams – putting some vehicles close to each other in a line. Ask a student to make a traffic jam with some of the toy cars and trucks. Then ask other students to talk to a partner for one minute about anything they notice about the vehicles' positions in the traffic jam. Take feedback

round the circle – repeat any position phrases such as **in front of**, **at the back of**, **next to**, **behind, between** to emphasise this vocabulary. Introduce any of the words and phrases students did not use, linking them to the vehicles' positions in the traffic jam, e.g. *The yellow car is between the red car and the purple truck.* Follow it with a corresponding question, e.g. *Where is the yellow car?*

Ask other students to make changes to the traffic jam. Ask questions like:

Can you put this truck behind the blue car / in front of the red truck / between the yellow taxi and the school bus?

Can you make a traffic jam with all the trucks at the front and the cars at the back?

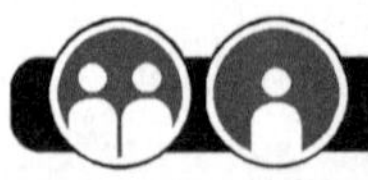 Main activity

Each pair of students should take four or five toy vehicles back to their table. They take it in turns to arrange the cars and trucks following instructions from their partner. The students should give one instruction each, one at a time. For example, the first student might say, *Put the blue car at the front.* Then the other student might say, *Put the yellow truck behind the blue car.* Circulate around the class modelling the key vocabulary.

Students complete the activities on Student Book page 151 individually, drawing on the practice they have had using the key vocabulary in the first part of the activity in pairs. Once they have completed their pattern for question 3, they should describe it to their partner and then listen to their partner's description of their own pattern, asking them questions about the pattern for further practice.

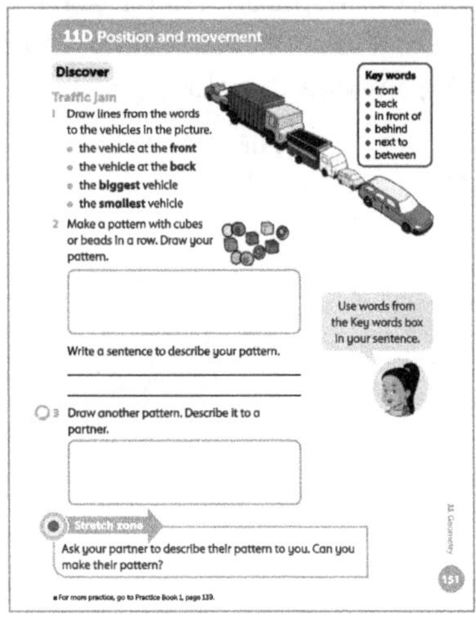

Differentiation

Supporting: Model the use of key language to develop students' fluency and help them to give instructions.

Consolidating: Follow students' instructions and make deliberate mistakes so that they have to be more specific in their instructions.

Extending: Ask questions such as, *How far in front? How far behind?*, as well as two-part instructions such as, *Which car is in front of the red car and behind the blue car?*

Stretch zone: *Ask your partner to describe their pattern to you. Can you make their pattern?*

Check that the vocabulary used is accurate and helps the student replicate their partner's pattern.

 ## Reflection time

Ask students to come to the front of the class and stand as if they were in a queue. Pick another student and ask for instructions from the class about where they should stand, e.g. *Stand in front of Yusuf*. Repeat this until there is a line of 10 students. The class then talk in pairs to say as many different things as they can about the relative positions of people in the queue.

Practice Book: Students can complete Practice Book page 139. This can be done directly after the main activity, as homework, or as the focus of a separate mathematics session to help students consolidate their learning and build fluency. Read through all the words at the top of the page and check that students know what each word or phrase means.

Differentiated outcomes	
All students	should be able to use the language of position to say something about the traffic jam and follow instructions.
Most students	will give precise instructions to change the traffic jam.
Some students	may use more complex language and include numbers in their instructions.

Answers

Student Book page 151

1 The vehicle at the front   red car
   The vehicle at the back   grey jeep
   The biggest vehicle   green truck
   The smallest vehicle   yellow car

2 Check that students' sentences accurately describe the pattern they have drawn.

3 Check that students' pattern descriptions are clear enough to follow.

Practice Book page 139

Check that students are using the key vocabulary accurately. Encourage them to use each of the words and phrases at least twice to describe things they see in the picture. Check that the two sentences reflect what is happening in the picture.

Stretch zone: Check that students' drawings match each instruction.

11D Position and movement

Explore   Student Book page 152 • Practice Book page 140

Specific learning focus

- Use everyday language of direction and distance to describe movement of objects.

Global skills

- **Creative skills:** problem solving
- **Real-world skills:** presenting information / interpreting information
- **Interpersonal skills:** communication / teamwork / leadership

Key vocabulary

- front, back, in front of, behind, next to, between, on top of, underneath, first, last

Resources

- photographs of local landmarks if possible – or labels such as 'Bridge', 'Department Store', 'Hospital'

Language support

Use the images of the local landmarks to make a large poster for display at the front of the class which describes a journey to school. This could be the journey that you take. Encourage students to talk about their own journey to school.

You can help by labelling their 'map' while they describe it to you.

 ## Introductory activity

Tell students they are going to describe a journey using images or labels of local places. Place the images of local landmarks (or labels) around the room. Ask a student to move to the place they pass first on their way to school.

Then they should move to other places in turn. For each part of the journey they should describe it using directional and positional language, for example, *I am next to the______, now I am passing ______.* They should also count the number of steps they take and describe any turns they make. Repeat with two more students who should take a different journey each time.

Main activity

Ask a student to come to the front of the classroom. Give them instructions so that they make a journey around the classroom. This should include the number of steps they take, and the direction they should turn. As they travel they should use phrases that begin with 'I am next to/ in front of/behind/between…' each time they stop to describe their position in relation to objects in the class.

If possible, select a student who is a confident English speaker. Ask them to give you instructions for a journey around the classroom. Repeat their instructions so that the whole class hears a good model of the vocabulary as necessary.

Ask students to work together in groups to form a line. Tell them to take turns to describe the position of one of the people in the group in several ways. Then ask students to sit down and answer the questions on page 152 in the Student Book, by recalling how they formed a line.

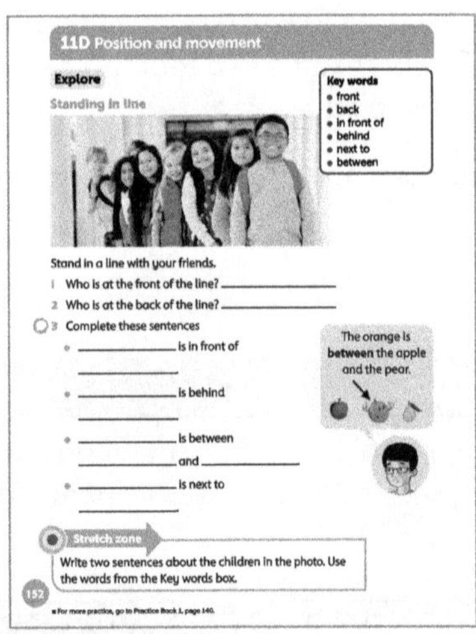

Differentiation

Supporting: For the Student Book activity, model the language as students stand in line and answer the questions about their positions in the line, e.g. *You are standing next to Benjy.*

Consolidating: Ask students to make up their own sentences about the positions they take in the line.

Extending: Use questions to get students to change their positions, e.g. *Can you move__________ to be the 4th student in the line? Can you describe where they are now?*

Stretch zone: *Write two sentences about the children in the photo. Use the words from the Key words box.*

Check that students are using the key vocabulary correctly. Help them with any additional vocabulary related to describing the appearance of the children as necessary.

 Reflection time

Ask each group to share some of their questions and answers about their positions in the lines they made.

Practice Book: Students can complete Practice Book page 140. This can be done directly after the main activity, as homework, or as the focus of a separate mathematics session to help students consolidate their learning and build fluency. Encourage students to keep their drawings simple and to focus on describing their picture to an adult.

Differentiated outcomes	
All students	should identify their position in the line and describe it using positional vocabulary.
Most students	will be able to describe their position in relation to other students in a complete sentence.
Some students	may change the line order by following directions or answering questions.

Answers

Student Book page 152

Check that students have accurately answered the positional questions about their line of friends.

Practice Book page 140

Check that students are using the key vocabulary accurately. Encourage them to use each of the words and phrases at least twice about different things in the picture.

Stretch zone: Students should continue on from the activity but use three of the phrases in their picture/ description.

11E Turns

Specific learning focus

- Use and understand the language of quarter and half turns.

Global skills

- **Creative skills:** problem solving / exploring / investigating
- **Real-world skills:** research / presenting information
- **Interpersonal skills:** communication

Key vocabulary

- full turn, half turn, quarter turn, three-quarter turn, clockwise, anti-clockwise

Resources

- simple drawing of a house or car on a piece of card
- cardboard 'L' shapes (one per student), or cardboard and scissors for students to make their own

Language support

Help students to make a physical connection to half, quarter and full turns and the vocabulary by asking them, for example, *Turn to face the board – that's a quarter turn. Now turn back a quarter turn. What are you facing now?*

 ## Introductory activity

Ask students to stand and face the front of the class. Facing the same way as students, say, *Copy what I do. Do a full turn.* Students copy you as you do a **full turn** slowly. *We did a full turn.* Repeat for a **half turn** and **quarter turn**, left and right. Repeat several times.

Ask students to stand and face the back of the class. Tell them that they're going to take a full turn. Ask, *Which way will you be facing after a full turn?* Then ask them to make the turn to see if they were correct. After doing that, ask them to think about which way they will face after a half turn, then make the half turn. Students should all end up facing the front of the class. Then ask them to make another half turn and return to facing the back. Now say that they are going to make a quarter turn to the left. *Which way will you be facing?* Repeat and mix up the instructions to do whole, half and quarter turns randomly.

 ## Main activity

Show students a picture drawn on a piece of card, such as a house or a car. Hold the picture the right way up and ask them to watch as you turn the picture a quarter turn clockwise as they are looking at it. *Think about what we did in the last activity with our bodies. How far did I turn the picture?* (a quarter turn). Write $\frac{1}{4}$ turn on the board.

Now turn the picture another quarter turn and ask students what they notice now (they might use the words 'upside down'). *Think about what we did in the activity with our bodies. How far have I turned the picture?* (half turn). Write $\frac{1}{2}$ turn on the board. Do another quarter turn and again ask students what they can notice. *How far have I turned the picture now?* (three-quarter turn) Write $\frac{3}{4}$ turn on the board.

After one more quarter turn, students will notice that the picture is the right way up again. *How many quarter turns did we make altogether?* (4)

Remind students of the difference between quarter and half turns, and discuss which way the picture was turned. *The direction we turned has a name: **clockwise,** because we turned the same direction as the hands on a clock face. Could we turn in the opposite direction?* (Yes) *It has a name: **anti-clockwise**.* Turn the picture, showing different fractions of a turn both clockwise and anti-clockwise and ask students to describe the turn.

Students can now complete the activities on page 153 of the Student Book individually. You can give students cardboard L shapes that you have prepared.

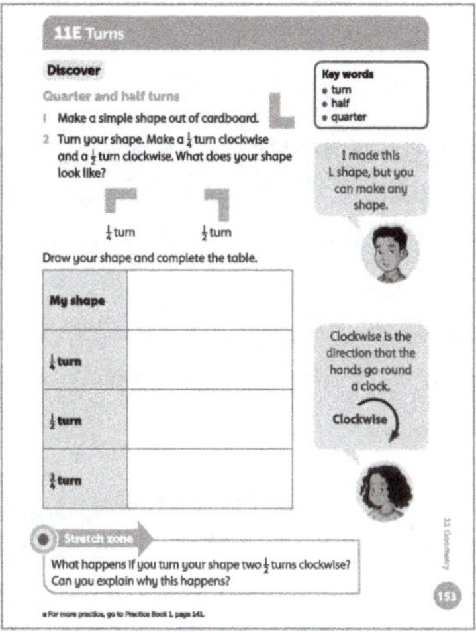

Differentiation

Supporting: Model the language as the students describe how the shape has been turned.

Consolidating: Ask students to describe the turn that has been used on the shapes and images.

Extending: Encourage the use of more complex language including clockwise and anti-clockwise.

Stretch zone: *What happens if you turn your shape two $\frac{1}{2}$ turns clockwise? Can you explain why this happens?*

Check that students understand that repeating a half turn brings the shape back to its starting position.

 **Reflection time**

Ask students to describe what happened to their shapes when they turned them by quarter, half and three-quarter turns. Did they notice anything as they made the turns? *How many quarter turns brings the shape back to where it started?* (4)

Practice Book: Students can then complete Practice Book page 141. This can be done directly after the main activity, as homework, or as the focus of a separate mathematics session to help students consolidate their learning and build fluency. Together, come up with suitable shapes to draw on card and cut out. Alternatively, provide students with pre-cut shapes to turn and then draw in their books.

Differentiated outcomes	
All students	should turn a shape by a half turn.
Most students	will turn shapes by quarter or half turns.
Some students	may turn shapes by a three-quarter turn and see that it is the same as a quarter turn in the other direction.

Answers

Student Book page 153

Check that students have drawn the shape in the correct orientation after a quarter turn, half turn and three-quarter turn.

Practice Book page 141

Check that students have drawn the shape in the correct orientation after a quarter turn, half turn and three-quarter turn.

Stretch zone: three-quarter turn clockwise

11E Turns

Explore   Student Book page 154 • Practice Book page 142

Specific learning focus

- Use and understand the language of quarter and half turns.

Global skills

- **Creative skills:** problem solving / exploring / investigating
- **Real-world skills:** research / presenting information / interpreting information
- **Interpersonal skills:** communication / teamwork

Key vocabulary

- full turn, half turn, quarter turn, three-quarter turn, clockwise, anti-clockwise

Resources

- shape (e.g. flag or arrow) drawn on squared paper, large pieces of paper or cloth, squared paper
- large clock face

Language support

Help students with the language of turning, reinforcing the vocabulary 'half turn', 'quarter turn', 'three-quarters turn', 'clockwise' and 'anti-clockwise'.

 **Introductory activity**

Show students a shape drawn on squared paper, perhaps a flag or some shape that is not symmetrical. Ask them to look closely at it and notice which way up it is.

Now hide it behind a piece of paper or cloth and tell students you are going to turn it a quarter turn clockwise. Ask them to draw in their book what they think it will look like when you reveal it. Check how many students could accurately picture what the turned flag looks like.

 **Main activity**

Continuing the theme from the introductory activity, ask students to work in pairs. One of them should draw a simple shape on squared paper and show it, then cover it and give it a turn, either a quarter turn clockwise or anti-clockwise, or a half turn. They tell their partner which turn they did, and the partner has to draw what the shape will look like when uncovered. They can take turns to do this with different turns or different shapes.

Students should complete page 154 in the Student Book individually. They can then compare their drawings with their partner's to see if they are correct.

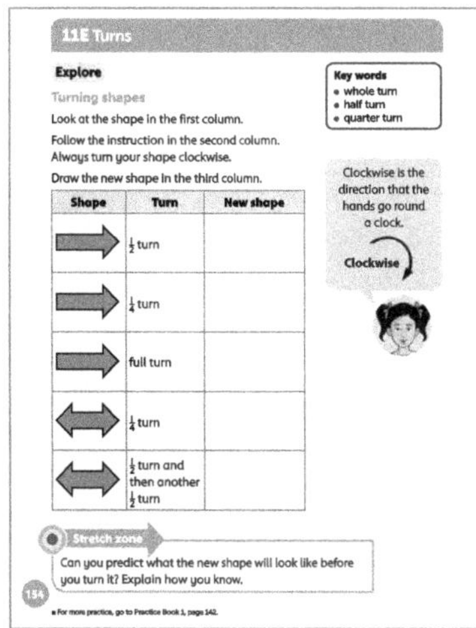

Differentiation

Supporting: Give students a piece of paper with a quarter turn, half turn and full turn drawn and labelled. Show them how to use this to help them turn their shape correctly.

Consolidating: Ask students to describe the turn that has been used on the shapes.

Extending: Ask students to follow two-part instructions including both anti-clockwise and clockwise.

Stretch zone: *Can you predict what the new shape will look like before you turn it? Explain how you know.*

Check that students can explain using the correct language and help them with visualising the new position if they need support.

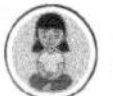 Reflection time

Show students a clock face which is not in the correct orientation, perhaps with the 12 where the 3 should be.

Ask students if they can tell you what turn needs to be done to get the clock back to the right way up. Repeat for different positions.

Practice Book: Students can complete Practice Book page 142. This can be done directly after the main activity, as homework, or as the focus of a separate mathematics session to help students consolidate their learning and build fluency. Students may find it useful to make their own arrow, which they can turn following the instructions, and then copy the new arrow in their books.

Differentiated outcomes	
All students	should draw a shape on squared paper, turn it and describe the turn, with support.
Most students	will work out what the shape looks like after a half turn.
Some students	may draw the shape when it is turned a three-quarter turn in either direction and recognise that a quarter turn in one direction is the same as a three-quarter turn in the opposite direction.

Answers

Student Book page 154

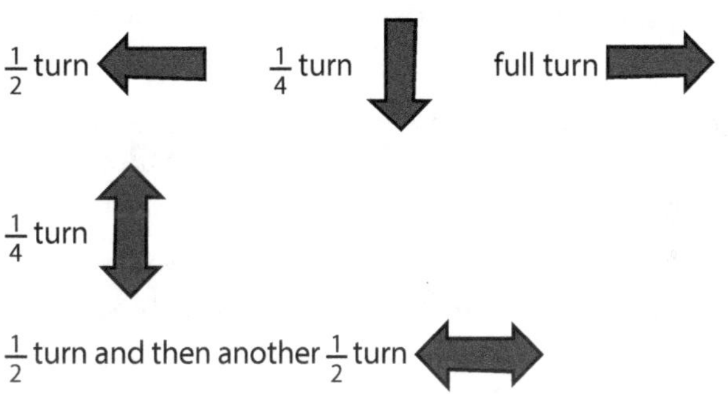

Practice Book page 142

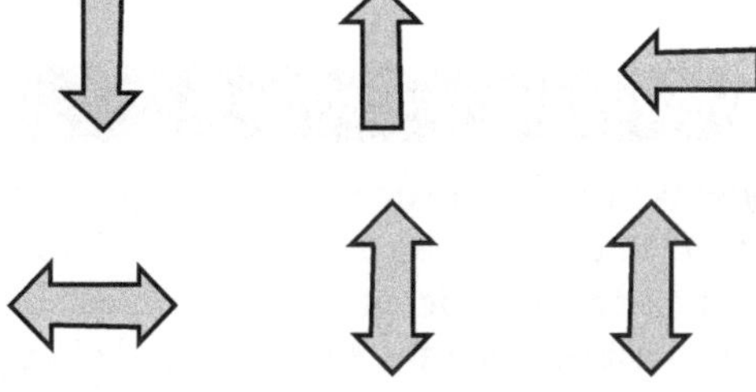

Stretch zone: Check that students have drawn a shape in the correct orientation after a quarter turn, half turn and three-quarter turn.

11 Geometry

Big idea

- I can describe the properties of shapes I see around me.
- I know some of the shape names.

Global skills

- **Creative skills:** exploring / investigating
- **Real-world skills:** research
- **Interpersonal skills:** communication / teamwork

Key vocabulary

- corner, circle, square, rectangle, triangle, side, symmetry, line of symmetry, mirror line, symmetrical, straight, curved, cube, pyramid, sphere

Resources

- pattern blocks (2D shapes), 3D shapes, mirrors

Language support

Working in pairs on the shape walk will allow those less confident in English to hear the language modelled by their peers.

 **Introductory activity**

Tell students that they are going to play a game called 'Think-tank'. To play this game the teacher thinks of a 2D or 3D shape and the students have to ask questions with a yes or no answer about the shape to try to discover what it is.

For example, *Does it have corners? Is it symmetrical? Does it roll?* If the answer to the question is 'yes', the teacher puts up a thumb. If the answer is 'no', the thumb is pointed down.

Whoever guesses the shape correctly takes the teacher's place for the next round. This is a good game to reveal students' knowledge and understanding of shape, and the language of shape.

 **Main activity**

Tell students they are going on a shape walk. Explain that they are going to look for some of the shapes they have been learning about. Look at page 155 in the Student Book together, review the names of the shapes in question 1 and agree on the things students need to look for on their walk.

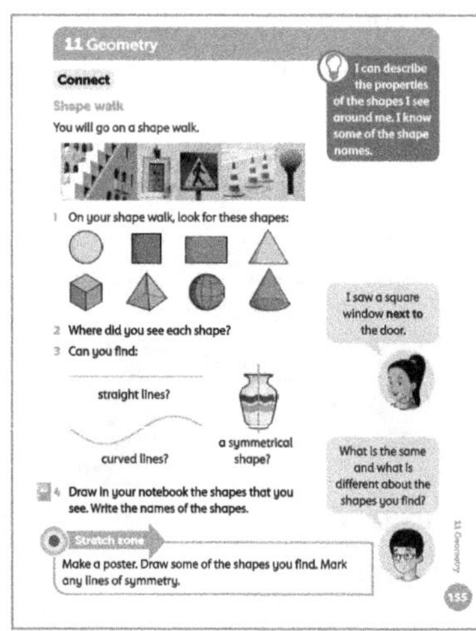

As students go along the walk, ask them questions such as, *What shape is this? Is it a 2D shape or a 3D shape? How many sides does it have? How many faces/edges/ vertices does it have? Is it symmetrical? We are looking for a **pyramid**. Where might we find one?*

Differentiation

Supporting: Use the vocabulary of 2D and 3D shapes to help students describe what they find on the shape walk.

Consolidating: Ask questions to develop vocabulary. For example, *How many faces/edges/vertices does this shape have?*

Extending: Ask students to identify lines of symmetry in their 2D shapes.

Stretch zone: *Make a poster. Draw some of the shapes you find. Mark any lines of symmetry.*

Check that the poster has symmetrical shapes with the lines of symmetry marked.

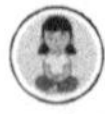 **Reflection time**

Ask students to make symmetrical patterns using 2D pattern blocks with exactly 8 black squares and 8 white squares. How many different patterns can they make? Invite students to share and describe their patterns with the class. Encourage them to use positional vocabulary, e.g. *This black square is next to another black square.*

Differentiated outcomes	
All students	should find and name common 2D and 3D shapes.
Most students	will find and name common 2D and 3D shapes and talk about their properties.
Some students	may find and name a wider range of 2D and 3D shapes and talk about their properties.

11 Geometry

Review Student Book page 156 • Practice Book page 143

Global skills

- **Creative skills:** problem solving / investigating
- **Real-world skills:** research / presenting information
- **Interpersonal skills:** communication
- **Self-development skills:** reflecting on learning

Student Book

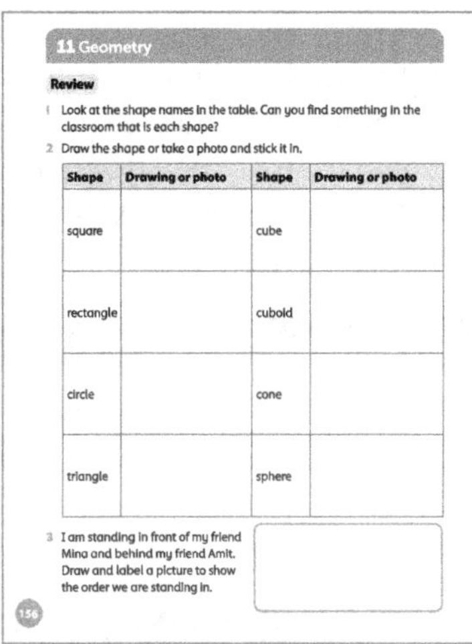

With young children, assessment activities are most effective when carried out as an everyday classroom activity. Students should have 2D and 3D shapes available so they can refer to them to support them.

Watch as students complete the table of images of shapes. Assess students as they work by asking, for example, *How many sides does a rectangle have? What do you know about the sides on a square? Which has more edges, a cube or a cuboid?*

Answers

Student Book page 156

Note who confidently identifies 2D and 3D shapes, and who finds it easier to recognise 2D shapes.

3 Check that students' labelled pictures match the description.

Practice Book

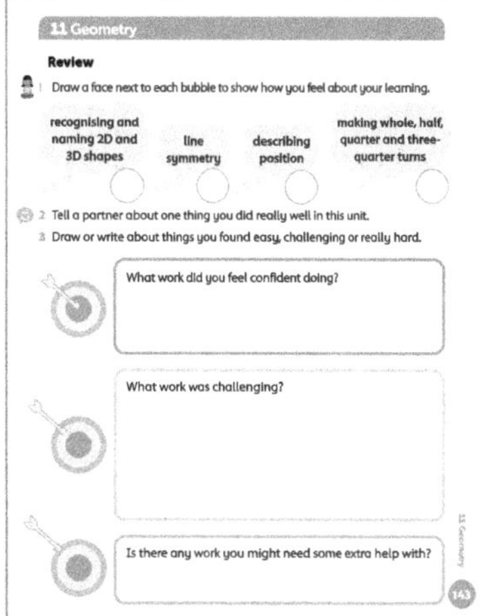

With young children it is appropriate to complete this as a whole-class discussion. You may choose to keep a record of the class discussion or a copy of the review page for your own records. Use the Student Book to briefly remind students of the areas of mathematics that they have worked on in this unit.

Give out the Student Book to support pairs of students as they discuss and answer the questions in the Practice Book.

Allow students plenty of time for discussion before asking them to share their responses with the rest of the class. If students complete this assessment at home, encourage them to discuss it with adults.

Make a note of areas that students still feel unsure about. You can build shape into everyday practice: for example, finding shapes and describing them around the class or the school grounds.

Additional material

There are additional end-of-unit assessments available on the *Oxford Owl* website.

12 Statistics

Overview

Big Idea

The Big idea for this section is that there are different ways of collecting, sorting and organising (representing) data or objects. Working with data goes through distinct stages. The starting point is a question or problem we want to find the answer to.

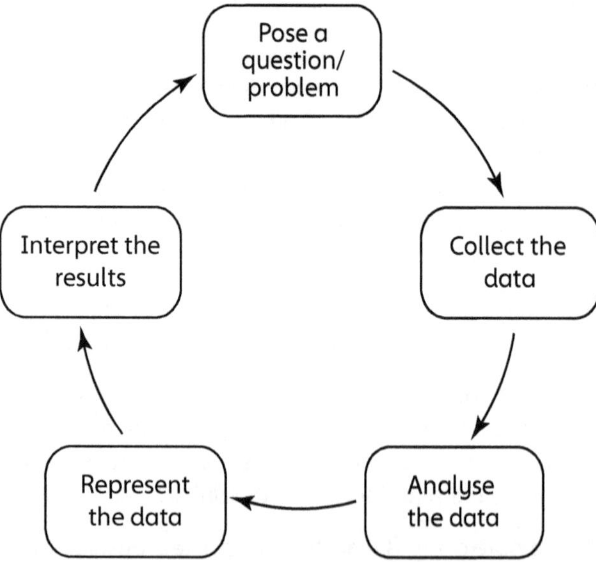

Asking the 'correct' question is important in order to make it clear what information (data) to collect.

Data can be represented visually using tables, charts and graphs. Students can gather, organise and display data to answer a question. Simple surveys with a limited number of responses help students to develop understanding of data presentation.

Students need lots of experience of sorting and classifying objects, making choices and justifying their decisions. They need to see how objects can share a particular property and then represent their ideas using pictures, drawings and numbers.

Look out for

- **Students who see data as a 'number'.** Seven is a number but 'seven students are wearing red' is data. Data are 'numbers' with a context. Data can be collected through surveys and through experiments, e.g. scores when throwing dice.
- **Students who do not count items accurately** before they represent the data.
- **Students who use inconsistent sizes** on block graphs or pictograms.
- **Students who place items incorrectly in Venn or Carroll diagrams** because they do not understand what each region means. These diagrams are difficult for young students to understand; they need lots of examples to help them.

Possible misconceptions

- **Students may think that pictures used in a key for a pictogram must represent exactly what they are recording**, e.g. Sadiq's vote for a football must be represented by a football.

Key vocabulary

- sort, groups, estimate, row, column, set
- collect, represent, diagram, list, data, key, pictogram, most, fewest, table, block diagram
- Venn diagram, criteria, intersection, Carroll diagram

Coverage in lessons

Learning focus	Learning outcomes (the ENC objectives)
Pictograms, lists and tables	Record, organise and represent categorical data in pictograms, lists and tables
Block diagrams	Record, organise and represent categorical data in block graphs.
Venn diagrams	Record, organise and represent categorical data in Venn diagrams.
Carroll diagrams	Record, organise and represent categorical data in Carroll diagrams.

12 Statistics

Engage   Student Book page 157

Big question

- How can we sort things? What is the best way to show how we have sorted them?

Global skills

- **Creative skills:** exploring / investigating
- **Interpersonal skills:** communication / teamwork
- **Self-development skills:** reflecting on learning

Key vocabulary

- sort, rows, group, estimate

Resources

- none needed

Language support

Listen to the ways in which students discuss and sort the information. Repeat key words to model pronunciation.

 **Introductory activity**

If you have access to an IWB, you could display the Student Book Engage page and discuss the Big idea of the unit with students. Ask students to talk in pairs about the picture on page 157 in the Student Book and to relate it to any previous experience they have of fruit and vegetables, either in a shop, a market or at home. Then ask them to share what they know with the class. Use this as an opportunity to review language of counting (cardinal and ordinal), position and comparison, for example:

How many baskets altogether? How many in each row?

Which basket is holding the biggest fruit? Which type of fruit is the smallest?

Can you describe the fruit that is in the middle basket in the second row?

What colour is the fruit in the basket above the basket of mangos?

How many bananas can you count?

 **Main activity**

Ask the class to look at the picture again and discuss how the fruit has been displayed. *How has the food been **sorted**? How many different **groups** are there?*

When you have several different ideas, such as by row, basket or type of fruit, ask, *How else could you sort the food?* Make a list of suggestions, e.g. by colour of fruit, colour of basket, what you like, what you don't like, size, shape.

Ask students to work in small groups to choose a way to sort the fruit into two groups and then sort it and record their choices.

*How could we **estimate** the numbers of fruit/vegetables in each group?* Give students time in their groups to discuss this. *Are some easier to count than others? Why is this? Which are the easy ones and which are the difficult ones? Why are they so difficult to count?* (They could be small, or there may be too many; larger, fewer items are easier to count.)

Differentiation

As this is a discussion-based activity, place students in mixed-attainment groups so that all students get to hear a wide range of vocabulary. While students are working, observe them and check that they can all sort objects into two groups. Ask students to explain their sorting and see who can describe accurately and fully the criteria they are using. Encourage these students to try sorting into more than two groups and to explain the criteria they use.

 Reflection time

Choose groups to share what they did with the rest of the class. *How did you sort the fruit? What did you find out? How many are there in each group? Does anyone want to ask this group some questions such as why did you do it that way, or did you find it easy/hard to sort his way?*

Discover 1 Student Book page 158 • Practice Book page 144

Specific learning focus

- Answer a question by collecting, sorting, organising and recording data or objects in a variety of ways, e.g. using block graphs and pictograms with practical resources; discuss the results.

Global skills

- **Creative skills:** investigating
- **Real-world skills:** research / presenting information / interpreting information
- **Interpersonal skills:** communication / teamwork

Key vocabulary

- collect, represent, list, data, sort, pictogram, key, diagram, most, fewest

Resources

- photo of each student's face or sticky notes for students to draw their face; a blank pictogram (see table on the right) prepared as a large poster
- interlocking cubes

Language support

Asking students to create their own graphs is very important as this helps them develop their knowledge and understanding and allows you to hear what misconceptions they may have.

 **Introductory activity**

Ask the students, *Who has a birthday this month?* Count how many there are in that month. Now ask, *Which month do you think has the most birthdays in our class? How can we find out?* Agree they could ask each student and keep track of what they find out. Write a **list** of the months of the year on the board, and ask in turn, *How many students have a birthday in January? February?… December?* Count the number of students for each month and record them on the board next to each month. *We have now **collected data** on the number of students who have a birthday in each month. Look at the data and say which month has the **most** students with birthdays, and which month has the **fewest**.*

 **Main activity**

Explain that you are now going to **represent** the birthday data on a **diagram**. Display the outline structure for a pictogram showing months in the left-hand column.

Month	Number of people
January	
February	
March	
April	
May	
June	
July	
August	
September	
October	
November	
December	

Explain that this diagram is called a **pictogram** because the data is going to be added to the diagram as pictures.

Now give each student a photo of their own face. If this is not possible, give each student a sticky note and ask them to draw a picture of their face. Then ask them to stick it next to their birth month on the board. Ask students to add their pictures in a row for each month. Make the point that the rows of pictures are easier to compare quickly than by looking at the numbers.

What does each face tell us? (That one person has their birthday in that month.) Explain that that a pictogram needs a **key** to explain what each picture represents. Add a key to the board:

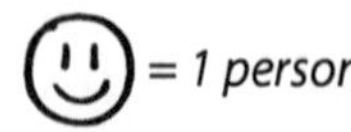 = 1 person

Ask a few questions about the pictogram to check understanding, e.g. *How many students have birthdays in April? Which month has the most students with birthdays?* Keep this pictogram displayed in the classroom. It is used in a later lesson.

Organise students into groups of six and give each group some interlocking cubes. Ask students to look at the Student Book page 158. Explain the diagram to students. Each student should ask the other five students in the group the question, *Do you prefer apples or grapes?* Each student records the answer to their question by putting a cube in the correct column until the pictogram is complete.

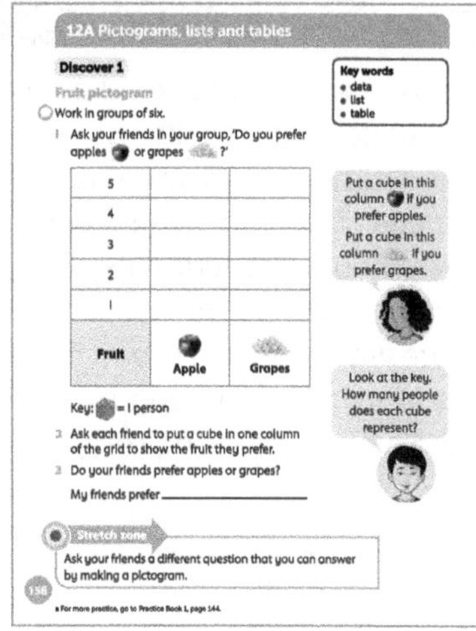

Differentiation

Supporting: Ask questions such as, *Which column is the shortest? What does that mean?*

Consolidating: Ask questions such as, *Can you explain what you did? Can you tell us what you found out? What information did you have? What did you need to find out?*

Extending: Ask students to make a range of statements about the data.

Stretch zone: *Ask your friends a different question that you can answer by making a pictogram.*

Check that students can formulate sensible questions that could result in data for a pictogram.

 ## Reflection time

Ask follow-up questions to discuss the outcomes from each group. For example, *Which fruit did most people prefer in your group? How many people in your group preferred grapes? How many people did you ask altogether? What is the difference between the number of people who preferred grapes and the number who preferred apples?*

Practice Book: Students can complete Practice Book page 144. This can be done directly after the main activity, as homework, or as the focus of a separate mathematics session to help students consolidate their learning and build fluency. There is no corresponding activity for Discover 2 so you may prefer to do this practice after the next lesson when students have dome more work with pictograms.

Differentiated outcomes	
All students	should understand the purpose of a pictogram and complete their own pictogram accurately.
Most students	will use the vocabulary with confidence.
Some students	may use the pictogram to make a range of statements about the data.

Answers

Student Book page 158

Assess the responses as you visit groups. The responses in the book should refer to the towers of cubes that students have built from asking friends whether they prefer apples or grapes.

Practice Book page 144

1 Margherita   6

2 Mushroom   4

3 Four cheese   2

4 Chicken   5

Stretch zone: Answers could include:

Four cheese was the least popular flavour.

Margherita was the most popular flavour.

17 pizzas were sold altogether.

12A Pictograms, lists and tables

Discover 2   Student Book page 159

Specific learning focus

- Answer a question by collecting, sorting, organising and recording data or objects in a pictogram.

Global skills

- **Creative skills:** problem solving/ investigating
- **Real-world skills:** research / presenting information / interpreting information
- **Interpersonal skills:** communication / teamwork

Key vocabulary

- collect, represent, list, data, sort, pictogram, key

Resources

- cubes or counters

Language support

Asking students to develop their own pictograms is very important as this helps them to develop their knowledge and understanding and allows you to hear what misconceptions they may have.

 ## Introductory activity

Carry out a short poll with the class, asking them to choose whether they prefer drinking water or orange juice. They can form two lines in the class, one for each choice. The lines have formed a 'human pictogram', with each student representing one student's choice. Discuss what they have found, and prompt with questions such as, *Which line is the longest? What does that mean? How many more are in the longest line than the shortest line?*

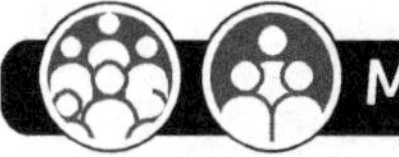 ## Main activity

Tell students they are going to work in groups of six to collect data to make a pictogram in their Student Books. Explain that they need to start with a question and today the question will be, *What do your friends like best?* Discuss different ideas for what they could ask about, such as, their favourite sandwich out of four options (e.g. meat, fish, cheese or egg) or their favourite season (summer, spring, winter or autumn). Tell students they should have no more than four possible answers (options).

Discuss how they could collect and record the data for making a pictogram.

Look at the partial pictogram on page 159 of the Student Book and discuss its features and where students need to record their data. *What does the green counter represent? In which box will you write each option? After each friend chooses, where should you put their counter?*

Give each group a set of counters and ask them to do the survey and complete their pictogram on page 159 of the Student Book. Each student in the group can choose a different question.

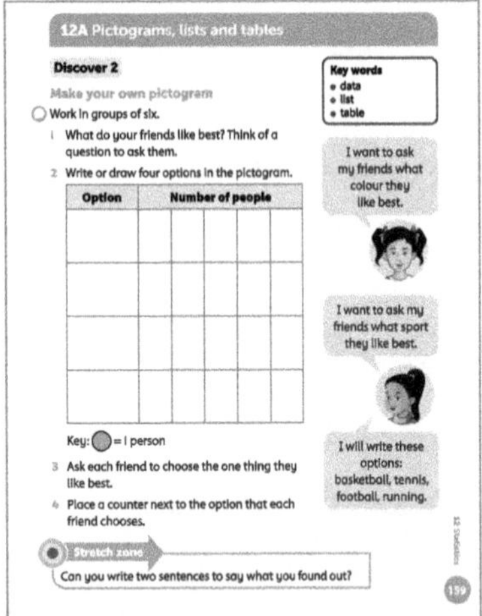

Differentiation

Supporting: Help students to write out the options in their pictogram and structure their question correctly.

Consolidating: Ensure that students are transferring the data onto the pictogram correctly and ask questions such as, *Which is the most popular sandwich option? How many students prefer the cheese sandwich?*

Extending: Encourage students to make a range of statements about the data.

Stretch zone: *Can you write two sentences to say what you found out?*

Check students' two sentences about what they did in completing the pictogram. Can they explain how they know they are accurate?

 **Reflection time**

Find out students' ideas by asking a group to feed back what they did, how they collected and represented the data and what they found out.

Differentiated outcomes	
All students	should understand the purpose of a pictogram and collect data for their own pictogram.
Most students	will use the vocabulary relating to data with confidence.
Some students	may use the pictograms to make a range of statements about the data.

Answers

Student Book page 159

Check that each group has produced an accurate table based on their preferences.

12A Pictograms, lists and tables

Explore 1 Student Book page 160 • Practice Book page 145

Specific learning focus

- Collect, sort and analyse data by using a table; discuss the results.

Global skills

- **Creative skills:** problem solving / investigating
- **Real-world skills:** research / presenting information / interpreting information
- **Interpersonal skills:** communication / teamwork

Key vocabulary

- collect, represent, list, table, data, sort, pictogram, key

Resources

- none necessary

Language support

The introductory activity and the beginning of the main part of this lesson can be used as formative activities, using questions to find out what the students already know, understand and can do.

Introductory activity

Remind the class of the last activity when they were asked to make their own pictogram. Tell them that this lesson is about pictograms, lists and tables, and this time they will be asking more students about different play equipment for the playground.

What question do you think you might ask? Why would you want to know? Who would you ask? How could you collect your data?

Main activity

Encourage students to look at the pictures of playground toys on page 160 of the Student Book. Tell them that they will be asking ten of their friends which playground toy they would like the school to buy, out of a choice of four items. Make it clear that each student can only choose one item.

Ask students to work in pairs within their group of ten to ask the question and to complete the **table** in their Student Books. Discuss how the table is organised, asking for example, *How many* **rows**? *What is shown in each row? What is each* **column** *used for?* Point out that every table needs a clear heading for each column. *What are the headings for the columns? If seven children choose skipping ropes where should I write this?*

Discuss how students will keep track of how many people choose each item: for example, making a list of the items and making a tick each time it is chosen and then counting how many ticks for each. Encourage students to record on separate paper before filling in the table in the Student Book.

Using the completed tables, ask each group to tell you what they found out about the choices of their group. Ask, for example, *Which toy is the most popular? Which toy is the least popular? What was the difference between the number of students who chose your most popular toy and the number who chose your least popular toy?*

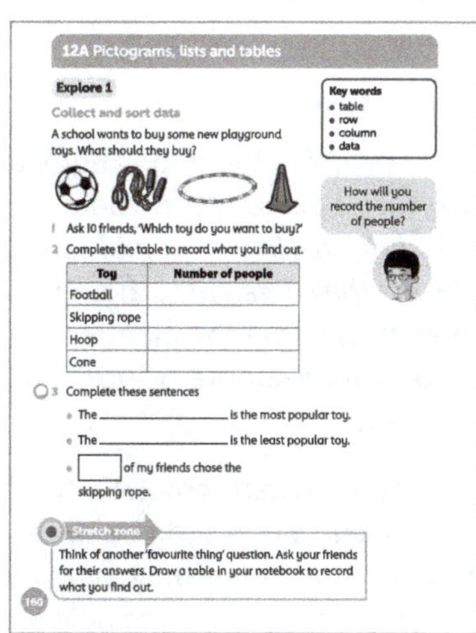

Differentiation

Supporting: Ask questions such as, *Which toy is the least popular? What does that mean?*

Consolidating: Ask questions such as, *Can you explain what you did? Can you tell us what you found out? What information did you have? What did you need to find out?*

Extending: Encourage students to make a range of statements about the data.

Stretch zone: *Think of another 'favourite thing' question. Ask your friends for their answers. Draw a table in your notebook to record what you find out.*

Check that students have collected suitable data and put it correctly in a table.

 Reflection time

Ask each group some 'How many' questions such as, *How many students did you ask? How many chose the most popular item?* Record the different results on the board so that all the class can see. Ask students to ask each other some questions about the tables, e.g. *What was the most popular toy in your group? Why do you think that is?*

Practice Book: Students can complete the first part of Practice Book page 145 at this point or work on both parts after completing Explore 2. This can be done directly after the main activity, as homework, or as the focus of a separate mathematics session to help students consolidate their learning and build fluency.

Differentiated outcomes	
All students	should understand the purpose of a table.
Most students	will complete the data collection and the table accurately.
Some students	may use the table to make a range of statements about the data.

Answers

Student Book page 160

Assess the responses as you visit groups. The responses in the book should refer to the data that they have collected about playground toys.

Practice Book page 145

Check that students have drawn the correct table to match the data they collected.

12A Pictograms, lists and tables

Explore 2   Student Book page 161 · Practice Book page 145

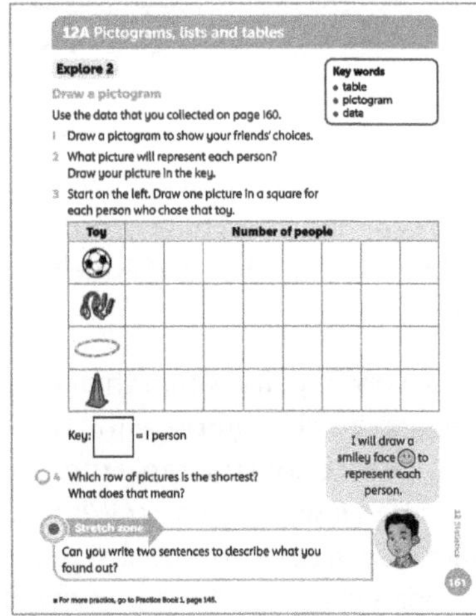

Specific learning focus

- Organise and record data in a pictogram and discuss the results.

Global skills

- **Real-world skills:** presenting information / interpreting information
- **Interpersonal skills:** communication / teamwork

Key vocabulary

- collect, represent, list, table, data, sort, pictogram

Resources

- labelled pictogram for display

Language support

Have a labelled pictogram for students to refer to so that they can remember the words or point to some element of a completed pictogram when they want to know the word.

 **Introductory activity**

Ask students to look at the table of data they collected about their preferences for new playground toys in the previous lesson. Explain that they are going to use the data they collected to draw a pictogram. First, ask students questions about the data in their tables, for example, *Which toy was the most popular? Which was least popular? Were any choices of playground toy equally popular?*

 **Main activity**

Ask questions about the data with a focus on how to display it in a pictogram. For example, *What information does the table have that we will use? In which row will you record the number of people who chose the hoop? Who can tell me what a key is? What picture could you use for your key? What does each picture tell us? Why might you prefer a pictogram to a table?*

Students can then work in the same groups as for the previous lesson groups and each produce a pictogram on page 161 of the Student Book, using their playground toy data.

Differentiation

Supporting: Ask questions such as, *Which playground toy is the least popular? What does that mean?*

Consolidating: Ask questions to assess how well students can interpret the data, such as, *Which row is the shortest? What does that tell us?*

Extending: Encourage students to make a range of statements about the data.

Stretch zone: *Can you write two sentences to describe what you found out?*

Check that students have written two sentences that describe something about the playground toy data.

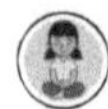 **Reflection time**

Students should be given the opportunity to discuss and share how they transferred the data from their table to the pictogram. *How are a table and pictogram the same? How are they different?* Students may wish to describe how they chose the pictures they used for each data item, for example, using a smiley face to represent 1 student.

Practice Book: Students can complete the second part of Practice Book page 145. This can be done directly after the main activity, as homework, or as the focus of a separate mathematics session to help students consolidate their learning and build fluency.

Differentiated outcomes	
All students	should understand the purpose of a pictogram and complete their own pictogram with support if needed.
Most students	will complete the pictogram accurately using the data collected.
Some students	may use the pictogram to make a range of statements about the data.

Student Book page 161

Check that students have produced an accurate pictogram based on the data they collected previously.

Practice Book page 145

Check that students have completed the pictogram correctly to match the data in their table.

Stretch zone: Ask students to share their easy and hard questions with you. *What made them hard? What made them easy?*

12B Block diagrams

Discover Student Book page 162

Specific learning focus

- Answer a question by sorting and organising data or objects using block graphs; discuss the results.

Global skills

- **Creative skills:** problem solving / exploring / investigating
- **Real-world skills:** research / presenting information / interpreting information
- **Interpersonal skills:** communication / teamwork

Key vocabulary

- collect, represent, data, sort, most, fewest

Resources

- interlocking cubes

Language support

Watch how students solve the problem of building the towers, listen to their discussions and correct use of vocabulary and language.

Ask questions such as, *Can you explain what you did? Can you tell us what you found out? Could you do it more quickly? What helped you? Was it something you already knew?*

Introductory activity

Pick up two handfuls of cubes using your two cupped hands. Ask the class how many they think you have picked up.

Take some answers and then leave the cubes in a randomly spread pile. Ask how many of each colour students think there are without counting them.

Which colour do you think I have most of? Which colour do you think I have fewest of?

How can we find out?

Main activity

Now sort the cubes into colours.

Ask students how they might work out which colour of cube there is most or fewest of. They may suggest building a tower of cubes for each colour and comparing the heights of the towers.

Build a tower for each colour. Ask questions about the height of each tower to compare and work out which colour of cube you have the most and fewest of. *Which tower is the tallest? Is there a tower that is shorter than the tower of [green] cubes? Which tower is the shortest? Which is taller, the [blue] tower or the [red] tower?*

Explain that students are going to work on page 162 of the Student Book to do the same as you have shown them. They collect as many cubes as they can in their cupped hands. Then they need to sort those cubes into colour groups and build towers of each colour. When all the towers have been built, ask students to put them next to each other in order of height to decide which is the tallest tower and which is the shortest one.

When students have completed the task once, ask them to work in pairs and do it again. Each student in the pair collects and builds their own towers, and discusses with their partner what they did and what they found out.

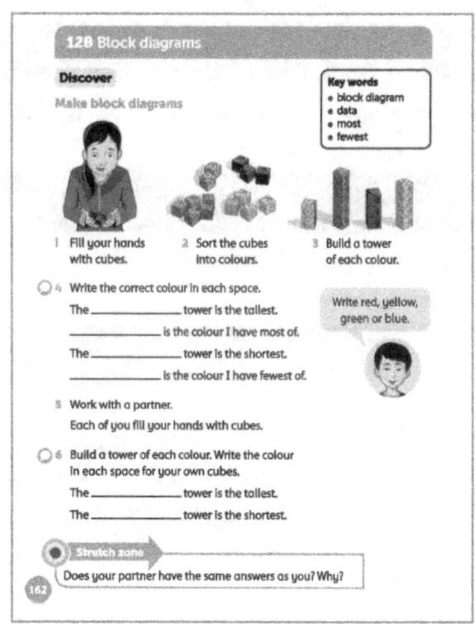

Differentiation

Supporting: Model the vocabulary carefully while students are engaged in the activity with a focus on shorter/shortest and taller/tallest.

Consolidating: Ask questions such as, *Can you explain what you did? Can you tell us what you found out? Could you do it more quickly? What helped you? Was it something you already knew?*

Extending: Ask students to write statements of comparison involving number, for example, 'I had 3 fewer cubes in my blue tower than my red tower', 'I had 13 cubes altogether', 'My green tower was twice as tall as my brown tower.'

Stretch zone: *Does your partner have the same answers? Why?*

Check that students understand that different-size hands will lead to different numbers of cubes being picked up, and hence different data.

 ## Reflection time

Ask the class to share with each other what they found out about the tallest and shortest towers and the most and fewest coloured cubes. *Was it the same for everyone? Why not? Who can explain why that happened?*

Differentiated outcomes	
All students	should collect cubes, build towers and order them accurately.
Most students	will use the vocabulary of counting and ordering appropriately.
Some students	may use larger containers to collect cubes and work with larger data.

Answers

Student Book page 162

Assess the responses as you visit groups. The responses in the Student Book should match the towers that students have built.

12B Block diagrams

Explore 1 Student Book page 163

Specific learning focus

- Answer a question by sorting and organising data using block diagrams; discuss the results.

Global skills

- **Creative skills:** problem solving / exploring / investigating
- **Interpersonal skills:** communication / teamwork

Key vocabulary

- collect, represent, table, data, sort, most, fewest

Resources

- interlocking cubes

Language support

Watch how the students solve the problem of building the towers, listen to their discussions and correct use of vocabulary and language.

Ask questions such as, *Can you explain what you did? Can you tell us what you found out? Could you do it more quickly? What helped you? Was it something you already knew?*

 ## Introductory activity

Remind students of the activity in the previous lesson when they collected cubes in their cupped hands. Ask if anyone can tell you what they did and what they found out. Encourage students to talk to the person next to them. Give some time for students to respond. Use their feedback to lead into the next part of the lesson.

 ## Main activity

Explain that, in this lesson, students are going to work through Student Book page 163 in groups of four. They will each collect their own cubes, sort them by colour and then build towers according to the colours.

Tell students that they will then ask each other questions to compare their towers with the towers of other students in the group to decide who has the most and fewest cubes of each colour.

As a class, come up with a list of questions that students can ask. Improve on the questions as necessary and model the correct question structure. For example, *How many blue cubes do you have? I have 5 green cubes. Do you have more or fewer green cubes than me? Who has the fewest/most yellow cubes?*

Make sure that everyone understands the task, modelling it for those who need more visual instructions. Ask students questions to check progress, for example, *How have you recorded the numbers of cube colours? How are you deciding who has the most cubes?*

For the final question in the Student Book, students work with the numbers of cubes collected by everyone in the group, recording the number of cubes each student in the group picked up. Discuss what this table might look like. *How many columns will you have? How many rows? What data will you put in each row? Where will you record the name of each person in your group? Where will you put how many red cubes a person collected?*

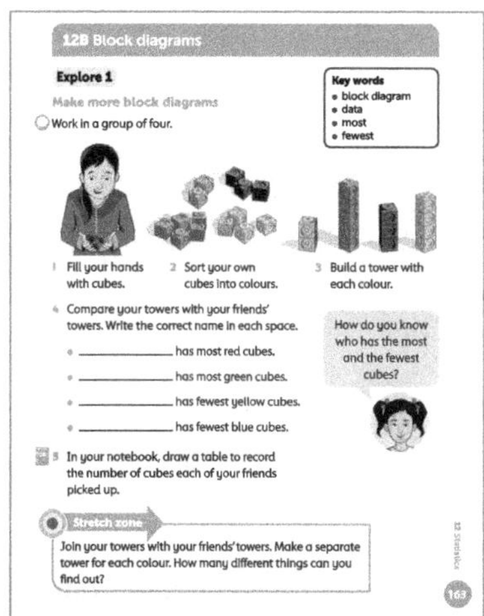

Differentiation

Supporting: Help students to create a table for question 5 in the Student Book and then ask them to fill it in with their data.

Consolidating: Ask questions such as, *Can you explain what you did? Can you tell us what you found out? Could you do it more quickly? What helped you? Was it something you already knew?*

Extending: Encourage students to write mathematical statements beyond comparing data, which relates to the different numbers of cubes: for example, 'The number of blue cubes is the same as the total of red and yellow cubes together'.

Stretch zone: *Join your towers with your friends' towers. Make a separate tower for each colour. How many different things can you find out?*

Check that students have combined their data accurately. Listen to the types of question they ask about their data.

 **Reflection time**

At the end of the lesson, ask questions so that groups can compare how many cubes of each colour they had. For example, ask, *Who had the most green cubes in your group? Did any other group have someone who had more green cubes than this? How many more? What was the fewest number of any colour of cubes you had?*

Differentiated outcomes	
All students	should collect cubes, build towers and order them accurately.
Most students	will use the vocabulary appropriately.
Some students	may analyse the numbers of different-coloured cubes in extended ways, for example, comparing totals of colours.

Answers

Student Book page 163

Assess the responses as you visit each group. The responses in the Student Book should match to the towers that students have built.

12B Block diagrams

Explore 2   Student Book page 164 • Practice Book page 146

Specific learning focus

- Answer a question by sorting and organising data or objects using a block diagram; discuss the results.

Global skills

- **Creative skills:** problem solving / investigating
- **Real-world skills:** presenting information / interpreting information
- **Interpersonal skills:** communication / teamwork

Key vocabulary

- collect, represent, table, data, sort, block diagram, title

Resources

- interlocking cubes

Language support

Include a simple block diagram as part of a class display. Add speech bubbles containing common questions, with answers in separate speech bubbles.

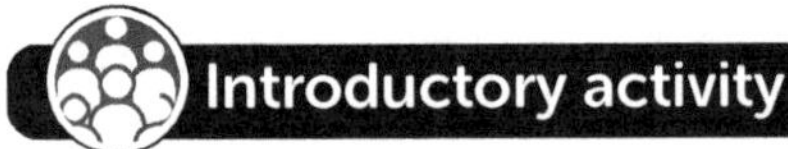
Introductory activity

Ask two students to the front of the class to gather handfuls of cubes. Ask two more children to join them to sort the cubes by colour and make them into towers. Stand the towers up in a row and ask children to describe what they see. Encourage them to use vocabulary such as more, most, fewer, fewest, tallest, shortest.

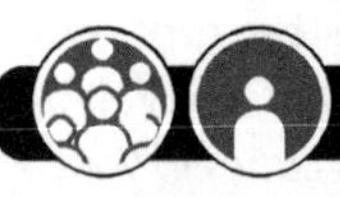
Main activity

Tell students that today they are going to gather data using coloured cubes, but instead of making towers to compare the colours as they just did, they will record the data in a **block diagram**. Explain that they will colour squares on the diagram to represent the cubes in the towers. The diagram will look like a pictogram but with coloured squares instead of pictures of cubes.

Use the cube data from the introductory activity to guide students through the steps of how to create a block diagram from towers. Begin by drawing a table on the board. Agree column headings for the table, e.g. 'Colour of cubes' and 'Number of cubes'. Then record the number of cubes in each colour from the introductory activity.

Then show students how to draw a block diagram using the data from the table.

Start by agreeing on a **title** for the diagram, such as 'Colours of cubes we collected' and write it on the board. On a grid background, draw the two axes and label the vertical axis 'Number of cubes' and the horizontal axis 'Colour of cubes'. Add numbers to the vertical axis. (If you think it is helpful, you can explain the word 'axis' to students.) *What do you think the numbers mean? (Number of cubes) Where should I record each colour name? Now we need to add our data. We have ____ blue cubes. We need to colour in squares to show that many. How many squares do we need to colour in?* Continue to complete the block diagram.

Ask students to talk to a partner about two things the block diagram tells them. Share these facts as a class.

Ask students to complete page 164 in the Student Book.

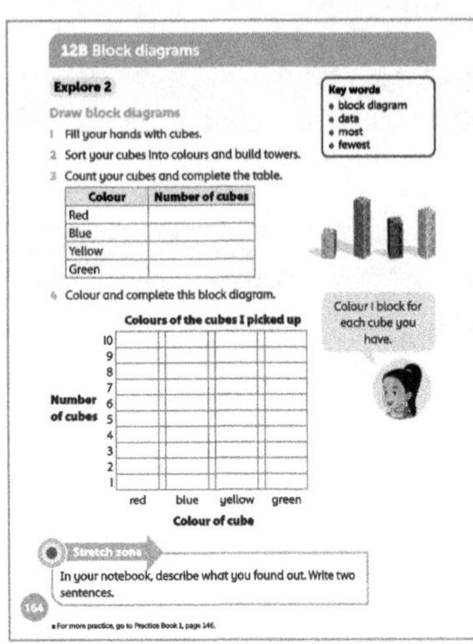

Differentiation

Supporting: Model the vocabulary carefully while students are engaged in the activity.

Consolidating: Ask questions such as, *Can you explain what you did? What did you find out? How is the block diagram the same as and different from your towers?*

Extending: Encourage students to compare the difference between numbers of coloured cubes, reading amounts from the vertical axis rather than counting blocks.

Stretch zone: *In your notebook, describe what you found out. Write two sentences.*

Check that students have described their data appropriately.

 Reflection time

Ask students to compare their towers and their block diagrams. They should see that, for example, the tallest tower is shown as the tallest block on the diagram. Ask students what is the same and what is different about the block diagram compared with the cube towers. Check that students can describe their cube data accurately. *What did you find out about the cubes from your diagrams?*

Practice Book: Students can complete Practice Book page 146. This can be done directly after the main activity, as homework, or as the focus of a separate mathematics session to help students consolidate their learning and build fluency. You may want to have students collect their data at school and then draw their block diagram as homework.

Differentiated outcomes	
All students	should sort cubes, build towers and count cubes.
Most students	will use the data in their table to complete a block diagram.
Some students	may interpret the data in the block diagram and use it to make statements about the data.

Answers

Student Book page 164

Check that the block diagram accurately shows the data collected in the table.

Practice Book page 146

Check that the data in the table matches the data shown in the block graph. The number of people for each sport should be the height of each column in the block diagram.

Stretch zone: Check that the statements students have written are appropriate for the information on the block diagram.

Explore 3
Student Book page 165 • Practice Book page 147

Specific learning focus

- Extract data from a pictorial diagram to construct a table and a block diagram.

Global skills

- **Creative skills:** investigating
- **Real-world skills:** research / presenting information / interpreting information
- **Interpersonal skills:** communication / teamwork

Key vocabulary

- collect, represent, table, data, sort, block diagram, title

Resources

- blank block diagrams prepared on square paper (one per pair) for students to complete

Language support

Continue to focus on the language of pictograms and block diagrams, modelling how to write titles and labels on diagrams.

 ## Introductory activity

Look at the large pictogram students made for their birthdays (12A Discover 2) and discuss what it tells them. *How many students have their birthday in February? Name me a month that has fewer birthdays.* Then talk about how the same data might look in a block diagram. *Would there still be the same number of students with their birthday in December? [January] has the most pictures. What does that tell us about how tall our block tower will be?*

 ## Main activity

Explain that students are going to prepare a block diagram to represent the birthday data that they collected earlier. Give each pair of students a copy of a blank block diagram you have prepared on squared paper and draw a similar one on the board. Discuss what the title for the block diagram could be and agree a suitable one. Then ask which labels should go along the bottom and up the side (on the axes). Together, write in the months of the year along the bottom axis and numbers up the side axis. Then use the data from the pictogram for the months of January and February to draw those block columns together. Students complete the rest of the diagram in their pairs.

Explain that they are now going to look at some birthday data in their Student Books, which was collected about other students. Ask students to complete the activities

on page 165 of the Student Book in pairs. Encourage all pairs to complete the Stretch zone activity.

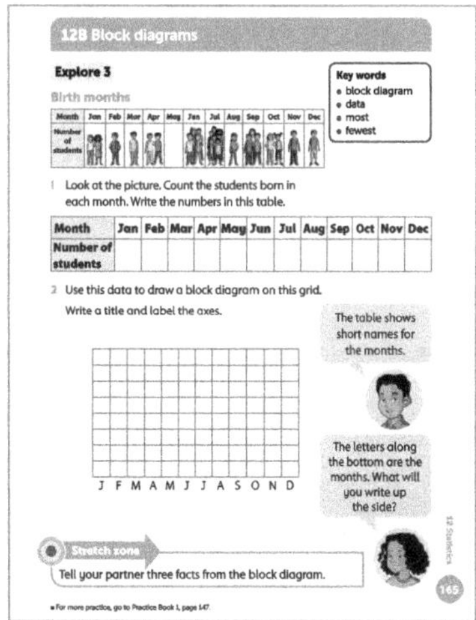

Differentiation

Supporting: Help students count the numbers for each month correctly and transfer the data to the table.

Consolidating: Ask questions such as, *Have you coloured the correct number of blocks for each month? How many blocks do you need to colour for May?*

Extending: Encourage students to write mathematical statements using the data in the table and block diagram, for example, 'There are more birthdays in June than any other month', 'There are fewer birthdays in the winter months than other seasons.'

Stretch zone: *Tell your partner three facts from the block diagram.*

Listen for students using accurate mathematical language and stating facts that show that they have interpreted the diagram correctly.

 ## Reflection time

Gather students to look at one of the students' completed block diagrams of the birthday month data in the Student Book. Ask a range of questions such as, *Which month has the most students with birthdays? Which month has the fewest? Do any months have exactly three students with birthdays?* Ask students to suggest questions about the block diagram for other students to answer.

Practice Book: Students can complete Practice Book page 147. This can be done directly after the main activity, as homework, or as the focus of a separate mathematics session to help students consolidate their learning and build fluency. Students could collect data at school and draw a block diagram as homework.

Differentiated outcomes	
All students	should begin to make a block diagram, including some key features.
Most students	will use data in a table to make a block diagram.
Some students	may ask suitable questions about the block diagram and answer them.

Student Book page 165

Check that students have correctly transferred the pictogram data onto the table and block diagram.

Practice Book page 147

Check that the data in the table matches the data shown in the block graph. The number of people for each animal should be the height of each column in the block graph.

Stretch zone: Ask students to share their easy and hard questions with you. *What made them hard? What made them easy?*

12C Venn diagrams

Discover 1  Student Book page 166 • Practice Book page 148

Specific learning focus

- Answer a question by collecting, sorting, organising and recording data in Venn diagrams, giving different criteria for grouping the same objects.

Global skills

- **Creative skills:** problem solving / exploring / investigating
- **Real-world skills:** research / presenting information / interpreting information
- **Interpersonal skills:** communication / teamwork

Key vocabulary

- represent, sort, Venn diagram, criteria, data, set

Resources

- large sheets of paper, images of animals (or similar) that can be sorted into two groups, sticky tack
- photographs or drawings of the fruits and vegetables from Student Book page 166 growing in the ground (if possible)

Language support

Ensure that all diagrams are clearly labelled, and that students understand what each of the sorting criteria means, e.g. underground, above the soil.

Introductory activity

On the board, show two large empty circles and have two **sets** of data ready which need to be sorted into the circles, such as images of animals with fur (e.g. rabbit, cat, bear, lion) and animals with scales (lizard, fish, snake). Also include some animals that don't fit either rule (e.g. bird, duck, elephant). Tell students that the circles are a way of sorting the animals. Each circle represents a different sorting rule for the animals. The circles are a type of diagram called a **Venn diagram**, and the different rules for sorting the animals are called **criteria**. Make labels 'Animals with fur' and 'Animals with scales' and stick them under the circles.

Choose students to come to the front of the class and, after discussing with the class, sort each animal image into the correct group. Some animals won't fit either rule, so they cannot go inside the circles. Ask students to stick these on the diagram, outside the circles. Once all the animal pictures have been sorted, ask questions about the data, such as *How many of the animals have scales?*

Main activity

Talk through the activity in the Student Book page 166, explaining that, instead of animals, students will be sorting different types of food. You may want to have pictures of these plants growing to help students identify whether they grow underground.

Ask students to look at the two circles. *What are the sorting criteria, or rules? Where do you find this information?* (Below the circles) *What type of fruit or vegetable should you put in the circle on the right?* Agree that there are two different types of fruit and vegetables shown, those that are orange and those that grow underground, and there are two circles labelled 'orange' and 'grows underground' and that these are the sorting criteria, or rules.

Ask students to complete page 166 of the Student Book in pairs.

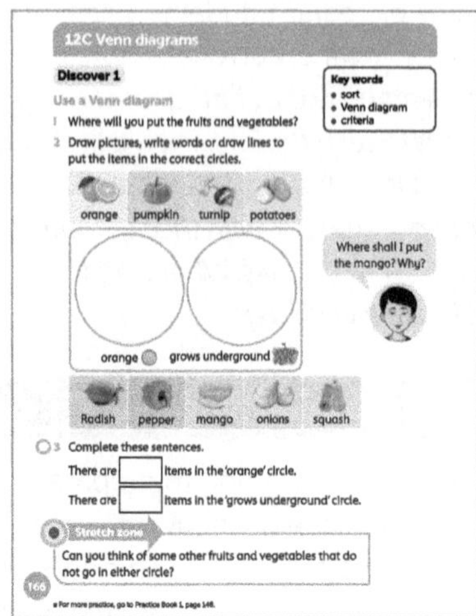

Differentiation

Supporting: Ask questions such as, *Where should I place the orange? Why?*

Consolidating: Ask questions such as, *Can you explain what you did? Can you tell us what you found out?*

Extending: Ask students if they can think of a fruit or vegetable that could be sorted into both sets.

Stretch zone: *Can you think of some other fruits and vegetables that do not go in either circle?*

Check that students have understood the question correctly and have identified fruits and vegetables that could not go in either circle. Ask them to explain their choices, e.g. 'I chose peas because they are green (not orange) and grow above ground.'

 ## Reflection time

Ask some pairs of students to share their own diagram. Ask them to tell the class what they did and what they found out. *What information did you have? What did you need to find out? How did you find out what you needed to know?*

Practice Book: Students can complete Practice Book page 148. This can be done directly after the main activity, as homework, or as the focus of a separate mathematics session to help students consolidate their learning and build fluency. You may choose to quickly review 2D shapes and their properties before students complete this activity. There is no practice activity for Discover 2, so you may prefer to complete this activity after that lesson.

Differentiated outcomes	
All students	should sort some items appropriately.
Most students	will sort most items appropriately and use appropriate vocabulary.
Some students	may make additional statements about the data.

Answers

Student Book page 166

Orange circle: orange, pumpkin, pepper, mango, squash

Grows underground: turnip, potato, onion, radish

There are 5 items in the 'orange' circle.

There are 4 items in the 'grows underground' circle.

Practice Book page 148

Check that students have placed their choice of shapes correctly. For example, 'rectangle' can go in 'has straight sides', and 'oval' can go in the 'curved sides'.

Stretch zone: Students should be able to explain that their chosen shape has straight or curved sides.

12C Venn diagrams

Discover 2 Student Book page 167

Specific learning focus

- Answer a question by collecting, sorting, organising and recording data or objects in Venn diagrams, giving different criteria for grouping the same objects.

Global skills

- **Creative skills:** problem solving / exploring / investigating
- **Real-world skills:** research / presenting information / interpreting information
- **Interpersonal skills:** communication / teamwork

Key vocabulary

- represent, sort, Venn diagram, criteria, data

Resources

- large sheets of paper
- photographs or drawings of the fruits and vegetables from Student Book page 167 growing in the ground (if possible)

Language support

Ensure that all diagrams are clearly labelled, and that students understand what each of the sorting criteria means, e.g. underground, above the soil.

 ## Introductory activity

Begin by asking students if they can tell you something about a Venn diagram. They can use the one they completed on page 166 of the Student Book for reference. Give some time for the students to respond and ask further questions. *Which fruits and vegetables do not belong in the set of things that grow underground? What do we use the circles for in this diagram? How do you decide which circle to sort something into? We sorted animals together as a class. Can you remember what the sorting rules were? What if an object does not fit either sorting rule? Where do you put it?*

Use students' feedback to lead into the next part of the lesson and encourage recall of key vocabulary.

 ## Main activity

Explain to students that a set of data such as fruits and vegetables can be sorted in different ways. Previously they sorted according to whether each item was either orange or grew underground, but today they are going to sort using different criteria. Show them the Venn diagram on page 167 of the Student Book. *Can you tell me what the sorting rules, or criteria, are on this diagram?*

How are these similar to and different from those in the last lesson? (They are still sorting vegetables and fruit but this time by whether they are red or whether they grow above the ground.) If possible, have pictures of these fruits and vegetables growing for the students to refer to so that they can find out whether a fruit or vegetable is grown above or below ground.

Now ask students to complete the activity on Student Book page 167 individually.

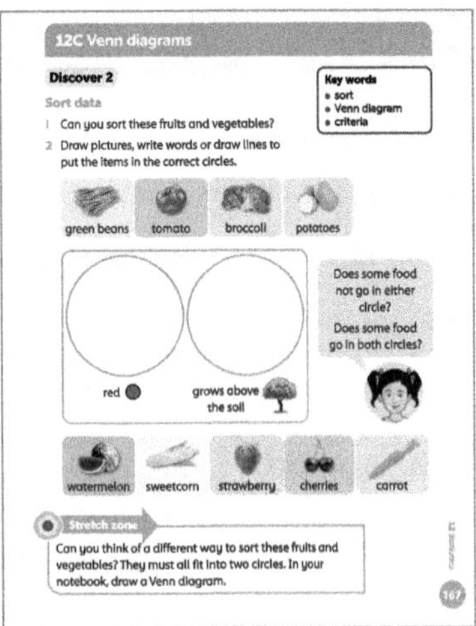

Differentiation

Supporting: Ask questions such as, *Where should I place the tomato? Why?*

Consolidating: Ask questions such as, *Can you explain what you did? Can you tell us what you found out?*

Extending: Encourage students to make a range of statements about the data.

Stretch zone: *Can you think of a different way to sort these fruits and vegetables? They must all fit into two circles. In your notebook, draw a Venn diagram.*

Check that students have sorted according to clear criteria and drawn an appropriate Venn diagram. You can suggest ways they could sort the fruit and vegetables, for example, having seeds, having peel, or fruit and vegetables that you may peel.

Reflection time

Look at the completed Venn diagrams. Ask students to describe why they placed each item where they did. Were any items difficult to place? Did they want to put some things in both circles? Draw out the fact that the potatoes and carrot cannot go in either circle, and all the items in the 'red' circle (cherries, strawberry and tomato) can also go in the 'grows above the soil' circle.

Differentiated outcomes	
All students	should sort most items appropriately.
Most students	will use the vocabulary with confidence.
Some students	may make statements about the data.

Answers

Student Book page 167

Red circle: tomato, strawberry, cherry

Grows above soil: green beans, tomato, broccoli, watermelon, sweetcorn, strawberry, cherry

Potato and carrot don't go in either circle.

Tomato, strawberry and cherry could go in both circles.

12C Venn diagrams

Explore 1 Student Book page 168 • Practice Book page 149

Specific learning focus

- Answer a question by collecting, sorting, organising and recording data or objects in Venn diagrams, giving different criteria for grouping the same objects.

Global skills

- **Creative skills:** problem solving / exploring / investigating
- **Real-world skills:** research / presenting information / interpreting information
- **Interpersonal skills:** communication / teamwork

Key vocabulary

- represent, sort, Venn diagram, criteria, intersection

Resources

- chalk or string
- images of animals, including some not featured in the Student Book, e.g. kangaroo, rabbit, duck, crocodile
- large hoops, large sheets of paper, toy animals (several for each group)

Language support

Have pictures of the animals that appear in the Student Book available, labelled with their names, so that students can refer to them easily when sorting for their Venn diagrams.

Explain that you are going to create a human Venn diagram by sorting students into groups and asking them to stand inside large circles (e.g. drawn in chalk on the floor). First of all choose criteria that mean each student will fit in one circle or the other (not both), e.g. one circle is for 5 year olds and the other circle is for 6 year olds. Then move on to using criteria which mean that some students will fit into both groups, e.g. 'students with a brother' and 'students with a sister'. *How can you be in both sets at the same time?* Students may suggest putting a leg in both circles, which then leads into the idea of 'intersection' as introduced in the main activity.

 Main activity

Display the first Venn diagram on page 168 of the Student Book. *What do you notice about the two sets?* (They have been filled in already with animals that live in water and animals that live on land.) *What about animals that live both on the land and in the water like frogs, turtles and seals? Where could we put them?* Give students a chance to think and respond. Accept all answers at this stage and follow their ideas. *I wonder what would happen if we move the two circles closer together so that they overlap?*

Ask students to look at the second Venn diagram in the Student Book. *What animals can you see? Where are they?* Agree that the seal, turtle and frog can live in both water and on land, so they are all in the overlapping part of the two circles – they belong to both circles. Explain that this is called the **intersection**.

Ask students to look again at the animals in the first Venn diagram. *Where should these animals go in the new Venn diagram. Should any of them go in the intersection? Why or why not?* Decide together where to sort all the animals and agree that none should go in the intersection because they do not live in water and on land.

Show students some pictures of other animals (e.g. kangaroo, rabbit, duck, crocodile) and discuss where you would place them in the Venn diagram.

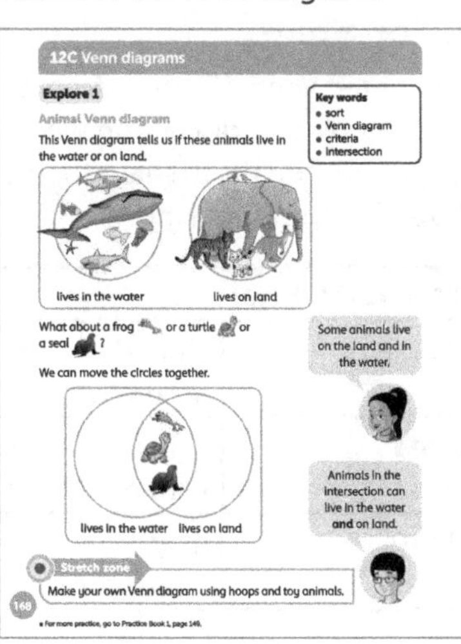

Give each group two large hoops and a selection of toy animals. Ask them to make their own Venn diagram, by choosing two criteria (sorting rules) and using them to sort the toy animals.

Differentiation

Supporting: Ask questions such as, *Where should I place the dog? Why?*

Consolidating: Ask questions such as, *Can you explain what you did? Can you tell us what you found out?*

Extending: Encourage students to make a range of statements about the data.

Stretch zone: *Make your own Venn diagram using hoops and toy animals.*

Check that students have placed the animals correctly according to the features they are sorting by.

 Reflection time

Ask students to share what they did with the stretch zone or, as a class, you could create a Venn diagram to represent the sort they did in the introductory activity when some students had both a brother and a sister and fitted both criteria. Students can create overlapping circles, and those students with both a brother and a sister can stand in the intersection.

Practice Book: Students can complete Practice Book page 149. This can be done directly after the main activity, as homework, or as the focus of a separate mathematics session to help students consolidate their learning and build fluency. There is no practice activity for Explore 2, so you may prefer to complete this activity after that lesson.

Differentiated outcomes	
All students	should sort items accurately.
Most students	will use the vocabulary with confidence.
Some students	may make statements about the data.

Answers

Practice Book page 149

Check that students have drawn lines from each vehicle to the correct part of the Venn diagram. Only the hovercraft should be in the intersection.

Stretch zone: The hovercraft should be in the intersection because it can travel on both land and water. Aircraft, helicopters and submarines would not fit in this diagram.

12C Venn diagrams

Specific learning focus

- Answer a question by collecting, sorting, organising and recording data or objects in a Venn diagram, giving different criteria for grouping the same objects.

Global skills

- **Creative skills:** problem solving / exploring / investigating
- **Real-world skills:** research / presenting information / interpreting information
- **Interpersonal skills:** communication / teamwork

Key vocabulary

- represent, sort, Venn diagram, criteria

Resources

- simple Venn diagram to display showing two intersecting sets (labelled with appropriate criteria) containing pictures of items, but with some items wrongly sorted
- large sheets of paper, images of things that fly or walk, or do both

Language support

Ask questions such as, *What do you know about a Venn diagram? Why do we use a Venn diagram? Can you draw a Venn diagram on the board?* Some students will need a picture and appropriate words to help them with explanations.

 **Introductory activity**

Display a Venn diagram on the board that has an intersection but includes some sorting errors. You may want to reuse some of the fruit and vegetable or animal pictures from earlier lessons. Ask students to work out which items have been sorted incorrectly.

 **Main activity**

Ask the class to look at the pictures in their Student Book on page 169. *What do you notice? Can you explain why each thing has been sorted that way?* Agree that the circles have been partly filled in already with things that fly and things that walk (and something that does both). *What if there are things that cannot fly or walk? Where could we put them?* Give students the chance to think and give their response. Accept all answers at this stage, and follow their ideas. *I wonder what we can do with the space outside the two circles?*

Agree that the box containing the circles can represent any item. Something that doesn't belong in either circle can be placed in the space between the circles and the outer box.

As a class, brainstorm some ideas of things that you could sort into each set (e.g. a kite, a bee, a monkey). Encourage students to explain how they know which set they would sort them into.

Students can complete the activity on Student Book page 169 individually.

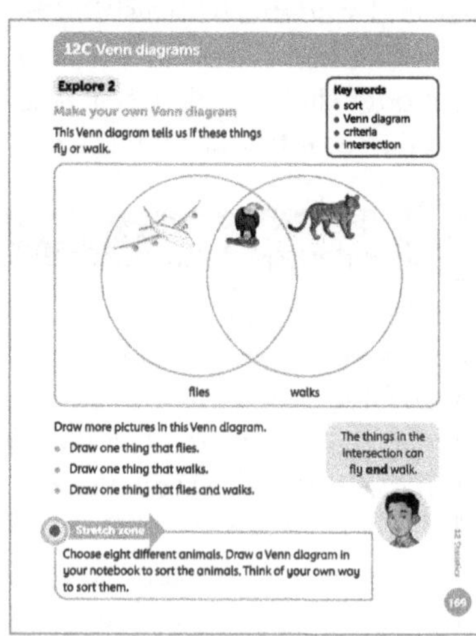

Differentiation

Supporting: Help students to give names to things that they draw if they do not know the name in English.

Consolidating: Ask questions such as: *Where would I place a snake? Why? What is the same and different about the things in the intersection?'*

Extending: Encourage students to make a range of statements about the data and sort more than one thing in each set and in the intersection.

Stretch zone: *Choose eight different animals. Draw a Venn diagram in your notebook to sort the animals. Think of your own way to sort them.*

Check that students have chosen suitable criteria for sorting and drawn the Venn diagram correctly.

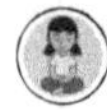 **Reflection time**

Use some completed Venn diagrams from the Stretch zone as a starting point for discussion with the whole class. Choose groups or pairs to share their own Venn diagrams about flying and walking.

Differentiated outcomes	
All students	should sort items accurately.
Most students	will use the vocabulary with confidence.
Some students	may make statements about the data.

Answers

Student Book page 169

Check the accuracy of the items in the Venn diagram.

12D Carroll diagrams

Discover 1 — Student Book page 170

Specific learning focus

- Answer a question by collecting, sorting, organising and recording data in Carroll diagrams.

Global skills

- **Creative skills:** problem solving / investigating
- **Real-world skills:** research / presenting information / interpreting information
- **Interpersonal skills:** communication / teamwork

Key vocabulary

- represent, sort, Carroll diagram, criteria, data

Resources

- shapes (for sorting by colour and shape), sticky tack

Language support

Ask questions such as, *What do you know about a Carroll diagram?* Help students to explain their ideas in a clear way. 'We use Carroll diagrams to sort things. You sort by putting things in different boxes. Each box has a different label. You ask a question about the objects and if the answer is yes they go in one box. If the answer is no they go in the other box.'

 **Introductory activity**

Draw two large empty boxes on the board. Write labels above the boxes, e.g. yellow, not yellow.

Show the class some blue, yellow, red and green shapes. *We are going to sort these shapes into the boxes.* One at a time, ask students which box they think each shape could go in, keeping to the sorting rule of the label: *Is this yellow?* (Yes/No).

Add each shape to the relevant box on board using sticky tack. Once the shapes have all been sorted, ask students to make statements about how the shapes have been sorted, e.g. *There are fewer shapes that are 'yellow' than 'not yellow'. There are [7] shapes that are not yellow.*

 **Main activity**

Tell students that they will be looking at a new way of sorting shapes. Explain that it is called a **Carroll diagram** and that, rather than sorting using circles, they will sort into boxes in a diagram. *Look at the Carroll diagram on Student Book page 170. Read the labels in the purple boxes. What do you think they tell us?.* (How to sort the fruit and vegetables) Agree that these are the sorting rules (criteria) for the diagram. You are sorting whether each item is or isn't a fruit. Then say, *In a Carroll diagram, we can ask a question for each thing we are sorting, and answer yes or no. Is it a fruit? Yes, so I sort it here. Is it a fruit? No, so I sort it there.*

Ask students which box they need to draw a line to if they sort an item that is a fruit. *What if it is not a fruit?*

Ask students to work on the problems on page 170 of the Student Book in pairs and discuss where to place the data. Provide support if the students are unfamiliar with some of the fruits and vegetables.

Once they have sorted the fruit and vegetables they count how many there are in each box.

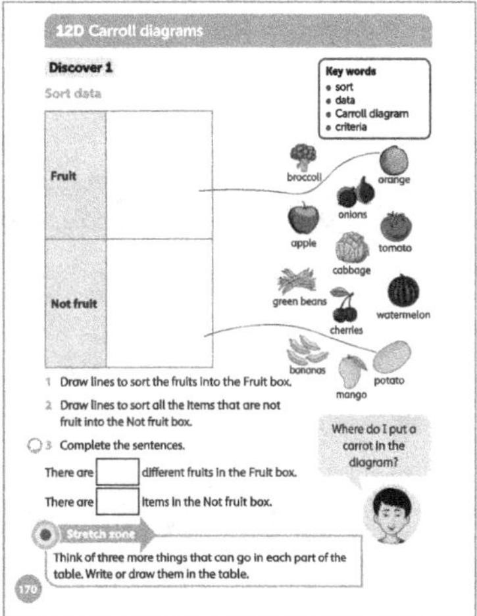

Differentiation

Supporting: Help students with identifying which are fruits and which are vegetables.

Consolidating: Ask questions such as, *Can you explain what you did? Where should I place the apple? Why?*

Extending: Ask students why every fruit or vegetable must sort into one box or the other.

Stretch zone: *Think of three more things that can go in each part of the table. Write or draw them in the table.*

Check that students have correctly placed additional items in the diagram.

 **Reflection time**

Bring the class back together and ask, *What was easy and what was difficult when you sorted the objects?*

Ask further questions such as, *How was this the same as sorting with a Venn diagram? How was it different?*

Differentiated outcomes	
All students	should sort familiar items accurately.
Most students	will use the vocabulary with confidence.
Some students	may make several statements about the data.

Answers

Student Book page 170

1 Fruit – orange, apple, cherries, watermelon, bananas, mango

2 Not fruit – broccoli, onions, green beans, potato, cabbage, tomato

3 There are 6 different fruits in the Fruit box.
There are 6 items in the Not fruit box.

12D Carroll diagrams

Discover 2 Student Book page 171 • Practice Book page 150

Discover 2 Student Book page 171 • Practice Book page 150

Specific learning focus

- Answer a question by collecting, sorting, organising and recording data or objects in Carroll diagrams, giving different criteria for grouping the same objects.

Global skills

- **Creative skills:** problem solving / investigating
- **Real-world skills:** research / presenting information / interpreting information
- **Interpersonal skills:** communication / teamwork

Key vocabulary

- represent, sort, Carroll diagram, criteria, data, row, column

Resources

- images of fruits and vegetables (for sorting by colour and fruit), sticky tack

Language support

Ask questions such as: *What do you know about a Carroll diagram?* Help students to explain their ideas in a clear way. 'We use Carroll diagrams to sort things. You sort by putting things in different boxes. Each box has a different label. You ask a question about the objects and if the answer is yes they go in one box. If the answer is no they go in the other box.'

Introductory activity

Recall with students the Carroll diagram from the last lesson where fruit and vegetables were sorted according to whether they were fruit or not. Draw a blank Carroll diagram on the board and label it with the same criteria, 'Fruit' / 'Not fruit'. Show students some pictures of other fruits and vegetables to sort into fruit/not fruit. Ask students to explain their choices and stick the pictures in the appropriate box in the diagram.

Can you think of other criteria we could use to sort these fruit and vegetables? Remind students that, for Carroll diagrams, the sorting rule should be of the form 'red' / 'not red', 'vegetable' / 'not vegetable', and they can sort by asking, e.g. 'is it red?' (Yes/No) Take suggestions (e.g. round/not round, has leaves/has no leaves, grows on trees/doesn't grow on trees). Choose new criteria from students' suggestions, relabel the diagram and sort the fruit and vegetables accordingly.

Main activity

What if I wanted to sort by more criteria? Could I add more boxes to my Carroll diagram?

Remove all the fruit and vegetables from the Carroll diagram used in the Introductory activity. Add two more boxes (an extra column) to the blank Carroll diagram so that there are four blank boxes (like the diagram on page 171 of the Student Book) and add new labels at the top of the grid for the two new criteria. These could be 'Red' and 'Not red', as in the Student Book, or one of the students' suggestions from the Introductory activity. Choose one of the vegetables and show students which section it should be placed in, describing how it fits the criteria for that row and that column. Ask a few students to sort other fruits and vegetables into the new Carroll diagram.

Ask students to complete the activities on page 171 in the Student Book individually.

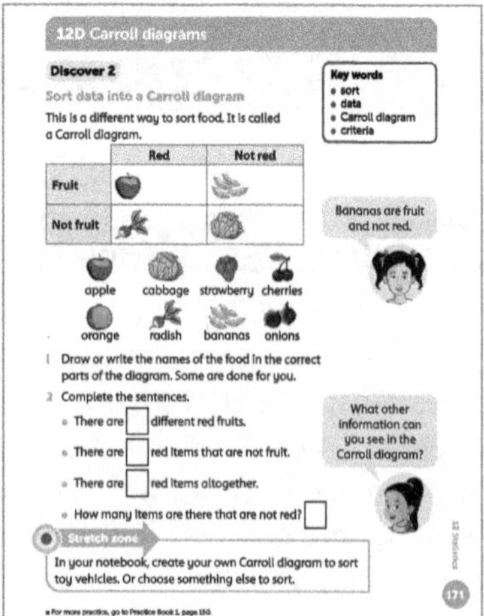

Differentiation

Supporting: Ask questions such as, *Where should I place the bananas? Why?*

Consolidating: Ask questions such as, *Can you explain what you did? Were there any fruits or vegetables that were difficult to sort? Why?*

Extending: Ask students to explain why every fruit or vegetable can be sorted into one of the boxes.

Stretch zone: *In your notebook, create your own Carroll diagram to sort toy vehicles. Or choose something else to sort.*

Help students choose suitable criteria for sorting their items and then let them sort and fill the Carroll diagram correctly.

 Reflection time

Discuss with students their completed Carroll diagrams. Have they placed all the items correctly? Were there any items they were not sure about? How does the diagram help them count, for example, how many red fruits there are?

Practice Book: Students can complete Practice Book page 150. This can be done directly after the main activity, as homework, or as the focus of a separate mathematics session to help students consolidate their learning and build fluency. Review the names of the 2D shapes to be sorted.

Differentiated outcomes	
All students	should sort familiar items accurately by two criteria.
Most students	will accurately place each fruit or vegetable into the correct box.
Some students	may reason why all fruit and vegetables will fit in one of the boxes.

Answers

Student Book page 171

1 Red, fruit: apple, strawberry, cherries
Red, not fruit: radish, onions
Not red, fruit: bananas, orange
Not red, not fruit: cabbage

2 There are 3 different red fruits.
There are 2 red items that are not fruit.
There are 5 red items altogether.
How many items are there that are not red? 3

Practice Book page 150

All sides straight, fewer than 5 sides: square, triangle, rectangle

All sides straight, not fewer than 5 sides: pentagon, hexagon, star, irregular pentagon

Not all sides straight, fewer than 5 sides: circle, semi-circle

Not all sides straight, not fewer than 5 sides: 'hexagon with one curved side'

Stretch zone: There are 7 shapes with straight sides.
There are 3 shapes with not all sides straight.
There are 5 shapes with fewer than 5 sides.
There are 5 shapes with 5 or more sides.

12D Carroll diagrams

Explore 1   Student Book page 172 • Practice Book page 151

Specific learning focus

- Answer a question by collecting, sorting, organising and recording data or objects in Carroll diagrams, giving different criteria for grouping the same objects.

Global skills

- **Creative skills:** problem solving / investigating
- **Real-world skills:** research / presenting information / interpreting information
- **Interpersonal skills:** communication / teamwork

Key vocabulary

- represent, sort, Carroll diagram, criteria

Resources

- images of foods or pictures of food cut out of magazines (if possible, the same selection of food items for each small group)

Language support

For students who need support, discuss what using 'not' means (the opposite of something). Ask students for examples: happy/not happy, big/not big.

Introductory activity

Give out a selection of images of food to each small group. They should discuss all the images on their table and sort them into two groups. When they have sorted them ask them to write down the sorting rule that they chose. Then ask each group to re-sort their group using a different sorting rule. Remind students that a Carroll diagram allows us to sort items using two sorting rules (criteria).

When all groups have chosen two sorting rules, ask for feedback and make a list of some of the criteria on the board. *Did each group choose the same sorting rules?* Agree that there are many different ways to sort the same items.

Main activity

Ask students, *What is healthy food? How do you know?* Take ideas, including examples of healthy food.

Ask students, working in the same small groups as previously, to now sort the same food items into 'healthy foods' and 'not healthy foods'. Then ask students to think about sorting the same foods into foods they like and foods they don't like.

Students should explain their choices to other students in their group, for example, 'I like apples and they are healthy so I would sort them into this pile. I don't like any vegetables and they are healthy. I like chocolate but it's not healthy, I don't like chips and they're not healthy'.

Once the small group has discussed how to sort the items, ask students to individually complete the activity in the Student Book page 172. Encourage students to think of other fruit and vegetables to sort. They may want to refer back to earlier pages in the Student Book for ideas.

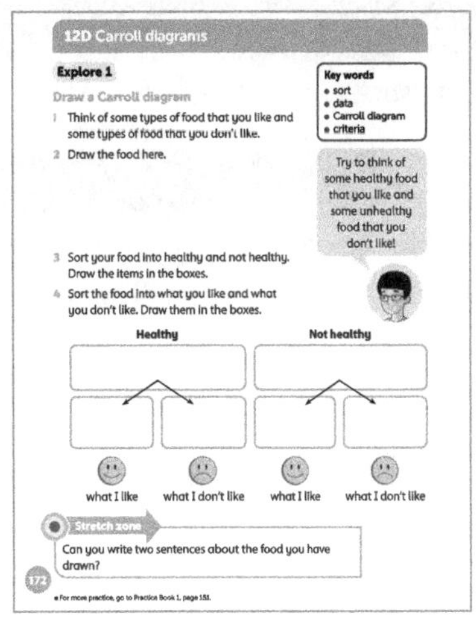

Differentiation

Supporting: Ask questions such as: *Why did you draw this food here?*

Consolidating: Ask questions such as: *Can you explain what you did? Can you tell us what you found out when you looked at someone else's diagram?*

Extending: Ask students to draw a Venn diagram to show the same data.

Stretch zone: *Can you write two sentences about the food you have drawn?*

Check that students have written correct sentences.

 **Reflection time**

Choose groups or individuals to share how they sorted the food.

Ask questions such as:

How did you decide what food to choose? How many healthy foods did you choose? How many unhealthy foods do you like?

Practice Book: Students can complete Practice Book page 151. This can be done directly after the main activity, as homework, or as the focus of a separate mathematics session to help students consolidate their learning and build fluency. There is no practice activity for Explore 2 so you may prefer to complete this activity after that lesson. Discuss possible criteria for sorting with students before they begin.

Differentiated outcomes	
All students	should sort items accurately.
Most students	will use the vocabulary with confidence.
Some students	may make statements about the data.

Answers

Student Book page 172

Check that students have written their data into the correct boxes.

Practice Book page 151

Check that students have written their data into the correct boxes in the Carroll diagram.

Stretch zone: Students' answers will vary, e.g. Venn diagrams use circles to show criteria. Carroll diagrams use a grid.

12D Carroll diagrams

Explore 2   Student Book page 173

- Answer a question by collecting, sorting, organising and recording data or objects in Carroll diagrams, giving different criteria for grouping the same objects.

Global skills

- **Creative skills:** problem solving
- **Real-world skills:** presenting information / interpreting information
- **Interpersonal skills:** communication / teamwork

Key vocabulary

- represent, sort, Carroll diagram, criteria

Resources

- images of foods or pictures of food cut out of magazines

Language support

For students who need extra support, discuss what using 'not' means (the opposite of something). Ask students to suggest examples, such as happy/not happy, big/not big.

 **Introductory activity**

Tell students that they are going to think about what they know about a Carroll diagram. Show them a completed diagram from yesterday and ask them questions such as, *Can you tell me what the criteria are for sorting in this Carroll diagram? What can you say about the two choices for each? Can there be any objects that can't be placed in the diagram? Why?*

 **Main activity**

Ask students to fill in the Carroll diagram on page 173 of the Student Book using the data from page 172. They can also sort the items that appear below the Carroll diagram. Remind them to follow the criteria carefully when placing each item.

hen they have completed the diagram individually, ask
em to work in pairs to compare diagrams. *Are there any
ferences? Do you agree with your partner about the placing
the items? Why or why not? Is it okay if they are sometimes
fferent?* Can they explain why they sorted some food
ms in a certain way? For example, 'I sorted pizza as food
ke that is healthy because I eat it with lots of vegetables
d cheese on top and they are both healthy types of food'.

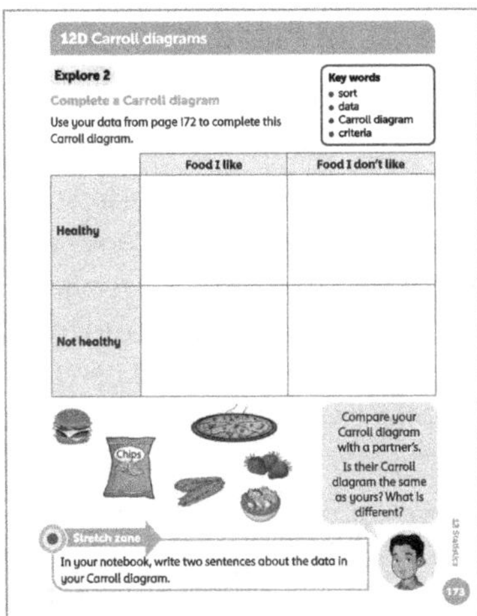

Differentiation

Supporting: Ask questions such as, *Why did you draw
this food here?*

Consolidating: Ask questions such as, *What did you
notice when you compared diagrams with your partner?
Why were they different?*

Extending: Ask students to draw a Carroll diagram to
show the same data, but change one of the criteria and
re-sort the data.

---

 *In your notebook, write two sentences about
the data in your Carroll diagram.*

Check that the sentences relate accurately to the diagram.

 ## Reflection time

Ask students to look at their Carroll diagram and make
a statement about the data to share with the class or
another student. For example, 'I like more healthy foods
than foods that are not healthy'. Then see if they can ask a
question which can be answered from the Carroll diagram.
For example, 'How many healthy foods do I not like'?

Students can ask each other their questions and use the
Carroll diagram to find the answer.

Differentiated outcomes	
All students	should sort items accurately according to their own preferences.
Most students	will describe their sorting choices to a partner.
Some students	may discuss and compare diagrams with a partner and be able to explain why they were the same and different.

Answers

Student Book page 173

Check that students have transferred the data sorted
from page 172 into the Carroll diagram correctly.

---

12 Statistics

Connect Student Book pages 174–5

Big idea

I know lots of ways to sort data. I can use
pictograms, block diagrams, Venn diagrams and
Carroll diagrams.

Global skills

- **Creative skills:** problem solving
- **Real-world skills:** research / presenting information / interpreting information
- **Interpersonal skills:** communication / teamwork
- **Self-development skills:** reflecting on learning

Key vocabulary

- pictogram, collect, represent, list, table, Venn diagram, data, Carroll diagram, sort, block diagram

Resources

- none needed

Language support

Listen to the ways in which the students discuss and
sort the information. Repeat key words to model
pronunciation.

 ## Introductory activity

Say to students, *We have been gathering data for different
types of diagrams. What did you find out and what ways
did you represent data?* You may want to use page 174 of
the Student Book to prompt students.

Give students a few minutes to discuss in pairs. Take
feedback and record their ideas on the board so that
all students can see them.

Recall the main features of and the differences between
tables, pictograms, block diagrams, Venn diagrams and
Carroll diagrams. Discuss how we used them.

Main activity

Explain to students that they are now going to collect some new data about the class and represent it one of the ways they have been learning about.

Place students in groups and explain that they need to talk to each other as a group to decide what data they want to collect. *Decide how you will collect the data* (list, table) *and how you will represent it* (block graph, pictogram, Venn or Carroll diagram). *You should use only one way of representing your data.* Make sure that all the groups understand the task by asking simple checking questions, such as, *Do you have any questions? Do you know what to do?*

As the students are working, move around the class to make sure that they all understand the task. Use questions such as, *How are you going to do this task? What information do you have? What do you need to find out? What diagram are you going to use? Why? What questions will you need to ask? How are you going to record what you are doing?*

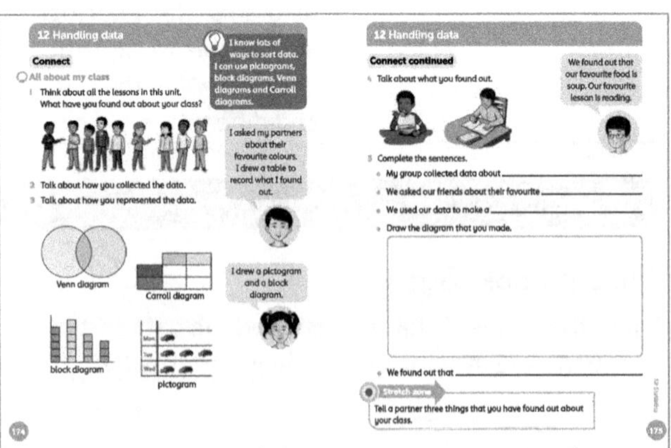

Differentiation

As this is a discussion-based activity, place students in mixed-attainment groups so that all students get to he a wide range of vocabulary.

Stretch zone: *Tell a partner three things you found out about your class.*

Walk around each group to get an understanding of th types of data students collected.

 **Reflection time**

Leave enough time at the end of the lesson so that eac group can talk about what they did and what they foun out. Ask questions such as, *How did you get your answer How did you organise collecting the data? Did you all ask questions or just one of you? Why? What have you learned or found out today? If you did it again, would you do it the same or differently? Why?*

Differentiated outcomes	
All students	should sort the information.
Most students	will sort data into groups and describe the criteria clearly.
Some students	may interpret the data and explain it clearly.

12 Statistics

Review Student Book page 176 • Practice Book page 152

Global skills

- **Creative skills:** problem solving
- **Real-world skills:** interpreting information
- **Interpersonal skills:** communication
- **Self-development skills:** reflecting on learning

Student Book

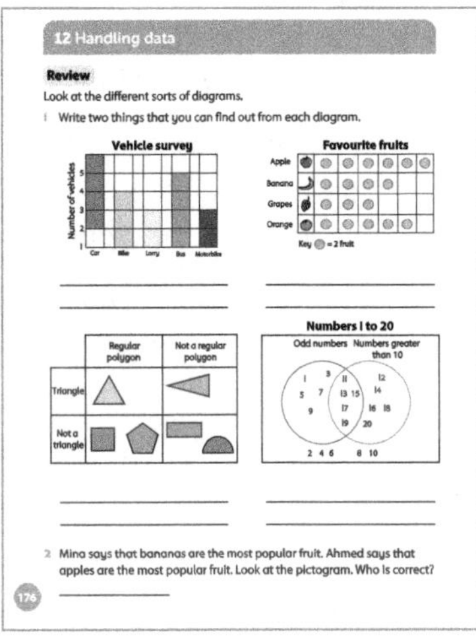

With young children, assessment activities are most effective when carried out as an everyday classroom activity. Students should be able to interpret block diagrams, pictograms, Carroll and Venn diagrams. Consider recording students' answers for them if writing will be a barrier to their answering.

It may help to remind students of the features of the different diagrams, for example, the criteria for sorting into Venn or Carroll diagrams.

Watch as children use the diagrams to make statements about the data.

Answers

Student Book page 176

1 Read through students' statements for each diagram. Ask for oral clarification as necessary.

2 Ahmed is correct.

Practice Book

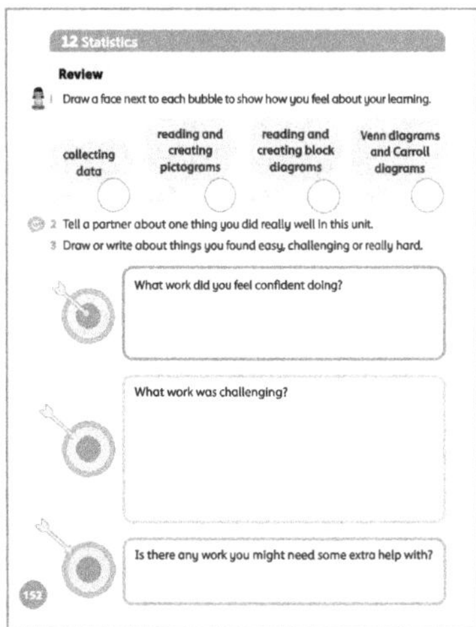

With young children it is appropriate to complete this as a whole-class discussion. You may choose to keep a record of the class discussion or a copy of the review page for your own records. Use the Student Book to briefly remind the students of the areas of mathematics that they have worked on in this unit.

Give out the Student Book to support pairs of students as they discuss and answer the questions in the Practice Book.

Allow students plenty of time for discussion before asking them to share their responses with the rest of the class. If students complete this assessment at home, encourage them to discuss this with adults.

Make a note of areas that students still feel unsure about.

Additional material

There are additional end-of-unit assessments available on the *Oxford Owl* website.

Glossary

add	combine two or more numbers or quantities to make a new number or amount. The sign for adding is +
answer	the solution to a problem
circle	a perfectly round 2D shape made by a single curved line
coin	a piece of money made from metal
cone	a 3D shape with a base that is a circle and a curved surface that comes to a point at its top
count	find the total number of something
cube	a 3D shape with 6 square faces, 8 vertices and 12 edges
cuboid	a 3D shape with 6 rectangular faces, 8 vertices and 12 edges
curved	a line or a surface that bends smoothly
cylinder	a 3D shape with 2 flat faces that are circles and one curved surface
day	1. a period of 24 hours 2. the hours of daylight when most people are awake and active
digit	a numerical symbol used to build numbers. The 10 digits are: 0, 1, 2, 3, 4, 5, 6, 7, 8, 9
double	multiply a number by 2 or make a number twice as many
edge	where two faces of a 3D shape meet. An edge can be curved or straight
equal	when one thing or number has the same value as another. The sign for 'is equal to' is =
estimate	make a reasonable guess at an amount, number or value without calculating
even	describes a number that can be divided exactly by 2
face	the flat or curved surface of a 3D shape
fast	describes something moving at high speed
group	a set of objects or numbers
half	divide something into 2 equal parts. The symbol for a half is $\frac{1}{2}$
heavy	describes the weight or mass of an object when the object has a great weight. It is the opposite of light
height	how tall something is or how far it is from the ground
length	how long something or a time is. You can find the length by measuring the distance or time between two points
light	describes the weight or mass of an object when the object has a low weight. It is the opposite of heavy
measure	find the size or amount of something using a measuring unit or device
metre	a metric unit of measurement used to measure lengths or distance
minus	the minus sign (−) tells you to take away, subtract or find the difference
money	a type of exchange in the form of coins and banknotes
month	a period of time: a month has 28, 29, 30 or 31 days. A year is divided into 12 months.
number	what we use to count or measure things
number sentence	a mathematical sentence that use numbers, signs and words to present a problem or statement
odd	describes a number that cannot be divided equally into two parts
ones	groups of 1. We use ones when we talk about place value in a number
operation	the process of changing a number by adding, subtracting, multiplying or dividing
plus	the plus sign (+) tells you to add
position	the place where something is
pyramid	a 3D shape with triangular faces that meet at a point

rectangle	a 2D shape with 4 straight sides and 4 right angles. Opposite sides are equal
ruler	an instrument for measuring the length of lines and objects and for drawing straight lines
season	part of the year: a year has four seasons: winter, spring, summer, autumn (fall)
shape	an area contained by sides and surfaces that can be 2D or 3D
sign	a short way of saying something. Some mathematical signs are: $+, -, \div$ and $\times$
size	how big something is
slow	describes something moving at low speed
sphere	a perfectly round 3D shape with one curved surface
square	a 2D shape with 4 equal straight sides and 4 right angles
subtract	take away one number from another or find the difference between two numbers
teen number	a 2-digit number that ends in 'teen'. The teen numbers are 13, 14, 15, 16, 17, 18 and 19
tens	groups of 10 ones. We use tens when we talk about the place value of a number
time	a measure of how long something lasts, e.g. in seconds, minutes, hours or days; a way to describe a fixed time in a day, e.g. it is 4 o'clock
total	the result of adding two or more numbers
triangle	a 2D shape with 3 straight sides and 3 angles
week	a time period of 7 days
width	how wide something is. The distance measurement across something or from side to side
year	a time period of 365 days (or 366 days in a leap year). The time it takes for the Earth to travel around the Sun
zero	another word for the digit 0. It means no amount, nothing or nought